Lecture Notes in Networks and Systems

Volume 56

Series editor

Janusz Kacprzyk, Systems Research Institute, Polish Academy of Sciences, Warsaw, Poland
e-mail: kacprzyk@ibspan.waw.pl

The series "Lecture Notes in Networks and Systems" publishes the latest developments in Networks and Systems—quickly, informally and with high quality. Original research reported in proceedings and post-proceedings represents the core of LNNS.

Volumes published in LNNS embrace all aspects and subfields of, as well as new challenges in, Networks and Systems.

The series contains proceedings and edited volumes in systems and networks, spanning the areas of Cyber-Physical Systems, Autonomous Systems, Sensor Networks, Control Systems, Energy Systems, Automotive Systems, Biological Systems, Vehicular Networking and Connected Vehicles, Aerospace Systems, Automation, Manufacturing, Smart Grids, Nonlinear Systems, Power Systems, Robotics, Social Systems, Economic Systems and other. Of particular value to both the contributors and the readership are the short publication timeframe and the world-wide distribution and exposure which enable both a wide and rapid dissemination of research output.

The series covers the theory, applications, and perspectives on the state of the art and future developments relevant to systems and networks, decision making, control, complex processes and related areas, as embedded in the fields of interdisciplinary and applied sciences, engineering, computer science, physics, economics, social, and life sciences, as well as the paradigms and methodologies behind them.

Advisory Board

More information about this series at http://www.springer.com/series/15179

Siddhartha Bhattacharyya
Aboul Ella Hassanien · Deepak Gupta
Ashish Khanna · Indrajit Pan
Editors

International Conference on Innovative Computing and Communications

Proceedings of ICICC-2018, Volume 2

 Springer

Editors
Siddhartha Bhattacharyya
Department of Computer Application
RCC Institute of Information Technology
Kolkata, West Bengal, India

Aboul Ella Hassanien
Faculty of Computers and Information
Cairo University
Giza, Egypt

Deepak Gupta
Department of Computer Science
and Engineering
Maharaja Agrasen Institute of Technology
New Delhi, Delhi, India

Ashish Khanna
Department of Computer Science
and Engineering
Maharaja Agrasen Institute of Technology
New Delhi, Delhi, India

Indrajit Pan
Department of Information Technology
RCC Institute of Information Technology
Kolkata, West Bengal, India

ISSN 2367-3370   ISSN 2367-3389 (electronic)
Lecture Notes in Networks and Systems
ISBN 978-981-13-2353-9   ISBN 978-981-13-2354-6 (eBook)
https://doi.org/10.1007/978-981-13-2354-6

Library of Congress Control Number: 2018952616

This Springer imprint is published by the registered company Springer Nature Singapore Pte Ltd.
The registered company address is: 152 Beach Road, #21-01/04 Gateway East, Singapore 189721, Singapore

*Prof. (Dr.) Siddhartha Bhattacharyya
would like to dedicate this book to his
father Late Ajit Kumar Bhattacharyya, his
mother Late Hashi Bhattacharyya, his
beloved wife Rashni, and his colleagues
Sayantan, Sandip, Chinmay, Pradip,
Santasil and Avijit, who has been beside
me through thick and thin.*

*Prof. (Dr.) Aboul Ella Hassanien would like
to dedicate this book to his beloved wife
Azza Hassan El-Saman.*

*Dr. Deepak Gupta would like to dedicate
this book to his father Sh. R. K. Gupta,
his mother Smt. Geeta Gupta, his mentors
Dr. Anil Kumar Ahlawat and
Dr. Arun Sharma for their constant
encouragement and his family members
including his wife, brothers, sisters, and kids.*

*Dr. Ashish Khanna would like to dedicate this
book to his mentors Dr. A. K. Singh and
Dr. Abhishek Swaroop for their constant
encouragement and guidance and his family*

members including his mother, wife, and kids. He would also like to dedicate this work to his (Late) father Sh. R. C. Khanna with folded hands for his constant blessings.

Dr. Indrajit Pan would like to dedicate this book to Dr. Sitansu Kumar Pan and Smt. Ratna Pan

ICICC-2018 Steering Committee Members

Chief Patron

S. Manjeet Singh, G. K. (President, DSGMC)

Patron-in-Chief

S. Manjinder Singh Sirsa, General Secretary, DSGMC

Patron

S. Gurudev Singh Gujural, Chairman, GNIM
S. Chaman Singh, Manager, GNIM
Prof. Dr. D. S. Jaggi, Director General, Higher Education, DSGMC

General Chair(s)

Prof. Dr. Valentina Emilia Balas, Aurel Vlaicu University of Arad, Romania
Prof. Dr. Vincenzo Piuri, The University of Milan, Italy
Prof. Dr. Siddhartha Bhattacharyya, Principal, RCC Institute of Information
Technology, Kolkata

Honorary Chairs

Prof. Dr. Aboul Ella Hassanien, Cairo University, Egypt
Prof. Dr. Subhaansu Bandyopadhyay, VC, Brainware University, West Bengal,
India
Dr. Ebrahim Aghajari, Azad University of Ahvaz (IAUA), Iran

Conference Chair(s)

Prof. Dr. A. K. Singh, Department of Computer Engineering, NIT Kurukshetra,
India
Prof. Dr. Anil Kumar Ahlawat, Dean, KIET Group of Institutions, India

Prof. Dr. Shubhra Saggar, Director, Guru Nanak Institute of Management, New Delhi

Technical Program Chair

Prof. Dr. Abhishek Swaroop, BPIT, Delhi, India
Dr. Arun Sharma, Associate Professor, IGDTUW, Delhi, India
Dr. Pradeep Kumar Mallick, VBIT, Hyderabad, India
Dr. Ahmed Faheem Zobaa, Brunel University London
Dr. Dac-Nhuong Le, Hai Phong University, Haiphong, Vietnam

Convener

Dr. Ashish Khanna, MAIT, Delhi, India
Dr. Deepak Gupta, MAIT, Delhi, India
Dr. Nidhi Khurana, HOD, Guru Nanak Institute of Management, New Delhi
Dr. Prasant Kumar Pattnaik, Professor, KIIT University, India
Dr. P. S. Bedi, HOD(IT), GTBIT, Delhi, India

Publicity Chair

Dr. Vishal Jain, Associate Professor, BVICAM, Delhi, India
Dr. Brojo Kishore Mishra, Associate Professor, C.V. Raman College of Engineering, India
Dr. Raghvendra Kumar, LNCT Group of College, MP, India
Dr. Anand Nayyar, KCL Institute of Management and Technology, Punjab, India
Dr. G. Suseendran, Assistant Professor, Vels University, Chennai

Publication Chair

Dr. D. Jude Hemanth, Associate Professor, Karunya University, Coimbatore, India

Co-convener

Dr. Brojo Kishore Mishra, C.V. Raman College of Engineering, Bhubaneswar, India
Mr. Nihar Ranjan Roy, GD Goenka University, Gurugram, India
Mr. Moolchand Sharma, MAIT, Delhi, India

ICICC-2018 Advisory Committee

Prof. Dr. Vincenzo Piuri, The University of Milan, Italy

Prof. Dr. Valentina Emilia Balas, Aurel Vlaicu University of Arad, Romania

Prof. Dr. Marius Balas, Aurel Vlaicu University of Arad, Romania

Prof. Dr. Mohamed Salim Bouhlel, University of Sfax, Tunisia

Prof. Dr. Aboul Ella Hassanien, Cairo University, Egypt

Prof. Dr. Cenap Ozel, King Abdulaziz University, Saudi Arabia

Prof. Dr. Ashiq Anjum, University of Derby, Bristol, UK

Prof. Dr. Mischa Dohler, King's College London, UK

Prof. Dr. Sanjeevikumar Padmanaban, University of Johannesburg, South Africa

Prof. Dr. Siddhartha Bhattacharyya, Principal, RCC Institute of Information Technology, Kolkata, India

Prof. Dr. David Camacho, Universidad Autonoma de Madrid, Spain

Prof. Dr. Parmanand, Dean, Galgotias University, UP, India

Dr. Abu Yousuf, Assistant Professor, University Malaysia Pahang, Gambang, Malaysia

Prof. Dr. Salah-ddine Krit, University Ibn Zohr, Agadir, Morocco

Dr. Sanjay Kumar, Biswash Research Scientist, INFOCOMM Lab, Russia

Prof. Dr. Maryna Yena S. Senior Lecturer, Kyiv Medical University of UAFM, Ukraine

Prof. Dr. Giorgos Karagiannidis Aristotle University of Thessaloniki, Greece

Prof. Dr. Tanuja Srivastava, Department of Mathematics, IIT Roorkee

Dr. D. Jude Hemanth, Associate Professor, Karunya University, Coimbatore

Prof. Dr. Tiziana Catarci, Sapienza University of Rome, Italy

Prof. Dr. Salvatore Gaglio, University Degli Studi di Palermo, Italy

Prof. Dr. Bozidar Klicek, University of Zagreb, Croatia

Dr. Marcin Paprzycki, Associate Professor, Polish Academy of Sciences, Poland

Prof. Dr. A. K. Singh, NIT Kurukshetra, India

Prof. Dr. Anil Kumar Ahlawat, KIET Group of Institutes, India

Prof. Dr. Chang-Shing Lee, National University of Tainan, Taiwan

Dr. Paolo Bellavista, Associate Professor, Alma Mater Studiorum–Università di Bologna

Preface

We hereby are delighted to announce that Guru Nanak Institute of Management, New Delhi, has hosted the eagerly awaited and much coveted International Conference on Innovative Computing and Communication (ICICC-2018). The first version of the conference was able to attract a diverse range of engineering practitioners, academicians, scholars and industry delegates, with the reception of abstracts including more than 1800 authors from different parts of the world. The committee of professionals dedicated toward the conference is striving to achieve a high-quality technical program with tracks on innovative computing, innovative communication network and security, and Internet of things. All the tracks chosen in the conference are interrelated and are very famous among the present-day research community. Therefore, a lot of research is happening in the above-mentioned tracks and their related subareas. As the name of the conference starts with the word "innovation," it has targeted out-of-box ideas, methodologies, applications, expositions, surveys, and presentations, helping to upgrade the current status of research. More than 460 full-length papers have been received, among which the contributions are focused on theoretical, computer simulation-based research, and laboratory-scale experiments. Among these manuscripts, 92 papers have been included in Springer proceedings after a thorough two-stage review and editing process. All the manuscripts submitted to ICICC-2018 were peer-reviewed by at least two independent reviewers, who were provided with a detailed review pro forma. The comments from the reviewers were communicated to the authors, who incorporated the suggestions in their revised manuscripts. The recommendations from the two reviewers were taken into consideration while selecting a manuscript for inclusion in the proceedings. The exhaustiveness of the review process is evident, given the large number of articles received addressing a wide range of research areas. The stringent review process ensured that each published manuscript met the rigorous academic and scientific standards. It is an exalting experience to finally see these elite contributions materialize into two book volumes as ICICC-2018 proceedings by Springer entitled *International Conference on Innovative Computing and Communications*. The accepted articles are broadly classified into two volumes according to the declared research domains. Whereas, the first volume covers networks and cryptography & security, further the second volume covers machine learning, data mining, and soft computing, big data and cloud computing.

ICICC-2018 invited nine keynote speakers, who are eminent researchers in the field of computer science and engineering, from different parts of the world. In addition to the plenary sessions on each day of the conference, five concurrent technical sessions are held every day to assure the oral presentation of around 92 accepted papers. Keynote speakers and session chair(s) for each of the concurrent sessions have been leading researchers from the thematic area of the session. A technical exhibition is held during all the 2 days of the conference, which has put on display the latest technologies, expositions, ideas, and presentations. The delegates were provided with a printed booklet of accepted abstracts and the scheduled program of the conference to quickly browse through the contents. The research part of the conference was organized in a total of 27 special sessions. These special sessions provided the opportunity for researchers conducting research in specific areas to present their results in a more focused environment.

An international conference of such magnitude and release of the ICICC-2018 proceedings by Springer has been the remarkable outcome of the untiring efforts of the entire organizing team. The success of an event undoubtedly involves the painstaking efforts of several contributors at different stages, dictated by their devotion and sincerity. Fortunately, since the beginning of its journey, ICICC-2018 has received support and contributions from every corner. We thank them all who have wished the best for ICICC-2018 and contributed by any means toward its success. The edited proceedings volumes by Springer would not have been possible without the perseverance of all the steering, advisory, and technical program committee members.

All the contributing authors owe thanks from the organizers of ICICC-2018 for their interest and exceptional articles. We would also like to thank the authors of the papers for adhering to the time schedule and for incorporating the review comments. We wish to extend my heartfelt acknowledgment to the authors, peer reviewers, committee members, and production staff whose diligent work put shape to the ICICC-2018 proceedings. We especially want to thank our dedicated team of peer reviewers who volunteered for the arduous and tedious step of quality checking and critique on the submitted manuscripts. We wish to thank my faculty colleagues Mr. Moolchand Sharma and Ms. Prerna Sharma for extending their enormous assistance during the conference. The time spent by them and the midnight oil burnt is greatly appreciated, for which we will ever remain indebted. The management, faculties, and administrative and support staff of the college have always been extending their services whenever needed, for which we remain thankful to them.

Lastly, we would like to thank Springer for accepting our proposal for publishing the ICICC-2018 conference proceedings. We are highly indebted to Mr. Aninda Bose, Acquisition– Senior Editor for the guidance in the entire process.

New Delhi, India   Ashish Khanna
                                                                               Deepak Gupta
                                                                   Organizers, ICICC-2018

About This Book (Volume 2)

International Conference on Innovative Computing and Communication (ICICC-2018) was held on 5–6 May at Guru Nanak Institute of Management, New Delhi. This conference was able to attract a diverse range of engineering practitioners, academicians, scholars, and industry delegates, with the reception of papers including more than 1,800 authors from different parts of the world. Only 92 papers have been accepted and registered with an acceptance ratio of 19% to be published in two volumes of prestigious Springer's Lecture Notes in Networks and Systems series. Volume 2 includes the accepted papers of machine learning, data mining, soft computing, big data, and cloud computing tracks. There are a total of 19 papers from machine learning track, a total of 18 papers from data mining track, and a total of 17 papers from soft computing, big data, and cloud computing track. This volume includes a total of 54 papers from these three tracks.

Contents

Editors and Contributors

About the Editors

Dr. Siddhartha Bhattacharyya [FIEI, FIETE, LFOSI, SMIEEE, SMACM, SMIETI, LMCSI, LMISTE, MIET (UK), MIAENG, MIRSS, MIAASSE, MCSTA, MIDES, MISSIP, MSDIWC] is currently Principal of RCC Institute of Information Technology, Kolkata, India. In addition, he is also serving as Professor of Computer Application and Dean (Research and Development and Academic Affairs) of the institute. He is the recipient of several coveted awards like **Adarsh Vidya Saraswati Rashtriya Puraskar**, **Distinguished HOD Award**, **Distinguished Professor Award**, **Bhartiya Shiksha Ratan Award**, **Best Faculty for Research**, and **Rashtriya Shiksha Gaurav Puraskar**. He received the **Honorary Doctorate Award (D. Litt.)** from the University of South America and the SEARCC International Digital Award **ICT Educator of the Year** in 2017. He has been appointed as **ACM Distinguished Speaker** for the tenure 2018–2020. He is a co-author of 4 books and the co-editor of 16 books and has more than 200 research publications in international journals and conference proceedings to his credit.

Dr. Aboul Ella Hassanien is Founder and Head of the Scientific Research Group in Egypt (SRGE) and Professor of Information Technology in the Faculty of Computers and Information, Cairo University. He is Ex-Dean in the Faculty of Computers and Information, Beni Suef University. He has more than 800 scientific research papers published in prestigious international journals and over 30 books covering such diverse topics as data mining, medical images, intelligent systems, social networks, and smart environment. He has won several awards including the Best Researcher of the Youth Award of Astronomy and Geophysics of the National Research Institute, Academy of Scientific Research, Egypt, in 1990. He was also granted Scientific Excellence Award in Humanities from the University of Kuwait in 2004 and received the superiority of scientific—University Award from the Cairo University in 2013. Also he was honored in Egypt as the best researcher at Cairo University in 2013. He also received the Islamic Educational, Scientific and

Cultural Organization (ISESCO) Prize on Technology (2014) and received the State Award for Excellence in Engineering Sciences in 2015. He was awarded the Medal of Sciences and Arts of the First Class by the President of the Arab Republic of Egypt in 2017.

Dr. Deepak Gupta received his Ph.D. in computer science and engineering from Dr. A. P. J. Abdul Kalam Technical University (AKTU), Master of Engineering (CTA) from University of Delhi, and B.Tech. (IT) from GGSIPU in 2017, 2010, and 2006, respectively. He is Postdoc Fellow in Internet of Things Lab at Inatel, Brazil. He is currently working as Assistant Professor in the Department of Computer Science and Engineering, Maharaja Agrasen Institute of Technology, GGSIPU, Delhi, India. He has published many scientific papers in SCI journals like JOCS, IAJIT, FGCS, IEEE Access, CAEE. In addition, he has authored/edited over 25 books (Elsevier, Springer, IOS Press, Katson) and 7 journal special issues including ASOC and CAEE. His research areas include human–computer interaction, intelligent data analysis, nature-inspired computing, machine learning, and soft computing. He is Convener and Organizer of "ICICC" Springer conference. He has also started a research unit under the banner of "Universal Innovator". He is also associated with various professional bodies like ISTE, IAENG, IACSIT, SCIEI, ICSES, UACEE, Internet Society, SMEI, IAOP, and IAOIP.

Dr. Ashish Khanna is from Maharaja Agrasen Institute of Technology, Delhi, India. He has around 15 years of experience in teaching, entrepreneurship, and R&D with a specialization in computer science engineering subjects. He received his Ph.D. from National Institute of Technology Kurukshetra in 2017. He has completed his M. Tech. in 2009 and B. Tech. from GGSIPU, Delhi, in 2004. He has published around 30 research papers in reputed SCI, Scopus journals, and conferences. He has papers in Springer, Elsevier, IEEE journals. He is co-author of 10 textbooks of engineering courses like distributed systems, Java programming. His research interests include distributed systems and its variants like MANET, FANET, VANET, IoT. He is Postdoc Fellow in Internet of Things Lab at Inatel, Brazil. He is currently working as Assistant Professor in the Department of Computer Science and Engineering, Maharaja Agrasen Institute of Technology, GGSIPU, Delhi, India. He is Convener and Organizer of ICICC-2018 Springer conference. He played a key role in the origination of a reputed publishing house "Bhavya Books" having 250 solution books and around 60 textbooks. He is also part of a research unit under the banner of "Universal Innovator". He also serves as a guest editor in various international journals of Inderscience, IGI Global, Bentham Science, and many more.

Dr. Indrajit Pan is from RCC Institute of Information Technology, Kolkata, India. He did his Bachelor of Engineering (BE) with honors in computer science and engineering from The University of Burdwan in 2005. Later, he completed Master of Technology (M.Tech.) in information technology from Bengal Engineering and Science University, Shibpur, in 2009, and stood first to receive university medal. He obtained Ph.D. in engineering from the Department of Information Technology, Indian Institute of Engineering Science and Technology, Shibpur, in 2015. Presently, he is serving as Associate Professor of Information Technology in RCC Institute of Information Technology, Kolkata. His total teaching service tenure is about 13 years. He has co-authored 3 edited books with reputed international publishers and 30 research publications in various international journals, conferences, and books. He also serves as a guest editor in different international journals of Taylor & Francis and Inderscience Publishers. His present research interests include cloud computing, community detection, and information propagation maximization. He is Professional Member of Institute of Electrical and Electronics Engineers (IEEE), USA; Association for Computing Machinery (ACM), USA; and Computer Society of India (CSI), India.

Contributors

S. M. Adikeshavamurthy Alvas Institute of Engineering and Technology, Mangalore, India

Aastha Aggarwal Department of Computer Science, MAIT, New Delhi, India

Mukul Aggarwal Department of Information Technology, KIET Group of Institutions, Ghaziabad, Uttar Pradesh, India

Shreya Agrawal Netaji Subhas Institute of Technology, Dwarka, New Delhi, Delhi, India

Savita Ahlawat Computer Science and Engineering Department, Maharaja Surajmal Institute of Technology, New Delhi, India

Hussain Ahmad Federal Urdu University of Arts Science and Technology, Islamabad, Pakistan

Udit Ahuja Department of Computer Science and Engineering, BPIT, GGSIPU, New Delhi, India

T. Amudha Department of Computer Applications, Bharathiar University, Coimbatore, India

M. Anjaneyulu Department of Computer Science, Dravidian University, Kuppam, India

P. Apurva Alvas Institute of Engineering and Technology, Mangalore, India

Monika Arora IT Department, Bhagwan Parshuram Institute of Technology, GGSIPU, New Delhi, India

D. R. Arun Kumar Department of Computer Science and Engineering, Sri Venkateshwara College of Engineering, Bengaluru, Karnataka, India

Pooja Asopa Banasthali Vidyapith, Newai, Rajasthan, India

Sneha Asopa Banasthali Vidyapith, Newai, Rajasthan, India

Arvind Bakshi National Institute of Technology Kurukshetra, Kurukshetra, India

Manju Bala Department of CS, IP College for Women, Delhi University, New Delhi, Delhi, India

Snehashish Banerjee Computer Science and Engineering Department, Maharaja Surajmal Institute of Technology, New Delhi, India

Abhay Bansal Department of CS, Amity School of Engineering Technology, Amity University Uttar Pradesh, Noida, Uttar Pradesh, India

Ankita Bansal Information Technology Division, Netaji Subhas Institute of Technology, Dwarka, New Delhi, India

Neha Bansal Department of Information Technology, Indira Gandhi Delhi Technical University for Women, New Delhi, Delhi, India

Vaibhav Batra Computer Science and Engineering Department, Maharaja Surajmal Institute of Technology, New Delhi, India

Neetika Bhandari Indira Gandhi Delhi Technical University for Women, New Delhi, Delhi, India

Hemankur Bhardwaj Department of Computer Science and Engineering, BPIT, GGSIPU, New Delhi, India

Balu Bhasuran DRDO-BU Center for Life Sciences, Bharathiar University, Coimbatore, Tamil Nadu, India

M. P. S. Bhatia NSIT – Delhi University, Dwarka, New Delhi, India

Neelam Bhatt Terna Engineering College, Navi Mumbai, India

Tanmay Bhattacharya Department of Information Technology, Techno India, Salt Lake, Kolkata, India

Priyanka Bhutani USICT, GGS Indraprastha University, Government of NCT of Delhi, Dwarka, New Delhi, India

Harlieen Bindra Department of Computer Science, Bharati Vidyapeeth's College of Engineering, New Delhi, Delhi, India

K. K. Biswas CSE Department, Bennett University, Greater Noida, India

Navya Chandra Delhi Technological University, New Delhi, Delhi, India

Ankur Chaurasia Amity University Uttar Pradesh, Noida, Uttar Pradesh, India

C. Chetan Department of Computer Science and Engineering, Sri Venkateshwara College of Engineering, Bengaluru, Karnataka, India

Vipul Chudasama Department of Computer Science Engineering, Institute of Technology, Nirma University, Ahmedabad, Gujarat, India

Smita Deshmukh IT Department, Mumbai University, Mumbai, India; Terna Engineering College, Navi Mumbai, India

Raghavendra Kumar Dwivedi Department of Information Technology, KIET Group of Institutions, Ghaziabad, Uttar Pradesh, India

Tanisha Gahlawat Amity School of Engineering and Technology, Amity University Uttar Pradesh, Noida, Uttar Pradesh, India

Akshit Garg Department of Computer Science, MAIT, New Delhi, India

Aman K. Garg Computer Science and Engineering Department, Maharaja Surajmal Institute of Technology, New Delhi, India

Suman Garg National Institute of Technology Kurukshetra, Kurukshetra, India

Vishakha Gautam Computer Science Engineering, Bhagwan Parshuram Institute of Technology, GGSIPU, New Delhi, India

Yogesh Gulhane Sipna COET, Amravati, Maharastra, India

Apoorva Gupta Department of Computer Science, MAIT, New Delhi, India

Deepak Gupta Maharaja Agarsen Institute of Technology, Rohini, Delhi, India

Prateek Gupta Department of Computer Science and Engineering, BPIT, GGSIPU, New Delhi, India

Vanyaa Gupta Department of Information Technology, Indira Gandhi Delhi Technical University for Women, New Delhi, Delhi, India

Thet Thet Htwe Cyber Security Research Lab, University of Computer Studies, Yangon, Myanmar

Aniket Jadhav IT Department, Mumbai University, Mumbai, India; Terna Engineering College, Navi Mumbai, India

Juhi Jain Computer Science and Engineering Department, Delhi Technological University, New Delhi, Delhi, India

Rachna Jain Department of Computer Science, Bharati Vidyapeeth's College of Engineering, New Delhi, Delhi, India

Rishabh Jain Maharaja Agarsen Institute of Technology, Rohini, Delhi, India

S. K. Jain National Institute of Technology Kurukshetra, Kurukshetra, India

Saksham Jain Netaji Subhas Institute of Technology, Dwarka, New Delhi, Delhi, India

Suruchi Jain Department of Computer Science & Engineering, Indira Gandhi Delhi Technical University for Women, New Delhi, Delhi, India

C. K. Jha Department of Computer Science, Banasthali Vidyapith, Jaipur, Rajasthan, India

S. K. Jha Division of MPAE, NSIT, Dwarka, New Delhi, India

Rahul Johari USIC&T, GGSIPU, New Delhi, India

Nisheeth Joshi Banasthali Vidyapith, Newai, Rajasthan, India

Parul Kalra Amity School of Engineering and Technology, Amity University Uttar Pradesh, Noida, Uttar Pradesh, India

Sheetika Kapoor Department of Computer Science and Engineering, USICT, GGSIPU, New Delhi, Delhi, India

Neeraj Kaushik Amity University Uttar Pradesh, Noida, Uttar Pradesh, India

Nidhi Kaushik Amity University Uttar Pradesh, Noida, Uttar Pradesh, India

Surendra Kr. Keshari Department of Information Technology, KIET Group of Institutions, Ghaziabad, Uttar Pradesh, India

Vaishali D. Khairnar Terna Engineering College, Navi Mumbai, India

Nang Saing Moon Kham Faculty of Information Science, University of Computer Studies, Yangon, Myanmar

Ashish Khanna Maharaja Agarsen Institute of Technology, Rohini, Delhi, India

Ashima Kukkar Department of Computer Science, Jaypee University of Information Technology, Waknaghat, India

Akshi Kumar Department of Computer Science & Engineering, Delhi Technological University, New Delhi, Delhi, India

Ashwini Kumar School of Computing Science & Engineering, Galgotias University, Greater Noida, Uttar Pradesh, India

Sandeep Kumar Amity University Rajasthan, Jaipur, India

Rajani Kumari JECRC University, Jaipur, India

S. A. Ladhake Sipna COET, Amravati, Maharastra, India

Shubham Lekhwar Amity School of Engineering and Technology, Amity University Uttar Pradesh, Noida, Uttar Pradesh, India

Nagaraj M. Lutimath Department of Computer Science and Engineering, Sri Venkateshwara College of Engineering, Bengaluru, Karnataka, India

Seema Maitrey Department of Computer Science and Engineering, KIET Group of Institutions, Ghaziabad, India

E. M. Malathy SSN College of Engineering, Chennai, Tamil Nadu, India

Pooja Malhotra USIC&T, GGSIPU, Dwarka, India

Ruchika Malhotra Computer Science and Engineering Department, Delhi Technological University, New Delhi, Delhi, India

Sanjay Kumar Malik USIC&T, GGSIPU, Dwarka, India

Iti Mathur Banasthali Vidyapith, Newai, Rajasthan, India

Deepti Mehrotra Department of IT, Amity School of Engineering Technology, Amity University Uttar Pradesh, Noida, Uttar Pradesh, India

Mamta Mittal G. B. Pant Government Engineering College, New Delhi, India

Rajni Mohana Department of Computer Science, Jaypee University of Information Technology, Waknaghat, India

A. V. Muley Division of MPAE, NSIT, Dwarka, New Delhi, India

Vijayalakshmi Muthuswamy College of Engineering, Anna University, Chennai, Chennai, Tamil Nadu, India

S. Nagaprasad Computer Science Department, S.R.R. Govterment Arts & Science College, Karimnagar, India

C. Namrata Mahender Department of Computer Science & IT, Dr. Babasaheb Ambedkar University, Aurangabad, Maharashtra, India

Jeyakumar Natarajan DRDO-BU Center for Life Sciences, Bharathiar University, Coimbatore, Tamil Nadu, India; Data Mining and Text Mining Laboratory, Department of BioInformatics, Bharathiar University, Coimbatore, Tamil Nadu, India

Shashwat Nayak Delhi Technological University, New Delhi, Delhi, India

A. Nayana Department of Computer Applications, Bharathiar University, Coimbatore, India

A. Nayeemulla Khan School of Computer Science and Engineering, Vellore Institute of Technology, Chennai, India

Anand Nayyar Graduate School, Duy Tan University, Da Nang, Vietnam

Anup Pachghare IT Department, Mumbai University, Mumbai, India; Terna Engineering College, Navi Mumbai, India

Payal Pahwa Bhagwan Parshuram Institute of Technology, Rohini, Delhi, India

Pallavi Pandey Indira Gandhi Delhi Technical University for Woman, New Delhi, Delhi, India

Sharad Panigrahi IT Department, Mumbai University, Mumbai, India; Terna Engineering College, Navi Mumbai, India

Arpit Paruthi Netaji Subhas Institute of Technology, Dwarka, New Delhi, Delhi, India

Anoop Kumar Patel National Institute of Technology Kurukshetra, Kurukshetra, India

Aayush Pipal Computer Science Engineering, Northern India Engineering College, GGSIPU, New Delhi, India

K. Prem Chander Department of Computer Science, Dravidian University, Kuppam, India

N. Radha Department of Information Technology, SSN College of Engineering, Chennai, India

Haji Rahman University of Buner, Buner, Pakistan

Sachet Rajbhandari Amity University Uttar Pradesh, Noida, Uttar Pradesh, India

Poonam Rana Department of Computer Science and Engineering, KIET Group of Institutions, Ghaziabad, India

Poonam Rani NSIT – Delhi University, Dwarka, New Delhi, India

B. V. R. Reddy USICT, GGSIPU, New Delhi, Delhi, India

R. Roopalakshmi Alvas Institute of Engineering and Technology, Mangalore, India

Bidisha Roy St. Francis Institute of Technology, University of Mumbai, Mumbai, India

Kalpna Sagar USICT, GGSIPU, New Delhi, Delhi, India

Anju Saha USICT, GGS Indraprastha University, Government of NCT of Delhi, Dwarka, New Delhi, India

Joydeep Saha Computer Science and Engineering Department, Maharaja Surajmal Institute of Technology, New Delhi, India

Suprativ Saha Department of Computer Science and Engineering, Brainware University, Barasat, Kolkata, India

M. S. Sandhya Alvas Institute of Engineering and Technology, Mangalore, India

Saurabh Raj Sangwan Department of Computer Science & Engineering, Delhi Technological University, New Delhi, Delhi, India

S. S. V. N. Sarma CSE Department, Vaagdevi Engineering College, Warangal, India

Archana Saroj Terna Engineering College, Navi Mumbai, India

Ankur Saxena Amity University Uttar Pradesh, Noida, Uttar Pradesh, India

K. R. Seeja Department of Computer Science & Engineering, Indira Gandhi Delhi Technical University for Women, New Delhi, Delhi, India

Krishna Sehgal Department of Computer Science, Bharati Vidyapeeth's College of Engineering, New Delhi, Delhi, India

Asim Kumar Sen Yadavrao Tasgaonkar College of Engineering and Management, University of Mumbai, Mumbai, India

Samridhi Seth USIC&T, GGSIPU, New Delhi, India

Muhammad Nouman Shafique Dongbei University of Finance and Economics, Dalian, China

A. Shahina Department of Information Technology, SSN College of Engineering, Chennai, India

Anjaneyulu Babu Shaik School of Computing Sciences, VISTAS, Chennai, India

Abhilasha Sharma Delhi Technological University, New Delhi, Delhi, India

Arun Sharma Department of Information Technology, Indira Gandhi Delhi Technical University for Women, New Delhi, Delhi, India

Neeraja Sharma Government Girls Inter College, Lalganj, Raebareli, Uttar Pradesh, India

Prerna Sharma Department of Computer Science, MAIT, New Delhi, India

Shraddha Shete Terna Engineering College, Navi Mumbai, India

Abhishek Kumar Singh Division of MPAE, NSIT, Dwarka, New Delhi, India

Archana Singh Amity University Uttar Pradesh, Noida, Uttar Pradesh, India

Umang Soni Netaji Subhas Institute of Technology, Dwarka, New Delhi, Delhi, India

Sujatha Srinivasan Department of Information Technology, School of Computing Sciences, VISTAS, Chennai, India

Awadhesh Kr Srivastava CSE Department, Uttarakhand Technical University, Dehradun, India; IT Department, KIET, Ghaziabad, India

Pankaj Srivastava Motilal Nehru National Institute of Technology, Allahabad, Uttar Pradesh, India

Bhawna Suri Department of Computer Science and Engineering, BPIT, GGSIPU, New Delhi, India

K. Swapnalaxmi Alvas Institute of Engineering and Technology, Mangalore, India

Shweta Taneja Department of Computer Science and Engineering, BPIT, GGSIPU, New Delhi, India

D. K. Tayal IGDTUW University, New Delhi, India

Narina Thakur Department of CS, Amity School of Engineering Technology, Amity University Uttar Pradesh, Noida, Uttar Pradesh, India

Surabhi Thorat Department of Computer Science & IT, Dr. Babasaheb Ambedkar University, Aurangabad, MS, India

Vikas Tripathi CSE Department, GEU, Dehradun, India

Ayush Trivedi Netaji Subhas Institute of Technology, Dwarka, New Delhi, Delhi, India

Vidhika Vasani Department of Computer Science Engineering, Institute of Technology, Nirma University, Ahmedabad, Gujarat, India

P. Vijaya Pal Reddy CSE Department, Matrusri Engineering College, Hyderabad, India

Mansi Yadav Department of Computer Science & Engineering, Indira Gandhi Delhi Technical University for Women, New Delhi, Delhi, India

Pooja Yadav Terna Engineering College, Navi Mumbai, India

An Optimization Technique for Unsupervised Automatic Extractive Bug Report Summarization

Ashima Kukkar and Rajni Mohana

Abstract Bug report summarization provides an outline of the present status of the bug to developers. The reason behind highlighting the solution of individual reported bug is to bring up the most appropriate solution and important data to resolve the bug. This technique basically limits the amount of time that the developer spent in a bug report maintenance activity. The previous researches show that till date the bug report summaries are not up to the developer expectations and they still have to study the whole bug report. So, in order to overcome this downside, bug report summarization method is proposed in light of collection of comments instead of single comment. The informative and phraseness feature are extracted from the bug reports to generate the all possible subsets of summary. These summary subsets are evaluated by Particle Swarm Optimization (PSO) to achieve the best subset. This approach is compared with the existing Bug Report Classifier (BRC) and Email Classifier (EC). For all approaches, the ROUGE score was calculated and compared with three human-generated summaries of 10 bug reports of Rastkar dataset. It was observed that the summary subset evaluated by PSO was more effective and generated less redundant, noise reduction summary and covered all the important points of bug reports due to its semantic base analysis.

Keywords Natural language processing · Feature weighting · Summarization
Unsupervised · Bug report · Particle swarm optimization

A. Kukkar · R. Mohana (✉)
Department of Computer Science, Jaypee University of Information Technology,
Waknaghat, India
e-mail: rajnivimalpaul@gmail.com

A. Kukkar
e-mail: ashi.chd92@gmail.com

© Springer Nature Singapore Pte Ltd. 2019
S. Bhattacharyya et al. (eds.), *International Conference on Innovative Computing and Communications*, Lecture Notes in Networks and Systems 56,
https://doi.org/10.1007/978-981-13-2354-6_1

1

1 Introduction

Other profession individuals have a myth that the software development includes only programming element; in contrast, this line of work has strong element of system information management. The software organization requires the management and formulation of artifacts like designs, source code with documentation, specifications, and bug reports. These bug reports are stored into the software bug repositories. Software bug repositories have extreme information to perform the work on project. Tester's needs to understand the artifacts related to project for recording the issues mentioned by customers and tracks the resolution of bugs. To find the main issue of a bug, a meaningful and huge conversation could happen between developer and reporter by bug reports. So the bug reports have a large amount of reported conversation in the form of messages from multiple peoples and these messages might contain few lines or multiple passages of unstructured text. For example, Mozilla bug #564,243 has 237 sentences and Mozilla bug #491,925 has 91 sentences as comments. To resolve the bug, the developer has to analyze all the comments in bug reports. This process can be tedious, time-consuming, burden, and very frustrating for the tester or developer. Sometimes, the amount of information might be overwhelming and sometimes, this information leads to duplicate, deserted, and non-optimized searches, just because the previous bug reports of the project has been ignored. In [1], the researcher suggested the way to reduce the developer effort and process time, consumed to identify the factual bug report is to provide bug summary. The ideal process of bug summarization is to develop a manually abstract of resolved bug by assigned developer. This summary is used by other developers for better understanding of the bugs. However, the ideal method is very difficult to use in practice because it demands excessive human effort. Therefore, it is need of automatic bug report summarization. The factual summary will save the developer efforts and time and helps in generating up to date summaries of the project on demand. This paper is focused on four challenges. The first challenge is extractive bug report summarization that has huge amount of information as comments and large search space cause NP-complete problem. When working with extensive comments containing a lot of words, sentence selection and sentence scoring become a difficult job and also affect the accuracy and speed of summarization. The second challenge is to increase the ROUGE score by selecting effective semantic text. For example, if there are 20 text lines to summarize and the user wants 20% summarization, then a total of 2^5 subsets are possible of semantic structure; if checked one by one, then this problem goes to intractable. The third challenge is sparsity of data. For example, suppose that one query term is not in the document, then the probability of occurrence of that term is zero to the document. If the probability of query term would be multiplied with the probability of every term of document, then the document would not be retrieved. The fourth challenge is reduction of information. Some high important features are assigned higher weight and low important features are assigned low weight. Feature Selection (FS) provides various heuristics methods which led to non-exhaustive search and reduced high-dimensional space by selecting the features. This drops the information and decreases the accuracy of summary by only selecting the higher weight features.

In such situations, the adoption of meta-heuristic, n-gram, and feature weighting (FW) methods is beneficial in accomplishing the optimal solution to resolve the problem in a productive way. Previously, there are two methods used for textual bug report summarization. First one is supervised learning-based method used by various researchers in [2–5]. The bottom line of this method is as follows: initially bug reports are manually summarized as training set, then text features are extracted from summarized bug reports, and statistical model are trained with these reports. Thereafter, this model is used to determine the features; further, these features are used to predict the manual summaries presented in training set. Later, the features are extracted from new bug report and its summary is predicted by using trained model. Second method is unsupervised learning-based method presented in [6–9]. The basic process of this method is as follows: In this, the centrality and diversity of the sentences are measured in a bug report to select the relevant sentences to put into the summary. In order to create effective, flexible bug report summary, reduce the time, effort of developer, and resolve above four challenges, we proposed an intelligent method for searching the effective semantic text by particle swarm optimization approach with concatenate scores of informativeness and phraseness methods for significant improvement in extractive bug report summarization.

Organization: This paper is divided into five sections: Sect. 2 presents an analysis by producing the summary of existing work related to this paper, Sect. 3 demonstrates the proposed system and algorithm in detail, and Sect. 4 explains the experiments and results thus exhibited along with an analysis of performance parameters. At last, the paper concludes and proposes the future work to be done in Sect. 5.

2 Related Work

Bug report summarization is the well-known research area in software industry, through the summarization dates back to 1958 when Luhn [10] generated the literature abstract. Text summarization process has been practiced in various domains like social media [11], videos [12], news articles [13], and audio [14]. From the past years, summarization process was also used by the incorporated world like Microsoft's Office Suite [15], Bug Triage Process [16], and IBM's Intelligent Miner for Text [17].

With the time, the summarization method is improved from simple term frequency methods to complex machine learning and natural language processing techniques.

Automatic bug report summarization is very difficult job; the vital challenge is how the sentences are picked to generate bug reports summary. To overcome this problem, the summarization methods are classified into supervised and unsupervised learning methods to select the appropriate sentences.

2.1 Supervised Learning Methods

In 2010, Rastkar et al. [1] used existing classifier like Email and Meeting Classifier (EMC), Bug Report Classifier (BRC), and Email Classifier (EC) to generate extractive summary. The human annotators generated the summary of 36 bug reports from (Mozilla, KDE, Eclipse, Gnome), and the bug report classifier performed better than other two classifiers by having <62% precision. Further in 2014, Murphy et al. extended the version of [18] by using the same classifier, to check whether the generated summaries helped in detecting the duplicates bugs. They performed task-based evaluation and found that summaries save the time as well as helped the developers in detecting the duplicates bug without the evidence of decreasing accuracy.

2.2 Unsupervised Learning Methods

The work by Mani et al. in 2012 [19] used four unsupervised learning methods to create the bug report summary. At first, noise reducer is developed to find out the inessential sentences from the reported bug reports after that Diverse Rank, Centroid, Grasshopper, and Maximum Marginal Relevance (MMR) approaches were applied on IBM DB2 bug report. These methods choose the sentences which were central to the bug reports. The author also compared this unsupervised summarization method to supervised method of [1] and found better results. Another work in 2012 by Lotufo et al. [20] modeled user reading process by three hypothetical model and applied heuristics using Markov chain and PageRank method to rank the sentences according to the probability of reading and generate summaries by choosing higher probability sentences. The precision of the model is improved up to 12% as compared to EC in [1] on Mozilla, Launchpad, Chrome and Debian projects. In 2016, Ferreira et al. [21] used Euclidean Distance (ED), Cosine Similarity (CS), PageRank, and Louvain community detection (LCD) algorithms to generate the summary from the comments instead of isolated sentences. It was observed that the ranking of the most appropriate sentences helped the developers in finding the relevant information as compared to manually generated summary of Angular, Bootstrap, and jQuery projects.

3 Proposed Summarization System

This section presents the proposed approach and proposed algorithm to generate the bug report summary.

3.1　Proposed Approach

The overall proposed system process is shown in Fig. 1; further, this is dived into summarization process by using particle swarm optimization. At first, the bug reports (free text) are taken as input, and this text is tokenized. It is then disintegrated into the sets of paragraphs, $D = \{PP1, PP2, PP3, ...\}$. Every paragraph is break down into the number of sentences. $P = \{SS1, SS2, SS3, ...\}$, each sentence contains many terms (features), $S = \{FF1, FF2, FF3, ...\}$. The output of tokenization task is segregated words, paragraph, and sentences. It has three main tasks: stop word removal, rooting extraction, and synonym replacement. The output of this phase is normalized text. The weights of all terms are concatenated to calculate the score of the terms in our system to provide the effective summary. These terms scores are added to sentences to gain the sentence score. After preprocessing, the two features scoring methods are used based on informativeness and phraseness principles [22]: TF-IDF and N-gram method. The bigram, trigram features are extracted using Eqs. 4–7 to preserve the semantics to bug summary, which reduced the sparsity of the text [23]. The TF-IDF features are extracted using Eqs. 1–3 from the normalized data to reflect how much the term provided the information about a bug report [21]. The matrix is considered for every bigram and trigram term. The number of rows and column showed the n-gram words (W) number that are extracted from these bug reports and their probabilities (P). After all the terms are extracted from bug reports dataset, the probability of each term is calculated by Eq. 8. This probability score and TF-IDF score are concatenated to get the combined score of each term. Instead of feature selection (FS), the feature weighting (FW) method is adopted. FW is a generalization from of FS which involves much larger searching space and have more flexibility to assign the continuous relative weight over FS. So these term weights (score) are added to the sentences to gain the sentence score, and the subsets of summary are produced by sentences according to user input, and subset score is calculated for each subset. The subsets of summary are chosen by using Particle Swarm Optimization (PSO) algorithm [24] due to its selecting nature of optimizing searching. It has global and local optimization to find effective semantic text and sentence, which increased the rogue score. This subset act as particles and subset score act as initial position.

Suppose S_i is the produced summary according to user input and has n subsets, $S = \{s_1, s_2, s_3, ..., s_n\}$. These sentences have corresponding weights W_k, $W = \{W_1, W_2, W_3, ..., W_n\}$. By assigning weight to each sentence, WS is produced, $WS = \{W_1S_1, W_2S_2, W_3S_3, ..., W_nS_n\}$. PSO algorithm is described as follows: At first, initialize the each particle's position to subset score. The swarm is updated at every cycle with the best value based on Eqs. 9–10. The position best (pb_i) is taken as the fitness value of summary and the global best (gb) is taken as the fitness value found by swarm best of personal bests. The new fitness value is compared to previous value of pb_i. If new fitness value is surpassed, then the previous value of pb_i is updated to new pb_i value. If the pbi value has not changed for any particle, then new pb_i is compared with previous gb value. If the new pb_i is surpassed than the previous gb, the value of pb_i is selected as new gb. At each iteration, the gb presented the position

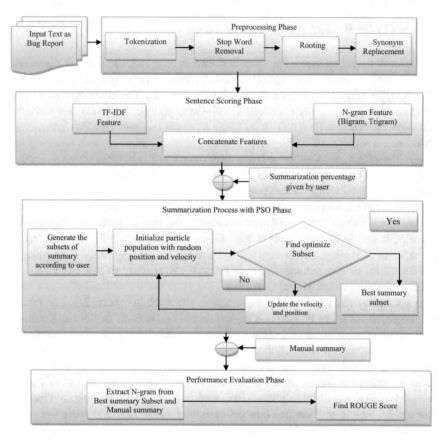

Fig. 1 Overall proposed system for bug report summarization

of the particle and showed a vector for best selected subset of the current bug report. The optimized summary is found after several iterations until the condition is not met. At last, ROUGE score [25] is calculated using Eq. 11.

3.2 Proposed Algorithm

Algorithm1: The summarization Process with PSO

Input: The XML files of bug reports

Output: ROUGE Score of optimized summary of bug reports

Step 1: Input text for preprocessing.

Step 2: Extract the TF-IDF Features

for (i = 1; i <=1;i++){

$$TF(t,D) = \frac{f_D(t)}{\min_{\omega \in D} f_D(\omega)} \tag{1}$$

$$IDF(t,D) = ln\left(\frac{|d|}{|\{D \in D : t \in D\}|}\right) \tag{2}$$

$$TFIDF(t,D,d) = tf(t,D) \cdot idf(t,d) \tag{3}$$

$F_D(t) :=$ frequency of term t in document D

$d :=$ Documment Corpus

}

i is number of feature.

Step 3: Extract the Bigram and Trigram Features

For bigram

$$P(Z_n|Z_0 \dots \dots \dots Z_{n-1}) \approx P(Z_n|Z_{n-1}) \tag{4}$$

$$P(Z_1^k) = \prod_{n=1}^{k} P(Z_n|Z_{n-1}) \tag{5}$$

For trigram

$$P(Z_n|Z_0 \dots \dots \dots Z_{n-1}) \approx P(Z_n|Z_{n-2}Z_{n-1}) \tag{6}$$

$$P(Z_1^k) = \prod_{n=1}^{k} P(Z_n|Z_{n-2}Z_{n-1}) \tag{7}$$

$P(Z_n) :=$ Probability of word n^{th} Z in document d

$Z_{1,2,3\dots k} =$ Word Sequence

Step 4: Concatenate both feature.

$$P_{total}(P_i) = TF - IDF(P_i) + bi - gram(P_i) + tri - gram(P_i) \tag{8}$$

Step 5: Make the subset of Summary produced by features and user input.

$$y = (x_1, x_2, x_3, \dots \dots \dots \dots \dots 2^n)$$

y:= is feature subset.

Step 6: Input these subset y in optimizer PSO as the particles form. Until an optimized subset is met, repeat the following:

Step 7: In PSO model for each particle j in S sentence , do for each dimension d in D

do

//Initialize each particle's position and velocity to random values.

$$X_{j,d} = Rnd(x_{max}, x_{min})$$

$$sv_{j,d} = Rnd(-v_{max}/3, v_{max}/3)$$

end for

//Initialize particle's best position and velocity

$$sv_j(l+1) = sv_j(l) + \gamma 1_i(pb_j - x_j(l)) + \gamma 2_i(gb - x_j(l)) \tag{9}$$

New velocity

$$x_j(l+1) = x_j(l) + sv_j(l+1) \tag{10}$$

Where,

j	Particle Index
1	Discrete time Index
sv_j	Velocity of jth Particle
x_j	Current Position of jth Particle
pb_j	Personal Best Position of jth Particle
gb	Global Best Position
$\gamma 1_i\ and \gamma 2_i$	Random number Between the [0,1]applied to the jth Particle

Step6: Set the global best position to its initial position
$$if f(x_i) > f(pb_i)$$
$$pb_i = x_i$$
// update global best position
$$if f\ (pb_i) > f(gb)$$
$$gb = pb_i$$
else
end if
end for

Step 7: Optimize subset is given by PSO, find ROUGE score.

Step 8:Optimizer summary and Manual Summary Concatenate for extraction of N-gram features

Step 9: Find the Rouge Score by the following formula

$$\frac{No\_of\_Overlapping\_Words}{total\_words\_in\_refernce\_summary} \qquad (11)$$

4 Experimental Result

In this paper, we used the benchmark Rastkar dataset [1, 18]. The ROUGE score is calculating based on three manual summaries of each bug report. Based on the above approach, two experiments are conducted to analyze the approach using ROUGE evaluation metric at 30% summary percentage given by user. If we decrease the summary percentage, then the important information is lost. If we increase the summary percentage, then unwanted data is included in the summary. In this experiment, the bug report summary is generated of every single bug report by using PSO optimization technique. The basic aim of this experiment is to overcome the four challenges which are discussed in the introduction section. This generated summary is compared with three human-generated summary [1, 18], and ROUGE score is calculated as demonstrated in Table 1.

It can be seen from Table 1. The range of ROUGE score based on PSO-generated summary of 10 bug reports was from 72 to 97%. The range of ROUGE score based on BRC and EC-generated summary of 10 bug reports was from 30 to 93% and 51 to 92%. The summary subsets evaluated by PSO were more effective and generated less redundant summary and covered all the important points of bug reports because PSO used n-gram approach for matching which is by default a semantic relation.

Table 1 Comparison of ROUGE score with manual summary using PSO, bug report classifier, and email classifier

Bug report no.	ROUGE score using PSO			ROUGE score using bug report classifier [1, 18]			ROUGE score using email classifier [1, 18]		
	Rouge-1	Rouge-2	Rouge-3	Rouge-1	Rouge-2	Rouge-3	Rouge-1	Rouge-2	Rouge-3
Bug 1	79.14	75.83	75.17	92.03	70.75	33.88	79.88	52.067	51.2
Bug 2	88.54	82.81	80.42	88.36	85.25	34.38	86.47	57.65	55.98
Bug 3	9.33	92.12	89.29	84.64	81.65	33.24	90.86	59.83	56.29
Bug 4	91.84	88.96	80.96	74.53	68.12	30.96	86.46	57.38	53.78
Bug 5	86.60	85.34	82.59	75.34	73.23	36.63	84.80	55.57	54.19
Bug 6	86.23	83.82	82.61	70.56	60.78	45.67	87.46	58.28	55.57
Bug 7	96.65	93.07	86.13	86.28	83.26	35.64	91.58	59.88	56.41
Bug 8	95.65	89.57	86.09	91.87	88.32	34.88	89.19	58.51	56.28
Bug 9	90.34	88.73	84.62	92.27	90.14	35.46	89.51	59.15	57.17
Bug 10	96.00	92.00	88.00	90.93	86.65	34.77	91.81	59.94	57.75

5   Conclusion

It can be concluded that the bug report summarization is very important task for bug triage process. The developer or user reported the report in BST after observing the unexpected behavior of software. With the passage of time, valuable information is accumulated of observed issues. These bug reports are not easy to read. They contained the free text as comments, discussions, and opinions about the bug fixing and bug resolving. So the reading task of these bug reports is time-consuming and tedious. To save the developer time and effort by not reading the entire bug report, need of an Extractive summary generation model of entire bug report. A novel approach is proposed to generate the bug report summary based on collection of comments. The PSO optimization technique is used to evaluate the appropriate summary subset. These subsets are generated by using informative and phraseness feature extraction methods. This method is compared with bug report classifier and email classifier. The summary subset evaluated by PSO was more effective and generated less redundant summary and covered all the important points of bug reports due to its semantic base analysis. In future work, some other optimized techniques will be applied with it to enhance the accuracy of the summarization.

Acknowledgements   We are mightily thankful to researchers Gail C. Murphy, Sarah Rastkar from the University of Columbia for proving the dataset. This helped us a lot in our research [1, 18].

References

1. Rastkar S, Murphy GC, Murray G (May 2010) Summarizing software artifacts: a case study of bug reports. In: Proceedings of the 32nd ACM/IEEE international conference on software engineering-volume 1. ACM, pp 505–514
2. Nomoto T, Matsumoto Y (July 2002) Supervised ranking in open-domain text summarization. In: Proceedings of the 40th annual meeting on association for computational linguistics. Association for Computational Linguistics, pp 465–472
3. Wong KF, Wu M, Li W (August 2008) Extractive summarization using supervised and semi-supervised learning. In: Proceedings of the 22nd international conference on computational linguistics-volume 1. Association for Computational Linguistics, pp 985–992
4. Yue Y, Joachims T (July 2008) Predicting diverse subsets using structural SVMs. In: Proceedings of the 25th international conference on machine learning. ACM, pp 1224–1231
5. Li L, Zhou K, Xue G.R, Zha H, Yu Y (April 2009) Enhancing diversity, coverage and balance for summarization through structure learning. In: Proceedings of the 18th international conference on World wide web. ACM, pp 71–80
6. Mihalcea R, Tarau P (2004) Textrank: bringing order into text. In: Proceedings of the 2004 conference on empirical methods in natural language processing
7. Erkan G, Radev DR (2004) Lexrank: Graph-based lexical centrality as salience in text summarization. J Artif Intell Res 22:457–479
8. Zhai C, Cohen WW, Lafferty J (June 2015) Beyond independent relevance: methods and evaluation metrics for subtopic retrieval. In: ACM SIGIR forum, vol 49, no 1. ACM, pp 2–9
9. Zhang B, Li H, Liu Y, Ji L, Xi W, Fan W, Ma WY (August 2005) Improving web search results using affinity graph. In: Proceedings of the 28th annual international ACM SIGIR conference on Research and development in information retrieval. ACM, pp 504–511

10. Luhn HP (1958) The automatic creation of literature abstracts. IBM J Res Dev 2(2):159–165
11. Lin YR, Sundaram H, Kelliher A (April 2009) Summarization of large scale social network activity. In: IEEE international conference on acoustics, speech and signal processing, 2009 ICASSP 2009. IEEE, pp 3481–3484
12. Han B, Hamm J, Sim J (January 2011) Personalized video summarization with human in the loop. In: IEEE workshop on applications of computer vision (WACV), 2011. IEEE, pp 51–57
13. Radev DR, Blair-Goldensohn S, Zhang Z, Raghavan RS (March 2001) Newsinessence: a system for domain-independent, real-time news clustering and multi-document summarization. In: Proceedings of the first international conference on Human language technology research. Association for Computational Linguistics, pp 1–4
14. Waibel A, Bett M, Metze F, Ries K, Schaaf T, Schultz T, Zechner K (2001) Advances in automatic meeting record creation and access. In: 2001 Proceedings of IEEE international conference on acoustics, speech, and signal processing, 2001 (ICASSP'01), vol 1. IEEE, pp 597–600
15. Microsoft office suite (2012). http://office.microsoft.com/en-us/word-help/automatically-summarize-a-document-HA010255206.aspx
16. Zhang T, Jiang H, Luo X, Chan AT (2016) A literature review of research in bug resolution: tasks, challenges and future directions. Comput J 59(5):741–773
17. Intelligent miner for text (2012). http://www-01.ibm.com/common/ssi/cgi-bin/ssialias? Subtype = ca\&infotype = an\&appname = iSource\&supplier = 897\&letternum = ENUS298-447
18. Rastkar S, Murphy GC, Murray G (2014) Automatic summarization of bug reports. IEEE Trans Softw Eng 40(4):366–380
19. Mani S, Catherine R, Sinha VS, Dubey A (November 2012) Ausum: approach for unsupervised bug report summarization. In: Proceedings of the ACM SIGSOFT 20th international symposium on the foundations of software engineering. ACM, p 11
20. Lotufo R, Malik Z, Czarnecki K Modelling the 'hurried' bug report reading process to summarize bug reports. Empir. Softw. Eng. 20(2):516–548
21. Ferreira I, Cirilo E, Vieira V, Mourao F (September 2016) Bug report summarization: an evaluation of ranking techniques. In: 2016 X Brazilian symposium on software components, architectures and reuse (SBCARS). IEEE, pp 101–110 (2015)
22. Han EHS, Karypis G, Kumar V (April 2001) Text categorization using weight adjusted k-nearest neighbor classification. In: Pacific-Asia conference on knowledge discovery and data mining. Springer, Berlin, Heidelberg, pp 53–65
23. Verberne S, Sappelli M, Hiemstra D, Kraaij W (2016) Evaluation and analysis of term scoring methods for term extraction. Inf Retr J 19(5):510–545
24. Kennedy J (2011) Particle swarm optimization. In: Encyclopedia of machine learning. Springer US, pp 760–766
25. Lin CY (2004) Rouge: a package for automatic evaluation of summaries. Text Summ Branches Out

Prediction of Air Quality Using Time Series Data Mining

Mansi Yadav, Suruchi Jain and K. R. Seeja

Abstract The recent bout of increased air pollution in Delhi has just made the task of identifying and controlling the causes of air pollution an extremely crucial task. In this paper, an Apriori-based association rule mining algorithm, which is a modified version of the Continuous Target Sequential Pattern Discovery (CTSPD), is used to generate a set of association rules that help in predicting the concentration of air pollutants. This algorithm considers the temporal aspect of the data and hence gives the rules with continuous events only as the result. The performance of the algorithm is evaluated by mining the air quality and meteorological data from Anand Vihar, New Delhi, over the period September 1, 2015 to August 31, 2016. The prediction of the proposed algorithm is compared with that of an existing prediction system, SAFAR and found that the proposed algorithm is more accurate than SAFAR.

Keywords Continuous target sequential pattern discovery
Time series data mining · Air pollution prediction

1 Introduction

Time series database consists of data points related to sequence of events measured in repeated time slots. Examples of time series data are the daily closing value of the Sensex and the hourly or daily concentration of air constituents of a particular location. The sequential data mining techniques do not take into account the temporal properties of the data. Many of the modern databases are temporal, which makes the

M. Yadav · S. Jain · K. R. Seeja (✉)
Department of Computer Science & Engineering, Indira Gandhi
Delhi Technical University for Women, New Delhi, Delhi, India
e-mail: seeja@igdtuw.ac.in; krseeja@gmail.com

M. Yadav
e-mail: mymansiyadav7@gmail.com

S. Jain
e-mail: suruchijain0508@gmail.com

© Springer Nature Singapore Pte Ltd. 2019
S. Bhattacharyya et al. (eds.), *International Conference on Innovative Computing and Communications*, Lecture Notes in Networks and Systems 56,
https://doi.org/10.1007/978-981-13-2354-6_2

13

task of studying and developing time series data mining techniques an important and much needed task. Time series data mining identifies time-dependent features from time series databases. These features are used for building predictive models. The algorithms that have been developed to tackle the problem of mining sequential patterns produce sequences that are not required to be continuous. In time series data mining, patterns that represent continuous sequences that imply a target event are identified for prediction. The air quality data is temporal data as the concentration of air pollutants of a particular location in next time frame (hourly or daily) depends on the current concentration of the air pollutant and metrological parameters like humidity, temperature, wind speed, etc.

In literature, both statistical and machine learning techniques [1, 2] are found to be used for air pollution prediction. Kurt et al. [3] proposed a neural network model for predicting of pollution levels of SO_2, PM_{10}, and CO up to 3 days. Statistical techniques [4] using past values of pollutant without considering meteorological conditions were also used for forecasting different pollutants. Peace et al. [5] proposed a prediction model based on wind speed and wind direction to predict CO. Kumar and Goel [6] studied the effect of wind in distribution of pollutants to predict the level of PM_{10}. Another research used Elman network [7] to forecast PM_{10} levels. Zhu et al. [8] proposed a coupled Gaussian model to mine air pollutant patterns at various locations from large quantity of air quality and metrological data. This paper proposes a temporal association rule mining algorithm, modified CTSPD [9], for predicting the air pollutant level.

2 Proposed Methodology

The proposed methodology is outlined in Fig. 1.

2.1 Data Collection

Delhi has always been counted among the most polluted cities in the world. Delhi Pollution Control Committee (DPCC) has sensors mounted at various locations in Delhi to sense and measure the components of the air quality. It posts these observed values along with the meteorological conditions on its website and updates them on an hourly basis. The ambient air quality data for the location Anand Vihar, New Delhi, obtained from DPCC [10] over the period of September 1st, 2015 to September 1st, 2016 has been selected as the dataset.

Fig. 1 Proposed methodology

2.2 Data Preprocessing

The collected data is then preprocessed to make it suitable for our algorithm implementation. The missing values in the collected data were replaced by the most probable values. The dataset is then sorted according to the ranges provided on the SAFAR website for different pollutants so that the continuous data is divided into discrete intervals.

2.3 Data Reduction

After preprocessing, the data is reduced and encoded to facilitate the implementation of the chosen data mining algorithm. The observed values of the air pollutants are divided into different ranges or bins. The bins have been designed in accordance with the air quality index provided by SAFAR [11].

2.4 Pattern Extraction Using CTSPD Algorithm

The proposed modified CTSPD [9] algorithm extracts different air quality patterns in the form of association rules. The algorithm works in two phases—frequent sequences generation phase and rules generation phase.

2.4.1 Frequent Sequences Generation Phase

Scan the database and calculate the Support of an event by counting the number of occurrence of the event. The events with support greater than the user-defined minimum support are frequent events. After finding the frequent events convert the given database to the collection of user sequences. Each sequence S is of the form *basic event → target event*. These are frequent 1-sequences. Then prune the frequent 1-sequences using the Apriori rule and find frequent 2-sequences. If a frequent sequence is followed by an infrequent sequence, there is a discontinuity. Store each continuous sequence in form of an ordered list of p levels, where each level depicts an incoming sequence that is not continuous with the previous sequence.

2.4.2 Rules Generation Phase

During the rule generation, calculate the confidence of each sequence and remove all sequences having confidence less than the user-defined minimum confidence. Then prioritize the obtained rules in the decreasing order of confidence.

3 Experiments and Results

The proposed methodology is implemented in MATLAB. The details of data and various parameter settings are as follows:

- Number of days in the training dataset $= 366$,
- Number of days in the test dataset $= 30$,
- Air Quality Constituents Predicted $= CO$, Ozone, NO_2, $PM_{2.5}$, PM_{10},
- Meteorological Features $=$ Temperature, humidity, and wind speed, and
- Experimental Value of Minimum Support $= 22\%$.

The severity of different pollutants is categorized into Good + Satisfactory, Moderate, Poor, Very Poor, and Severe as shown in Table 1.

The graph in Fig. 2 shows how the number of air quality patterns obtained differs according to the minimum support chosen for formulating the frequent rules.

Few of the air quality patterns obtained when minimum support was set as 22% is given in Table 2.

In order to test the effectiveness of our solution, the rules were applied to the Ambient Air Quality Data recorded by SAFAR-India and the results were compared against the predictions made by SAFAR. The predictions of SAFAR and CTSPD are compared with the actual values and are shown in Figs. 3, 4, 5, 6 and 7.

According to the above graphs, the accuracy of both SAFAR and CTSPD is calculated and is as follows:

Table 1 Severity classification

Description	Good + Satisfactory	Moderate	Poor	Very poor	Severe
CO (ppm)	0–1.7	1.8–8.7	8.7–14.8	14.9–29.7	29.8–40
Ozone (ppb)	0–50	51–84	85–104	105–374	375–450
NO_2 (ppb)	0–43	44–96	97–149	150–213	214–750
$PM_{2.5}$ ($\mu g/m^3$)	0–60	61–90	91–120	121–250	251–350
PM_{10} ($\mu g/m^3$)	0–100	101–250	251–350	351–430	431–550

Fig. 2 Minimum support versus number of patterns

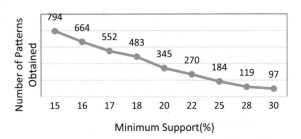

Table 2 Sample air quality patterns obtained

Rule	Support (%)	Confidence (%)
PM_{10} is satisfactory, humidity is moderate → CO (next day) is moderate	25	92
Ozone is satisfactory, NO_2 is moderate, temperature is high → ozone (next day) is satisfactory	26	93
Ozone is satisfactory, NO_2 is moderate, PM_{10} is moderate → NO_2 (next day) is moderate	25	81
$PM_{2.5}$ is satisfactory, temperature is high → $PM_{2.5}$ (next day) is satisfactory	22	85
Ozone is satisfactory, $PM_{2.5}$ is satisfactory, PM_{10} is moderate → $PM_{2.5}$ (next day) is satisfactory	22	83
Ozone is satisfactory, PM_{10} is very poor → PM_{10} (next day) is very poor	30	84
CO is moderate, PM_{10} is satisfactory, humidity is moderate → PM_{10} (next day) is very poor	22	84

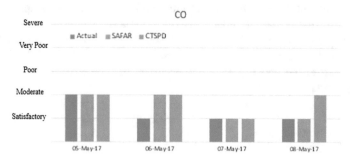

Fig. 3 Prediction of CO

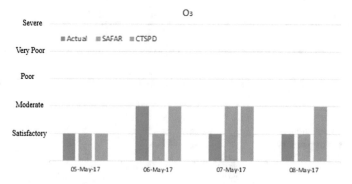

Fig. 4 Prediction of ozone

- Accuracy of SAFAR predictions = 50% and
- Accuracy of CTSPD predictions = 60%.

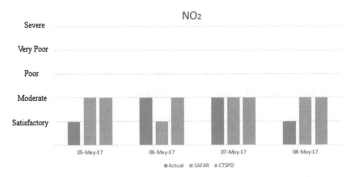

Fig. 5 Predictions of NO₂

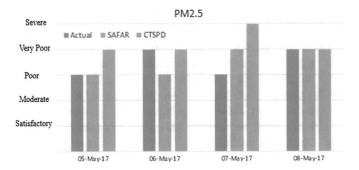

Fig. 6 Predictions of PM$_{2.5}$

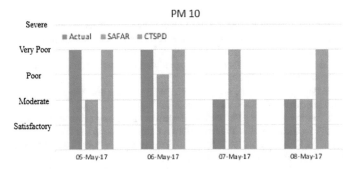

Fig. 7 Predictions of PM$_{10}$

By studying the obtained air quality patterns, it was found that the concentration of a pollutant need not depend on all the other pollutants. Table 3 illustrates the relationship between a pollutant and its affecting factors.

According to the obtained patterns, the concentration of CO and ozone depends on all the other air quality constituents. The concentration of NO$_2$, PM$_{2.5}$, and PM$_{10}$ depends on ozone, PM$_{10}$, temperature, and humidity.

Table 3 Factors affecting the concentration of different pollutants

Pollutant	Factors affecting the pollutant
CO	CO, ozone, NO_2, $PM_{2.5}$, PM_{10}, temperature, humidity, wind speed
Ozone	CO, ozone, NO_2, $PM_{2.5}$, PM_{10}, temperature, humidity, wind speed
NO_2	Ozone, NO_2, PM_{10}, temperature, humidity
$PM_{2.5}$	Ozone, $PM_{2.5}$, PM_{10}, temperature
PM_{10}	CO, Ozone, $PM_{2.5}$, PM_{10}, temperature, humidity

4 Conclusion

This paper proposes an efficient algorithm to predict the concentration of the various air pollutants by using time series data mining techniques. The time series data mining algorithm CTSPD or Continuous Target Sequence Pattern Discovery has been used for the prediction of air pollutants. The predictions made by the proposed solution are compared with the predictions made by SAFAR-India and found that the proposed solution provides more accurate results. By studying the obtained air quality patterns, it was found that the concentration of a pollutant need not depend on all the other pollutants. This problem can be further extended to include the spatial characteristics of the data by using the air quality data of different locations as the data source.

References

1. Athanasiadis IN, Karatzas KD, Mitkas PA (2006) Classification techniques for air quality forecasting. In: Fifth ECAI workshop on binding environmental sciences and artificial intelligence, 17th European conference on artificial intelligence
2. Niharika, Venkatadri M, Rao PS (2014) A survey on air quality forecasting techniques. Int J Comput Sci Inf Technol 5(1):103–107
3. Kurt A, Gulbagci B, Karaca F, Alagha O (2008) An online air pollution forecasting system using neural networks. Environ Int 34(5):592–598
4. Kandya A, Mohan M (2009) Forecasting the urban air quality using various statistical techniques. In: Proceedings of the 7th international conference on urban climate
5. Peace M, Dirks K, Austin G (2005) 5.24 The prediction of air pollution using a site optimized model and mesoscale model wind forecasts
6. Kumar A, Goyal P (2014) Air quality prediction of PM_{10} through an analytical dispersion model for Delhi. Aerosol Air Qual Res 14:1487–1499
7. Ababneh MF, Ala'a O, Btoush MH (2014) PM_{10} forecasting using soft computing techniques. Res J Appl Sci Eng Technol 7(16):3253–3265
8. Zhu JY, Zheng Y, Yi X, Li VO (2016) A Gaussian Bayesian model to identify spatio-temporal causalities for air pollution based on urban big data. In: 2016 IEEE conference on computer communications workshops (INFOCOM WKSHPS). IEEE, pp 3–8
9. Gudes E, Marina L (November 2004) Discovering target event rules based on time-consecutive pattern mining. In: 4th IEEE conference on data mining, Brighton, UK
10. DPCC Real Time Ambient Air Quality Data. http://www.dpccairdata.com/dpccairdata/display/index.php
11. SAFAR India. http://safar.tropmet.res.in/index.php?menu_id=1

Optimisation of C5.0 Using Association Rules and Prediction of Employee Attrition

Harlieen Bindra, Krishna Sehgal and Rachna Jain

Abstract IBM Watson Human Resource Employee Attrition Dataset is analysed to predict the employee attrition based on five selected attributes which are Gender, Distance from Home, Environment Satisfaction, Work Life Balance and Education Field out of 36 variables present in the dataset. Association Rule Algorithm 'Apriori' along with Decision Tree Algorithm 'C5.0' is used. The processing time taken to predict an attrition using the selected attributes using C5.0 with association is 0.02 ms while using traditional C5.0 is 2 ms. RAM consumption for C5.0 with association is 30.89 MB while for traditional C5.0, it is 48 MB. This is a new approach to predict the employee attrition which is better in efficiency than simply applying decision tree algorithms.

Keywords Apriori · Association technique · C5.0 · Data mining · Decision tree
Employee attrition · Entropy · IBM Watson HR · Information gain

1 Introduction

One of the biggest challenges that companies face is employee attrition. There are many factors leading to attrition and in these factors, though everything cannot be controlled, we can try to look into those factors that seem controllable and can help us improve any company's human resource [1]. Factors such as promotions, number of projects, average number of hours spend per month by the employees, salary and job rotation are a few which are easier to manage.

H. Bindra (✉) · K. Sehgal · R. Jain
Department of Computer Science, Bharati Vidyapeeth's College
of Engineering, New Delhi, Delhi, India
e-mail: harlieenbindra@gmail.com

K. Sehgal
e-mail: krishnasehgal2108@gmail.com

R. Jain
e-mail: rachna.jain@bharatividyapeeth.edu

© Springer Nature Singapore Pte Ltd. 2019
S. Bhattacharyya et al. (eds.), *International Conference on Innovative Computing and Communications*, Lecture Notes in Networks and Systems 56,
https://doi.org/10.1007/978-981-13-2354-6_3

21

Being able to extract possible thresholds for the controllable factors, we can understand the factors that are responsible for the premature leaving of employees. We are going to achieve our objective using data mining [2].

2 Literature Survey

In the year 2006, Lee along with other Chia-Chu [3] Chiang studied about frequency of occurrence of a data value in an item set using Apriori algorithm and frequently occurring data value in an item set using Apriori algorithm. They proposed that several implementation techniques involving Apriori Algorithm could be used to find frequently occurring data items. They worked on the improvement of Bodon's work and partitioned the dataset into disjoint partitions and analysed its performance on a parallel computer. Bodon [4] optimized the dataset using trie structure which performed better as compared to hash tree.

In the year 2014, Bashir et al. [5] with other research scholar used ensemble classifiers on two diabetes datasets from UCI repository. They found that ensemble of classifiers gave better results as compared to using single classifiers.

In the year 2014, Dongre and Prajapati [6] gave a holistic view of the Apriori algorithm for finding rules of association and frequent items from transaction dataset. The paper further describes the associative rule mining and its two attributes: support and confidence. The paper uses interesting examples to explain the concept and gives formula for finding support and confidence. It shows how Apriori algorithm is used in associative rule mining.

In the year 2015, Mehta and Shukla [7] used the naïve Bayes algorithm for the optimization of C5.0 classifier. Naïve Bayes is a probabilistic classifier based on the application of naïve (strong) Bayes theorem which assumes that there is strong independence between different features. Intensive studies have been done on this. It is used for text categorization and helps in deciding which document belongs to which category. The main advantage of using this algorithm is that it uses a small dataset for training purpose. It considers each feature to contribute independently regardless of any contribution among them. This algorithm is therefore used with C5.0 for optimization for post pruning of decision tree.

3 Materials and Methods

We have analysed IBM Watson Human Resource Employee Attrition Data (source—Kaggle [8]) set to predict the employee attrition based on five selected bases classes, which are Gender, Distance from Home, Environment Satisfaction, Work Life Balance and Education Field variables out of the set of 36 variables.

Tools used are Microsoft Visual Studio [9] and Microsoft SQL Server [10] on core i7, 7th Generation processor with 16 GB RAM.

3.1 Association Technique

In data mining, association rule [11] is used to find the relationship between different variables that belong to a large database. These relationships are used to render important decisions in medical diagnostics, marketing activities such as promotion and discount, and performance of an employee.

3.1.1 Apriori Algorithm

This algorithm is used to extract item sets that are frequently present in a dataset which is further used to determine association rule to culminate important trends in a database. After finding item sets that occur frequently in database, Apriori [12] algorithm elongates these item sets to make them as long as item sets that occur frequently in database.

3.2 Decision Tree

It is a predictive modelling technique that makes use of historical data to make important decisions within an organization. In a decision tree [13], each node represents a test case and each branch indicates the outcome of test cases. The leaf node represents the decision taken after evaluating all the attributes.

3.2.1 C5.0 Algorithm

It is a data mining tool [14] which is used to make predictions on a dataset, it first discovers pattern in a dataset. Afterwards, it classifies the data items into categories based on which it makes predictions.

Table 1 The improved efficiency when C5.0 algorithm is used along with association rule mining as compared to the traditional C5.0 algorithm

Basis	Improved C5.0 with association	Traditional C5.0
Processing time consumption (ms)	0.02	2
RAM consumption (MB)	30.89	48

4 Proposed Research

The dataset used has been acquired from Kaggle. After acquiring the dataset, we selected base classes for the employee attrition which are: Gender, Distance from Home, Environment Satisfaction, Work Life Balance and Education Field. Then we have cleaned and transformed the dataset according to the selected attributes. During cleaning, we have removed entries that contain redundant values and incomplete tuples.

After transforming the dataset, we have applied C5.0 algorithm to predict employee attrition and compare its efficiency with the new approach.

Then to use our new approach that is C5.0 with association, we have applied association rule mining using Apriori algorithm to form association rules using selected attributes. Then, using these association rules we have trained the C5.0 decision tree. Using this optimized model, we have then predicted the attrition of employee and then matched the predicted results with the actual attrition to evaluate the efficiency of the proposed algorithm on this dataset.

5 Performance Comparison and Result

The dataset used has been acquired from Kaggle. After acquiring the dataset, we are using the improved approach that is C5.0 with association; we have observed that out of 1151 instances 1047 instances were correctly predicted whereas 104 instances were incorrectly predicted.

Therefore, C5.0 with association is more efficient in time and memory consumption as compared to traditional C5.0 algorithm as shown in Table 1 (Figs. 1, 2, 3, 4, 5, 6 and 7).

Hardware Specification:
Processor: Intel Core i5-7200U CPU @2.50 GHZ,
System type: 64-bit Operating System, and
RAM: 8 GB.

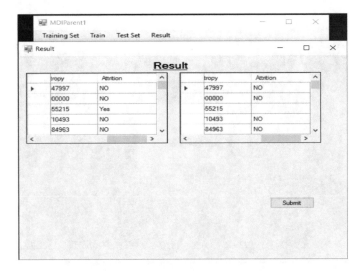

Fig. 1 Result of prediction made by optimized C5.0 algorithm. The grid view on the left of the image shows the actual values of attrition while the grid view on the right shows the predicted values of attrition. Blank cell in the column shows that the prediction was incorrect

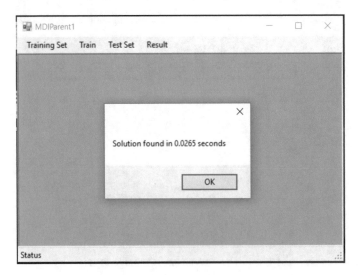

Fig. 2 Processing time consumption by the optimized C5.0 algorithm

6 Future Work

The future work includes the application of CURE clustering [15] for further improvement in the algorithm. CURE clustering, i.e. clustering using representatives, works better than other clustering algorithms with non-spherical shaped clusters and

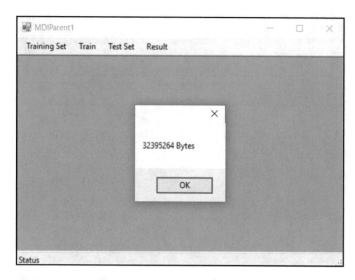

Fig. 3 RAM consumption by the optimized C5.0 algorithm

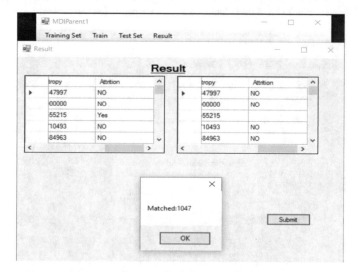

Fig. 4 Correctly predicted instances by the optimized C5.0 algorithm

clusters with variance. It also helps in preventing outliers. It uses hierarchical clustering to avoid the problems of variant clusters. In this algorithm, a constant c of all the points scattered around the cluster is chosen, and the points around the centre are shrunk towards the centroid. The shrunken points towards the centroid are chosen as representative, and these representatives are later merged together if they are close enough. This is how CURE clustering prevents outliers and uses hierarchical clustering.

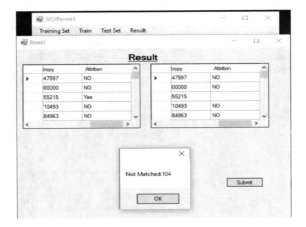

Fig. 5 Incorrectly predicted instances by the optimized C5.0 algorithm

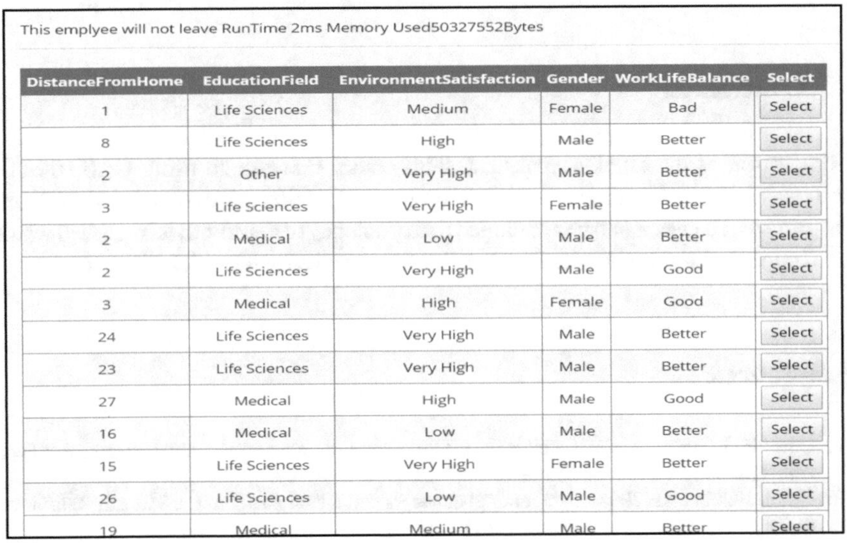

This emplyee will not leave RunTime 2ms Memory Used50327552Bytes

DistanceFromHome	EducationField	EnvironmentSatisfaction	Gender	WorkLifeBalance	Select
1	Life Sciences	Medium	Female	Bad	Select
8	Life Sciences	High	Male	Better	Select
2	Other	Very High	Male	Better	Select
3	Life Sciences	Very High	Female	Better	Select
2	Medical	Low	Male	Better	Select
2	Life Sciences	Very High	Male	Good	Select
3	Medical	High	Female	Good	Select
24	Life Sciences	Very High	Male	Better	Select
23	Life Sciences	Very High	Male	Better	Select
27	Medical	High	Male	Good	Select
16	Medical	Low	Male	Better	Select
15	Life Sciences	Very High	Female	Better	Select
26	Life Sciences	Low	Male	Good	Select
19	Medical	Medium	Male	Better	Select

Fig. 6 Processing time and RAM consumption by the traditional

Fig. 7 Graphical
representation of processing
time and RAM consumption
by the traditional and
optimized algorithm

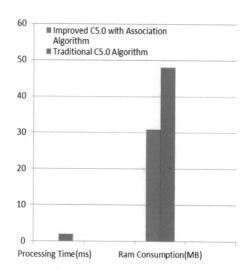

7 Conclusion

Comparison of the efficiency of C5.0 with association rule to using C5.0 without
association shows that time to predict an employee attrition and consumption of RAM
is optimized; hence, it can be concluded that efficiency of algorithm is improved when
C5.0 is used with association rule algorithm as compared to using C5.0 independently.

References

1. Core JE, Holthausen RW, Larcker DF (1998) Corporate governance, chief executive officer
compensation, and firm performance. J Financ Econ 371–406
2. Han J, Kamber M (2006) Data mining: concepts and techniques. In: The Morgan Kaufmann
series in data management systems, 2nd edn
3. Lee Y, Chia-Chu (2006) A parallel apriori algorithm for frequent item-sets mining. In: Proceedings of the fourth international conference on software engineering research. IEEE, p
93
4. Bodon F (2003) A fast apriori implementation. In: Proceedings of the IEEE ICDM workshop
on frequent item-set mining implementations, vol. 90 of CEUR workshop proceedings
5. Bashir S, Qamar U, Khan FH, Younus Javed M (2014) An efficient rule-based classification
of diabetes using ID3, C4.5 & CART ensembles. In: Proceedings of the 2014 IEEE 12th
international conference on frontiers of information technology, pp 226–231
6. Dongre J, Prajapati GL (2014) The role of apriori algorithm for finding the association rules in
data mining. In: Proceedings of international conference on issues and challenges in intelligent
computing techniques (ICICT), pp 657–660
7. Mehta S, Shukla D (2015) Optimization of C5.0 classifier using Bayesian theory. In: IEEE
international conference on computer, communication and control
8. https://www.ibm.com/communities/analytics/watson-analytics-blog/hr-employee-attrition/,
Accessed on 30 Oct 2017

9. https://en.wikipedia.org/wiki/Microsoft_Visual_Studio, Accessed on 30 Oct 17
10. https://en.wikipedia.org/wiki/Microsoft_SQL_Server, Accessed on 30 Oct 17
11. Sethi M, Jindal R (2016) Distributed data association rule mining: tools and techniques. In: 3rd international conference on computing for sustainable global development INDIACom
12. Yuan X (2017) An improved apriori algorithm for mining association rules. In: AIP conference proceedings, vol 1820, issue 1
13. Du W, Zhan Z (2002) Building decision tree classifier on private data. In: CRPIT '14 Proceedings of the IEEE international conference on privacy, security and data mining, vol 14, pp 1–8
14. Bujlow T, Riaz T, Pedersen JM (2012) A method for classification of network traffic based on C5.0 machine learning algorithm. In: International conference on computing, networking and communications (ICNC). IEEE
15. Guha S, Rastogi R, Shim K (2001) Cure: an efficient clustering algorithm for large databases. Inf Syst 26(1):35–58

Stress Analysis Using Speech Signal

Yogesh Gulhane and S. A. Ladhake

Abstract In the area of human–computer interaction, a number of methods have been developed. The recent popular theme is "emotional intelligence". For research, our main target objective is to observe and analyze the effect of emotions on the performance of persons while performing the tasks. In this paper, we are imposing a new approach of stress detection and classification for students during the examination period. We used mel-frequency cepstral coefficients (MFCC) for feature extraction and support vector machine (SVM) classifier for better performance. In this system, three types of corpora have been tested and classified. Support vector machine combines with the rule-based approach with energy and fundamental frequency rules. Indian dataset is created by 50 students, including male and female both. Testing of corpus proved that native area, nationality, and living place have an effect on speech frequencies. At the end of result analysis, we can see that Indians' normal speech frequency is nearly equal to the Mongolian's angry frequency. And as per our target view, the results show that emotions affect the performance at an average rate of 20–30%. That is, if the person is with positive emotions, then his task will achieve 20–30% better result with high speed and opposite to this person with negative emotions will move towards the failure or will get a reduced rate in his performance about the task. The accuracy of the system achieved more than 90% for depressive stress and aggressive stress. The result proved that in the examination period, the performance of students increases in excited state and decreases in a depressive state.

Keywords Classification · Emotion detection · MFCC · Stress · SVM
Performance analyzer

Y. Gulhane (✉) · S. A. Ladhake
Sipna COET, Badnera Road, Amravati, Maharastra, India
e-mail: yogesh1_gulhane@yahoo.co.in

S. A. Ladhake
e-mail: saladhake@gmail.com

© Springer Nature Singapore Pte Ltd. 2019
S. Bhattacharyya et al. (eds.), *International Conference on Innovative Computing
and Communications*, Lecture Notes in Networks and Systems 56,
https://doi.org/10.1007/978-981-13-2354-6_4

31

1 Introduction

Emotions of the living beings are their energy-in-motion. Globalization and digitization are the part of today's life traditional ways are going away peace of mind by YOGA was played an important role to keep balance physical and mental health. Due to competitive view, the youths live in a fast and changing modern life and correlates of stress. Students are facing problems by a rapid change in lifestyle in society and academic challenges in the race of competition. A number of factors are important in daily life for consideration of "Stress of Study". Main factors giving Stress is an increase in competition. From the past decade, researcher communities working in the area of signal processing introduced various approaches related with emotion detection and classification and on the other hand, some Communities bi-forget the emotion and gives the various methods for the problem of stress detection and classification. Those methods were based on facial expression, i.e., image processing based, some was base on brain signal simulation and some was based on voice, i.e., signal processing.

From the various researches, it comes to know that stress or emotion affects not only physically but also psychologically. In the area of communication, it is found that voice or speech is a good effective medium for capturing the any emotions or feelings. These speech signals consist of various parameters and features due to this formulation of the speech signal using digital signal processing can easily possibly for emotion or stress detection and classification. The principal of emotion recognition is depending on the acoustic difference with respective uttering. Emotions and speech are directly propositional to each other. Speech is an effective medium of nonverbal communication and it carries important information about mental as well as the physical condition of the Human or target speaker. The uttering of the speech signals shows the various features that can be classified as the various moods. This paper addresses the problem of stress on the student during the examination period. Researchers define stress as a physical, mental, or emotional response to events that causes body or mental tension. When considering the interpretation of this study, certain limitations should consider. However, there is a lot of scope for research in the same area which could address some of the challenges discussed in the literature review section. According to some research, mental stress can change the physiological balance. This is to contribute in human–machine interaction by enhancing the power efficiency with the help of introducing new approaches motivate me for conducting research on impact of emotion on the performance of students. The aim of the research is to develop a system for the students and exploring dependencies the nature of utterance at the level of negative emotion for classification as depressive stress and aggressive stress.

This paper is organized into five sections, and lastly, we have discussed the results. The first section gives the brief introduction and from the second section (Literature Survey), we are able to find the contribution of different researchers toward this area. Along with this, we can point out the strength and weakness of the different previous of the model. This survey motivates us to contribute data and work to this field. In

Sect. 3 (Proposed Model), we build the model for stress analysis and performance analyzer. The idea of building model is coming from literature review loopholes. Section 4 contains the mathematical formulas and application in our model. Finally, in Sect. 5, we describe the various types of inputs in the form of speech (.wav format). Speeches are collected from the standard data sets. And this input speech has been applied for testing and analysis of our model. At the end, we have shown various results obtained in the observation, which strongly support our targeted goal that "negative emotions (stress)" affects the performance.

2 Literature Survey

For the last few decades, a number of researches and methodologies have been introduced concern with the speech and emotions. Different approaches like pattern matching, template matching, and feature extraction have been used. An important part of the speech processing is the extraction of the features from the speech signal. A lot of feature extraction techniques are available such as filter banking, cepstral technique, and linear predictive coding.

In [1], researchers studied the emotion recognition system for human speeches and they implemented it with segmentation and classifier. The performance was based on hit and false positive rate. The different emotions such as anger, fear, sadness, natural, and happy are taken for classification of the speech by using the utterance segment. They had chosen a simple segmentation method of speech, sentence, or speech to analyze the effect of emotional state. At the end, they observed that the selection of the classifier and the recognizer is most important in case of classification method.

In [2], the fuzzy-based algorithmic approach is used for recognizing the emotions along with this artificial intelligence classification of the speech by using the SVM. They have chosen a multilevel classifier and find a confusion matrix of emotional state 1. At the end, they observed and concluded that vocal signals of the persons are directed.

Influenced by the emotional state of the person, they had analyzed normal, angry, and other emotions. Few researcher communities studied the compact feature vector and the researcher tried to give a better performance for recognition of different emotions that include different states. At the medical side, the researchers used Stroop test and the sensors to recognize the stress. This is helpful to measure that in which pressure distribution sensors are a most feasible sensor to achieve a higher accuracy rate? Artificial neural network was used to classify the feature in many cases. Somewhere used of DWT and MFCC for recoding speech input to get better performance of the classification. Parameters of voice which affect the emotion are stated as voice quality, utterance timing, and utterance pitch contour. In addition to this, troops have been used with approach of fuzzy SVM classification and this is able to classify the stress and relaxation of the patients. ECG was used to compute the performance. Use an SVM and neural Network is mostly popular technique. In [3], feature extraction, data mining, and classification are used. PCA-based classification was proposed.

In [4], the researchers tried to reduce the misclassification rate. For this purpose, they used the MFCC algorithm approach; for the classification, researcher used classifier and supervised learning approach. We found that there is a discussion about deep neural network for eliciting the emotional state in wild. In the result, it has been shown that there we can use the combination of NN and Bayesian process for the recognition of emotional state [5], to detect the emotion. For the performance measurement, they used segment-level output and SVM classifier.

In [6], the author wrote about the nature of the different stages in the education system and their relation with the emotional state of students. He observed that academic or professional circumstances affect on the health of the person, and hence, performance can affect with emotional intelligence.

Recently [7], the author contributes his work respective to controlling multiple emotions in DNN-based speech synthesis. His model was based on information input process. He proved the emotional speech modeling accuracy by adding the particular information to the listener. He used the supervised training model for the implementation. Another researcher in [8] tried to detect the stress in the speech signal using the spectral slope measurement. His method was based on probability density function. In his result, he showed that by using the random forest classifier, he is able to obtain the accuracy of stress analysis up to 92.06%.

Authors of [9] consider speech as a center to represent the features of audio data. On the basis of features obtained, they tried to classify the emotions in the speech. In the result, the researcher stated hierarchical sparse coding (HSC) scheme give the feature classification set of audio signals.

3   Proposed System Model

In this section, we describe the approach toward the detection and classification of stress among the students during the examination period. The prepared dataset had different attributes including energy and f0 excursions. The f0 excursions attribute indicates the emotional state of the person and A and D for aggressive stress and Depressive stress, respectively. Instead of including all of the different human emotions, we have used only one emotion as it can clearly reveal the impact of stress on the student performance. Our algorithm follows some important stages that are shown in Fig. 1, including data capturing, background noise filtering, windowing, framing, normalization, feature extraction, and classification. Data capture that is an input speech signal can be chosen by two ways one either by recorded dataset which has been already stored in system or you can choose real-time input speech. This input is recorded by using a microphone. All voice inputs will get tested in the form of .wav signals.

Then, preprocessing is performed to improve the quality of input speech by using filters. In the filtering process, background noise is eliminated to extract the original

Fig. 1 Block
diagram—stress analysis
with speech signals

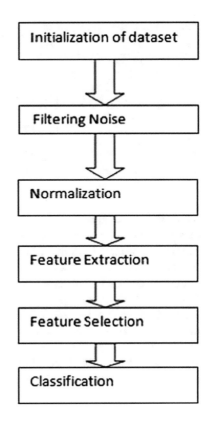

Fig. 1 Block
diagram—stress analysis
with speech signals

speech because the collected emotional data usually get degraded due to external noise (background and "hiss" of the recording machine). This filtered signal may contain a sort of hiss which will affect the process of feature extraction and classification in terms of accuracy. Hence, the normalization process is necessary for the signal processing. Normally, two types of technique of normalization are preferred in case of speech signal processing; they are pitch normalization and energy normalization.

After the normalization process of emotional speech signal, we have divided the segments of the speech containing meaningful units. These small units represent different emotion in a speech signal. The next step is the extraction of relevant features. These emotional speech features can be classified into different categories. Long-term features appear for a long period of time and short-term features those are occurring for short time period characteristics. The formants in the speech appear just once. Hence, it comes in the category of short-term feature similarly pitch, and energy follows the same. The statistical approach is used in case of long-term features for computing digitized speech signal. For the feature extraction, we can extract the meaning, the standard deviation from the set of inputs. As we increase the set of feature extraction for evaluation, the chance of more improvement increases

Table 1 List of some features used to compute classification

Input	Features	Label
Speech in .wav format	Energy	E
Speech in .wav format	Fundamental frequency	Fo
Speech in .wav format	Standard deviation	Sd
Speech in .wav format	Mean	Mo
Speech in .wav format	Minimum speech frequency	Fmin
Speech in .wav format	Maximum frequency in speech	Fmax
Speech in .wav format	Cepstrum	Co

the classification process for the next inputs. For the measurement of frequency of a tone, mel unit is used. We are focusing on negative energy to find out the stress in the speech. Feature Extraction: In this section, the feature is extracted from the speech signal; most relevant feature mean, standard deviation, peak value, mel-frequency, and energy are extracted from the signal. For the spectral analysis of two frequency bands, low-frequency band and high-frequency band are important. We use the discrete wavelet and mel-frequency cepstral coefficients (MFCC) to extract the features. Table 1 shows the list of some features used to compute classification.

Classification: For the classification purpose, SVM classifier is used. To classify the stress from the detected negative energy speech signal. SVM has general optimistic performance and hence it has been forefront in the race of classifiers.

4 Mathematical Modulation

Every Speech signal contains the energy either at high level or at low level. This energy plays an important role in the classification of zero-crossing speech. The energy parameter is calculated using the short-term signal energy, which is defined in Eq. 1.

$$E = \frac{1}{N} \sum_{n}^{N-1} [x(n)]^2 \tag{1}$$

where

N   number of samples represented and
n   particular sample.

There was overlap between two adjacent frames to ensure stationary between frames, Eq. (2)

$$w(n) = 0.54 + 0.46 \left\lfloor (\frac{1}{2}N - n)\frac{2\prod}{N} \right\rfloor \tag{2}$$

Emotions	Number of samples	Age group
Anger	50	19–25 year
Hot anger	50	
Neutrals	50	
Sad	50	
Happy	50	
Fear	50	

Table 2 Indian corpus collection

The important frequency is the fundamental frequency. The fundamental frequency is the "lowest frequency of object"; final calculation of the fundamental frequency is carried out using a simple formula

$$F_0 = \frac{F_s}{K} \qquad (3)$$

Hamming windowing method and formulate the signal with (4). Hamming window is used as window shape just for considering the next block in feature extraction, processing chain and integrates all the closest frequency lines.

$$W(N) = \alpha - \beta \cos \frac{2\pi N}{N-1} \qquad (4)$$

where, $\alpha = 0.54$ and $\beta = 0.46$

Mel-frequency scale transformation by using the equation (Eq. 5). As pitch level is not linear in the speech, mel frequency is used to decide the range of pitch.

$$mel(f) = 259 \log_{10}(1 - \frac{F}{700}) \qquad (5)$$

This $mel(f)$ will obtain in Hz and then further get converted into the time domain using the discrete cosine transform (DCT).

5 Corpus Selections

Indian dataset is formed using 50 students called as the subject of the age group 19–25. Subjects consist of males and females. We recorded the voice using the sound recorder installed in the system and microphones attached to that system. While the data recording we make sure that only two frequencies that is low and high required for classification. Table 2 shows the collected number of samples with respective emotions. Samples are collected from the college students those are preparing for examination before a paper hour. All the student is among (19–25) age group as shown in Table 1.

Table 3 Berlin database count

Emotions	Number of samples
Anger	127
Boredom	81
Disgust	46
Fear	69
Happy	71
Neutral	79
Sad	62

Berlin database of Emotional Speech (EMO-DB): It is collected from the Technical University of Berlin, Whole information regarding Berlin speech dataset of emotions are available on "emodb.bilderbar.info". This database was composed of 10 different actors expressing different emotions. Table 3 shows the emotions and the related number of samples of the Berlin emotional Database.

6   Result and Discussion

For the performance measurement of the stress analysis system, key term was its accuracy of recognition. The analysis is performed against various methods. All the setups and experiments were performed on the same platform of Matlab. Trained and tested databases are used for analyzing the performance accuracy of the stress analysis system. Anger was again classified as aggressive anger and depressive anger. Similarly, sadness was also classified as aggressive sadness and depressive sadness. For the final testing of the system, subjects were asked to identify the stress during the examination period.

In Table 4, three types of emotional impact are shown. In column 1 (normal), we can see that whenever objects (persons) work with natural emotions, then it is not a big deal because the outcome of the work achieved is approximately equal that we expected. But from column 3 (depressive) and column 4 (aggressive), it shows that depression or hypertension affects negatively on the achievements of goals. From Table 4, it is clear that when students are stress free, they appear in normal state. Due to this in normal state, the result gets increased many times. But when a student in depression or in any stress, then they lack their performance.

Table 4 Stress impact on performance

No/type	Normal (%)	Depressive (%)	Aggressive (%)
1	−3	10	5
2	1	5	0
3	0	15	3
4	0	5	5
5	1	6	3
Average	0	8	3

7 Conclusion

Almost all nationalities were having their uniqueness geographically and culturally. It is observed that in case of emotion, vibration, the range of very high frequency is 30–300 MHz with the wavelength 1–10 m which gets captured in aggressive stress, and the range of very low frequency is 3–30 kHz with the wavelength 100–10 km which is captured in the depressive stress. During the testing, it has been seen that for the Mongolian dataset high frequency 1.0 is classified as the aggressive stress in early stage, on the other hand same frequency 1.0 indicates the normal speech in case of Indian dataset. From this observation, we can easily understand that the extent of F0-excursions in speech increases slightly with age, and also geography and culture affect the F0. While comparing the approach with dynamic programming, we obtain the better satisfactory results. Studies have found that stress affects the performance of the student at an average percentage of 10% as compared to normal performance. This system can extend to the medical application in case of physiological testing.

Acknowledgments The authors would like to thank Principal of Sipna College of Engineering Amravati for perception testing for the study.

References

1. Datcu D et al (2009) Multimodel recognition of emotions. Thesis, Datcu uuid:1867c5a9-8043-427e-bc6d-c7a40013c006
2. Bhakteyari K et al. Fuzzy model on human emotion recognition. Recent Adv Electr Comput Eng
3. Mikhail M. Real time emotion detection using EEG. Seminar The American University in Cairo-09
4. Shah R, Hewlett M (2007) Emotion detection from speech. Final projects cs 229 machine learning autumn
5. Lee J, Tashev I (2015) High-level feature representation using recurrent neural network for speech emotion recognition. In: Interspeech
6. Surace L et al (2017) Emotion recognition in the wild using deep neural networks and Bayesian classifiers, arXiv:1709.03820v1 [cs.CV]

7. Lorenzo-Trueba et al (2018) Investigating different representations for modeling and controlling multiple emotions in DNN-based speech synthesis. Speech Commun. 99 (undefined)
8. Simantiraki O et al (2018) Stress detection from speech using spectral slop mesurement. In: Nuria et al (eds) Pervasive computing paradigms for mental health: selected papers from mind care
9. Torres-Boza D et al (2018) Hierarchical sparse coding framework for speech emotion recognition. Speech Commun 99:80–89 © 2018 Elsevier B.V. All rights reserved

Subject-Independent Emotion Detection from EEG Signals Using Deep Neural Network

Pallavi Pandey and K. R. Seeja

Abstract There is a large variation in EEG signals from human to human. Therefore, it is a tough task to create a subject-independent emotion recognition system using EEG. EEG is reliable than facial expression or speech signal to recognize emotions, since this cannot be self-produced. The proposed study aims to develop a subject-independent emotion recognition system with a benchmark database DEAP. In this work, deep neural network with simple architecture is used to classify low–high valence and similarly low–high arousal. EEG signals are nonstationary signals. In this, the stochastic properties as well as spectrum changes over time. For these types of signals, the wavelet transform would be suitable as features, hence, wavelet transform is used to obtain different frequency bands of EEG signals.

Keywords Electroencephalogram · Affective computing · Deep neural network

1 Introduction

Emotion recognition is the task which comes under the field of affective computing. Affective computing relates with the HCI (Human–Computer interaction) which involves the study of computers to process and recognize emotions. Emotions recognized from facial features or speech may not be accurate because they may be fake therefore EEG signals can be a good choice. Since the recording of EEG signal is a noninvasive method, researchers are using EEG to study neural activity of the brain relating to emotional responses. The main challenge in developing subject-independent emotion recognition system is that the EEG varies from person to person. Moreover, on the same stimulus, one person may be less reactive than other.

P. Pandey (✉) · K. R. Seeja
Indira Gandhi Delhi Technical University for Woman, New Delhi, Delhi, India
e-mail: jipallavi@gmail.com

K. R. Seeja
e-mail: seeja@igdtuw.ac.in

© Springer Nature Singapore Pte Ltd. 2019
S. Bhattacharyya et al. (eds.), *International Conference on Innovative Computing and Communications*, Lecture Notes in Networks and Systems 56,
https://doi.org/10.1007/978-981-13-2354-6_5

41

Emotion recognition system is needed for the medical purpose also, as it may be helpful to detect the emotional state of a person and if he is in stress or depress state, a cure may be started. For mental health and behavioral sciences, emotions study is required. Hindustan Times, December 10, 2017, Delhi, quotes an incident in which a boy at the age of 16 years murders his mother as well sister because of some metal state imbalance. These types of cases can be stopped, if emotional states can be monitored and if there is any abnormality, necessary actions could be taken. Emotion extraction process falls into two types of framework: Categorical and dimensional. Paul et al. [1] suggested six basic emotions: Happy, sad, surprise, anger, disgust, and fear. These emotions are called "Basic" because by birth, these emotions are present. In dimensional model, the emotions are modeled on valence arousal space.

2 Related Work

In EEG signal analysis, the main difficulty occurs to obtain a good feature set which can be capable to identify emotions correctly. These signals highly vary from person to person therefore identifying good feature set is a difficult task. EEG features used in the literature can be characterized as statistical features, time domain features, frequency domain features, or both. Fourier transforms (FT), STFT (short time FT), continuous wavelet transform, and discrete wavelet transforms are few signal processing techniques used for EEG signal analysis.

Petrantonakis [2] have used statistical features, band powers of frequency bands, Hjorth parameters, and fractal dimensions as features. They have selected the most relevant features using mRMR method. Classifier used was SVM with RBF(radial basis function) kernel. Zhuang et al. [3] have compared self-induced emotions with movie-induced emotions and reported that self-induced emotions are nearly the same as movie-induced emotions. For six discrete emotions, they have got classification accuracy of 54.52% by using self-induced emotions and 55.65% accuracy by using movie-induced emotions. To classify positive and negative emotions classification, 87.36% accuracy is achieved for self-induced emotions as compared to 87.20 for movie-induced emotions.

3 Proposed Methodology

The process of emotional state recognition from EEG signals consists of several stages. In the first stage, EEG data will be collected. To collect EEG for emotion recognition, the subject will watch a stimulus, wearing an electrode cap in the controlled environment, and the EEG signals will be recorded using some software. Here, the stimuli can be either video or picture. Emotion occurs spontaneously by watching the stimuli. Then, the recorded signals will be preprocessed to remove noises. Since the size of EEG data is very large, feature selection procedures are applied to extract features. Then, the extracted features are fed to a classifier for classifying various emotions like happy, sad, fear, and satisfaction.

3.1 DEAP Database

DEAP database [4] for emotion recognition is used in the proposed work. It is a multimodal database which contains EEG recordings of 32 subjects. Participants have watched 1-min long video and rated it on the scale of valence, arousal, liking/disliking, and dominance. Emotions are modeled on valence arousal scale given by Russell [5] and rated using self-assessment manikin given by Bradley and Lang [6].

Data is collected using electrode cap with 10–20 international electrode placement system and uses 40 electrodes to record various data among which 32 electrodes are there to record EEG and rest are for other peripheral physiological signals. There are 32 .mat files, one for each subject. There are total 40 videos out of which 20 videos data is selected. For each video, this database contains readings for 40 electrodes and each electrode contains 8064 voltage values. That means for one subject data is of the form $40 \times 40 \times 8064$, in which, the first dimension of the data is for video, the second for electrode position, and the third for EEG. From the selected 20 videos for this experimentation, 10 videos falls in the happy quadrant of valence arousal model and other 10 videos corresponds to sad quadrant of valence arousal model. The authors of the database have taken 40 videos corresponding to 4 quadrants. They have already preprocessed the data.

3.2 Wavelet Coefficients as Features

For nonstationary data, the wavelet transform is more suitable because it contains time as well as frequency information both. The discrete wavelet transform is used to find out different coefficients of the signal which corresponds to different frequency bands of EEG. The wavelet function used was db8. The discrete wavelet transform represents the degree of correlation between the signal under analysis and the wavelet function at different instances of time. In the wavelet transform, high-pass filter is designed first for wavelet function and then, low-pass filter is designed for scaling function. Low-pass and high-pass filters should have the same frequency to cover the entire signal. In this, the wavelet provides details and scaling gives trend (average). The size obtained of Gamma band is 1931, Beta band is 973, Alpha band is 494, theta is 254, and Delta is 254.

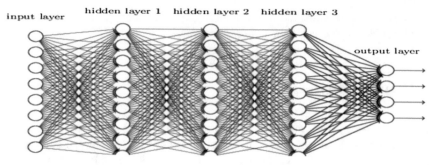

Fig. 1 Deep neural network model

3.3 Classification

Deep neural network (DNN) is selected to classify low/high valence and arousal states. The No. of hidden layers as well as the No. of neurons at each hidden layer may vary. The model used is given in Fig. 1. Optimization algorithm used in the module of DNN utilized for the proposed work is gradient descendent and rectified linear unit (ReLU) function is used as activation function. Rectifier activation function is defined as the maximum of input value or zero as given in Eq. 1.

$$f(n) = \max(0, n) \tag{1}$$

where n is the input to the neuron. The input neurons correspond to the features (wavelet coefficients). The output neurons correspond to the valence/Arousal level (one at a time) whether high or low.

4 Implementation

A quick start module of DNN implemented in Python with tensor flow is used for implementation.

The data unit in the tensor flow is a tensor. It uses Visual C++ as back end. The proposed work is purely subject (user) independent and test data is completely different from train data. The data of those subjects are used for testing which was not used for training. DEAP database used in this research contains EEG signals of 32 subjects. These signals are already downsampled to 128 Hz. At a time, data for all the 32 subjects corresponds to 1 electrode position are considered. Then using wavelets of various frequency bands of EEG are extracted. Theta band of 1 electrode for all 32 subjects is taken as input features. Features for 30 subjects were used to train the classifier and the other 2 subject data are used to test the classifier. For 20 videos and 1 subject, data is of the form 20 × (254 + 494) that is 20 × 748, where

Table 1 Accuracy obtained by combining Alpha of Fp1 and Theta of F4 as features for classifying low–high valence

Number of neurons at three hidden layers	No. of iterations involved	Accuracy (%)
(10, 20, 10)	3400	62.50
(9, 8, 9)	3400	61.25
(9, 8, 9)	3000	60.00

Table 2 Accuracy obtained by combining Alpha of Fp1 and Theta of F4 as features for classifying low–high arousal

Number of neurons at three hidden layers	No. of iterations involved	Accuracy (%)
(10, 20, 10)	3400	64.25
(9, 8, 9)	3400	60
(9, 8, 9)	3000	62.50

Table 3 Performance comparison

Article	Database	DNN classifier accuracy
Zhang et al. [7]	DEAP	57.67
Jirayucharoensak et al. [8]	DEAP	Valence 53, arousal 52
Proposed	DEAP	Valence 62.5, arousal 64.25

254 correspond to theta coefficients obtained from 1 electrode (F4) data and 494 correspond to Alpha coefficients of a single electrode (Fp1). For 30 subjects, data will be of size 600×748 (i.e., $30 \times 20 \times 748$) which is used to train the classifier. So, the training data size with Alpha band of electrode Fp1 and Theta band of electrode F4 were 600×748. The test data was of size 40×748.

From Table 1, it can be seen that the highest accuracy obtained for valence classification is 62.50% with combination of electrode F4 (frontal) of Theta band and Fp1 of Alpha band. Similarly, Table 2 gives the accuracy for arousal classification with the varying number of neurons at three hidden layers and number of iterations.

The result obtained by the proposed method is also compared with that of the other research in this field by using DEAP database and is summarized in Table 3.

5 Conclusion

This paper proposes a subject-independent emotion recognition system using DWT and deep neural network. The previously reported works are either subject dependent or gender specific and mainly used self-created databases. This proposed method is better than the earlier methods as it has used publically available benchmark database which contains data from 32 subjects from both the genders. Moreover, it has employed data from two channels only to recognize valence and arousal. In future, the work may be extended to explore other models of deep neural network like RNN since RNN (Recurrent Neural Network) are best suited for time series data.

Acknowledgements This study uses publically available DEAP database.

References

1. Paul S, Mazumder A, Ghosh P, Tibarewala DN, Vimalarani G (2015) EEG based emotion recognition system using MFDFA as feature extractor. In: International conference on robotics, automation, control and embedded systems (RACE). IEEE, pp 1–5
2. Petrantonakis PC, Hadjileontiadis LJ (2010) Emotion recognition from EEG using higher order crossings. IEEE Trans Inf Technol Biomed 14(2)
3. Zhuang N, Zeng Y, Yang K, Zhang C, Tong L, Yan B (2018) Investigating patterns for self-induced emotion recognition from EEG signals. Sensors 18(3):841
4. Koelstra S, Muhl C, Soleymani M, Lee J, Yazdani A, Ebrahimi T, Pun T, NIjhilt A, Patras I (2012) Deap: a database for emotion analysis; using physiological signals. IEEE Trans Affect Comput 3(1):18–31
5. Russell JA (1980) A circumplex model of affect. J Pers Soc Psychol 39(6):1161–1178
6. Bradley MM, Lang PJ (1994) Measuring emotion: the self-assessment manikin and the semantic differential. J Behav Ther Exp Psychiatry 25(1):49–59
7. Zhang J, Chen M, Zhao S, Hu S, Shi Z, Cao Y (2016) ReliefF-based EEG sensor selection methods for emotion recognition. Sensors 16(10):1558
8. Jirayucharoensak S, Pan-Ngum S, Israsena P (2014) EEG-based emotion recognition using deep learning network with principal component based covariate Shift adaption. World Sci J 2014(627892):10

SoundEx Algorithm Revisited for Indian Language

Vishakha Gautam, Aayush Pipal and Monika Arora

Abstract Writing a noun word with different spellings leads to inconsistency in the database while retrieving the records. This discrepancy occurs because the retrieval of records is done through exact match of the spelling of words. Thus, to improve this, phonetic algorithms were introduced to retrieve the records from the database where matching is done through phonetics of the word instead of spelling match. In this paper, we have tried to implement the SoundEx phonetic algorithm for Hindi language to retrieve the noun words from the database. Here, we have tried to include all the vowels and consonants of Hindi language which were not included in the earlier version of the similar algorithm.

Keywords Phonetic algorithms · SoundEx · Edit distance · Database
Noun words

1 Introduction

Phonetic matching is a technique used for information retrieval by giving indexes to words (generally proper nouns) in the way they are pronounced. It comes into play when people from different cultures and backgrounds come together and want to retrieve data from an enormous database. People have variations in writing styles and

V. Gautam (✉)
Computer Science Engineering, Bhagwan Parshuram Institute of Technology,
GGSIPU, New Delhi, India
e-mail: vishakha1203@gmail.com

A. Pipal
Computer Science Engineering, Northern India Engineering College,
GGSIPU, New Delhi, India
e-mail: aayush.pipal@yahoo.in

M. Arora
IT Department, Bhagwan Parshuram Institute of Technology,
GGSIPU, New Delhi, India
e-mail: monikaarora@bpitindia.com

© Springer Nature Singapore Pte Ltd. 2019
S. Bhattacharyya et al. (eds.), *International Conference on Innovative Computing
and Communications*, Lecture Notes in Networks and Systems 56,
https://doi.org/10.1007/978-981-13-2354-6_6

pronunciation styles for many languages [1], which have the same meaning; thus, phonetic matching resolves this issue by dealing with the pronunciation of given strings and it does not require for the spellings of the strings to match. Phonetic matching is used for information retrieval in India because more than 60% people reside in rural areas and as it is known that in rural areas, people spell and pronounce words in a different and wrong manner. Thus, phonetic matching helps in retrieving the data in such cases.

Many researches have been carried out for information retrieval under data mining for languages like English, German, etc., but in this paper, we have proposed an improved approach for the Indian language—Hindi.

2   Related Work

A lot of interesting researches exist for information retrieval under data mining. The first-ever technique for phonetic matching was given by Robert C. Russell et al. patented in 1918 and 1922 [2] and is known as SoundEx algorithm. Later, many other techniques like Daitch Mokotoff, Metaphone, Double Metaphone, etc. were also given but in this paper SoundEx approach is used. The researches related to this have been finely divided into two categories: The first is working on applications based on SoundEx and the second is finding an accurate SoundEx algorithm for different languages. Our work is mostly related to the work in the latter category.

In [1], the authors have given a SoundEx algorithm for two Indian languages: Gujarati and Hindi, and have shown if two strings which maybe in Gujarati or Hindi are same or not. In [3], the authors have proposed techniques for Hindi and Marathi and they have worked not just on SoundEx but they have also worked on Q-gram and Indic-Phonetic algorithms by developing various states based on length-of-string (LOS) or differ in vowels, for Marathi and Hindi languages and have concluded that Indic-Phonetic approach has more efficacy among the three algorithms used in the paper. In [4], the researchers have found few of the restrictions in using the available approaches for Hindi and Marathi languages and thus they have put forth solutions for phonetic matching. In [5], the authors have given an improved SoundEx algorithm which reflects the similarity between the texts received through SMS and its equivalent text in English or Spanish language. Thus, [5] works on the application of SoundEx algorithm for SMS.

In [6], the researchers have studied the existing phonetic matching algorithms and have created an algorithm and contrasted it with the existing ones. In [7], the authors have given the description of basic algorithms which are a success in dealing with name variations.

In [8], the work gives a combinational phonetic algorithm for developing a Sindhi spell checker. Bhatti et al. [8] give an application for Sindhi language using SoundEx and ShapeEx algorithms. Contrary to the above-stated researches, our work is based completely on Hindi language and we have proposed an improved version of algorithm using SoundEx outlines to detect the similar sounding Hindi names even if they have variations in their spellings.

3 Phonetic Algorithms

Numerous approaches have been proposed for finding phonetic matching of string and they are SoundEx, Caverphone, Kstring, Metaphone coding, Daitch Mokotoff, Double Metaphone, Edit distance, etc. [9, 10].

Daitch Mokotoff and Metaphone are both phonetic matching systems which consider sequences of letters to determine how the name might be pronounced, unlike the original SoundEx. Daitch Mokotoff and Metaphone are used for Eastern European names and English names, respectively [11].

Edit distance is a technique which judges how dissimilar two strings are to one another by counting the minimum number of operations required to convert one string into the other.

The phonetic matching approach used in this paper is SoundEx algorithm because SoundEx being the base algorithm does not generate a code for sequences of letters in order to determine how the name is pronounced instead it gives an efficient and pellucid method of matching the strings unlike the other techniques mentioned above.

SoundEx algorithm [Robert C. Russell and Margaret King Odell, 1918] is one of the phonetic matching techniques which works on string matching based on their pronunciation. The algorithm tells that similar sounding words can be retrieved from a large database by converting each word into a code which comprises the initial letter of the given string followed by three numbers. The numbers are assigned to letters of the given string according to the algorithm until a four-letter code called SoundEx encoding is generated. SoundEx algorithm is a prominent feature of well-known database software, namely, MySQL, SQLite, and Oracle.

4 Proposed Approach: Improved SoundEx Algorithm for Hindi Language

Phonetic matching works on the pronunciation of words rather than their spellings. For understanding the matching function, let us consider a large database which has names in Hindi language like विशाखा, विसाखा, विशाका, विसाका, वैशाली, and विशाकशा.

Let us assume that the name we want to retrieve is विशाखा. The matches that are returned as results are called positives, and the ones that it rejects are called negatives. The positives that are significant are called as true positives, and the rest are known as false positives [11].

Taking an example of the names in the database above, the matches for विशाखा are विशाखा, विशाका are relevant and are the outcomes that we wanted, and thus these

Table 1 SoundEx encoding given in [3]

Code:	0	1	2	3	4	5	6
Letters:	अ, इ, ओ, उ, व, ह	ब, प	च, ग, ज, क, स	ड, ट, त	ल, ळ	म, न	र

are true positives; विशाकशा is not what we wanted, thus it is false positive. From the remaining names, विसाका and विसाखा are the two that we desired as our result too so they are false negatives and वैशाली is the one that we were not interested in so it is a true negative [11].

A lot of researches have been carried out for information retrieval under data mining but our technique uses SoundEx algorithm as the phonetic matching approach for Indian language—Hindi.

Name retrieval has always been an issue in a large database. This particular problem was put to an end by Robert Russell and Margaret King Odell when they proposed SoundEx algorithm and patented it in 1918 and 1922, respectively [2]. SoundEx is the best phonetic matching approach as it is simple. SoundEx is a system that assigns codes to strings on the basis of their sounds and translates the string into an encoding with four alphanumeric characters where the first letter of the string is retained. SoundEx assigns same value to words which sound similar. These values are known as SoundEx encodings. If the four-letter code is not formed, then the remaining code is assigned to zeros on the right side [12]. American SoundEx works well for English language but there are problems still associated with retrieving names from a database having Hindi names.

The purpose behind our approach is to augment the throughput of SoundEx algorithm for Hindi language. SoundEx works well for English language but, in India, spelling alone is not the problem but words are translated from one Indian language to another. For example, in a railway reservation chart, names are written in both English and Hindi. If a person from Bengal sees that the name is translated to some other language, then the sole factor that remains same is its pronunciation. Thus, using SoundEx, similar sounding words can be retrieved from the database, regardless of their spellings.

The encodings for present phonetic matching algorithm proposed by S. Chaware and S. Rao which are used for Hindi language are shown in Table 1, whereas Table 2 shows the encodings for our improved SoundEx for Hindi language.

In our algorithm, we have included all those vowels and consonants also which were absent in the present algorithm. The improved algorithm gives 0–9 encodings covering most of the Hindi vowels and consonants.

Table 2 Proposed improved SoundEx encoding

Codes:	0	1	2	3	4	5	6	7	8	9
Letters:	अ आ इ ई उ ऊ ए ऐ ओ औ अँ अः व य ह	क ख ग घ ङ	च छ ज झ ज़	ट ठ ड ढ ड़	त थ द ध त	न ण म	प फ ब भ फ़	ल	श ष स क्ष	र ऋ

1. The first letter of the string remains same.
2. Change all the occurrences of the following letters to zero: अआइईउऊएऐओऔअँअःवयह
3. Assign the following numbers to the remaining letters (after the first) as follows:
कखगघङ= 1
चछजझज़ = 2
टठडढड़= 3
तथदधत्र= 4
नणम = 5
पफबभफ़= 6
ल=7
शषसक्ष=8
रऋ=9
4. Remove all the pairs of digits with same encoding which occur adjacent to each other from the string that resulted after the above step.
5. Remove all the zeros from the string that occur in the code.
6. The first four characters of the code are given as output and if the string doesn't generate four letter code then zero is assigned on the right side of the code till the four letter code is generated.

Fig. 1 Improved SoundEx for Hindi language

Figure 1 shows the outlines for the improved SoundEx algorithm for Indian language—Hindi. It exhibits the steps required for obtaining the encodings for different strings to check if they are pronounced similarly even if they have slight differences in their spellings or for checking if they match phonetically or not.

The first step of the algorithm demands the first letter of the string to remain same, and steps 2 and 3 require changing the occurrences of the consonants and vowels in the given strings to the codes shown in Fig. 1. Steps 4, 5, and 6 entail that all the pairs of digits with same encoding which occur adjacent to each other in the string that resulted after the above steps to be removed and all the zeros should be removed too; leaving a four-letter code as output. If the string does not generate a four-letter code, then zeros are assigned on the right side of the code until a four-letter code is generated. These steps sum up the SoundEx algorithm.

The flowchart in Fig. 2 manifests the working of the SoundEx algorithm for Hindi language in which we have compared SoundEx codes generated using improved SoundEx algorithm for Hindi language (Fig. 1) and analyzed whether the inputted strings are similar sounding or not.

5 Result

The implementation of SoundEx algorithm for Indian language was carried out in Python programming language. Python 3.6.4 Shell was used for the same. The code for implementing SoundEx algorithm for Hindi language was kept without any complex loops and recursions.

The designed code gives the SoundEx encodings for the strings provided to it and states "The strings are similar sounding" as a result if the SoundEx encodings are same and vice versa if the SoundEx encodings are distinct. The code works as required for the Hindi strings given to it.

The encodings given in both the approaches, shown in Tables 1 and 2, were used to get the results for certain words, and the results obtained are shown in Figs. 3 and 4, respectively. Figure 3 represents the comparison of the inputted strings, and the result shows that by applying the SoundEx codes for Hindi as per algorithm in [3], the similar sounding strings are not showing the correct results and this problem is rectified in our algorithm which is shown in Fig. 4.

Figure 4 shows that the names विशाखा and विसाखा have different spellings but are pronounced similarly and they are true positives and thus have same SoundEx encodings; same is the case for the names आयुष and आयुस, spelled distinctly but pronounced likewise as they are true positives and have same SoundEx encodings. Contrary to the results produced in Fig. 4, the results in Fig. 3 are not acceptable as it shows that the strings विशाखा, विसाखा and आयुष, आयुस are not similar sounding names, which shows that according to [3] they are false positives which is not true in the real world. The results generated by algorithm in [3] produce incorrect results due to the deficiency of few vowels and consonants in their algorithm like ए, अँ, ऋ, ष, and many more letters which are shown in Table 2 but due to the presence of all these letters in our approach, it helped us retrieve correct and appropriate results as displayed in Fig. 4. The addition of the missing vowels and consonants to the algorithm in [3] made it possible for us to refine it and fetch the results that are accurate in the actual world.

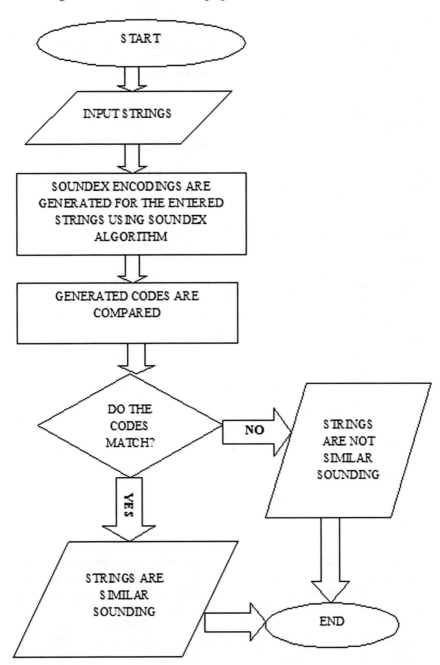

Fig. 2 Flowchart showing the flow of the SoundEx code for Hindi

```
Python 3.6.4 Shell
File  Edit  Shell  Debug  Options  Window  Help
Python 3.6.4 (v3.6.4:d48eceb, Dec 19 2017, 06:04:45) [MSC v.1900 32 bit (Intel)] on win32
Type "copyright", "credits" or "license()" for more information.
>>>
============== RESTART: C:\Users\HP\Desktop\research\marath.py ==============
Enter name 1 = विशाखा
Enter name 2 = विसाखा
NAME              SOUNDEX
विशाखा             व000
विसाखा             व200
 The strings are NOT Similar Sounding.
>>>
============== RESTART: C:\Users\HP\Desktop\research\marath.py ==============
Enter name 1 = आयुष
Enter name 2 = आयुस
NAME              SOUNDEX
आयुष               आ000
आयुस               आ200
 The strings are NOT Similar Sounding.
>>>
```

Fig. 3 Results from SoundEx encodings in [3]

```
Python 3.6.4 Shell
File  Edit  Shell  Debug  Options  Window  Help
Python 3.6.4 (v3.6.4:d48eceb, Dec 19 2017, 06:04:45) [MSC v.1900 32 bit (Intel)] on win32
Type "copyright", "credits" or "license()" for more information.
>>>
================ RESTART: C:\Users\HP\Desktop\research\av.py ================
Enter name 1 = विशाखा
Enter name 2 = विसाखा
NAME              SOUNDEX
विशाखा             व810
विसाखा             व810
The strings are Similar Sounding.
>>>
================ RESTART: C:\Users\HP\Desktop\research\av.py ================
Enter name 1 = आयुस
Enter name 2 = आयुष
NAME              SOUNDEX
आयुस               आ800
आयुष               आ800
The strings are Similar Sounding.
>>> |
```

Fig. 4 Results from the proposed approach

6 Conclusion

Phonetic algorithms play a vital role in the field of information retrieval. These algorithms help in retrieving noun words written with different spellings. The SoundEx, a phonetic algorithm, was designed to retrieve similar sounding noun words in English language. The variation of the same algorithm was used for Hindi language in Analysis of Phonetic Matching Approaches for Indic Language [3] in which they divided the Hindi letters into seven categories (0–6) which does not include all the vowels and consonants like ए, ॲं, ऋ, and ष due to which their algorithm does not yield correct results. Thus, we redefined the SoundEx algorithm for Hindi language which covers all those letters also which are not used for providing SoundEx codes in [3]. As it is shown in Figs. 3 and 4 that the words विशाखा, विसाखा, and आयुष, आयुस are similar sounding although they have different spellings, these same words result as different soundings when implemented using the approach in [3] as some of the letters used in these words are not defined in the algorithm given by Chaware and Srikanth Rao [3].

Therefore, concluding this paper was initialized with the objective of upgrading the available primitive SoundEx algorithm for Hindi language. Considering the results and the implementation, the proposed approach was successfully programmed, hence, resulting in an efficient and effective measure to use SoundEx for Hindi language for retrieving similar sounding names explicitly from the database. Talking about the standards of efficiency and effectiveness of the approach, it is observed that the upgraded version has successfully yielded appropriate results as compared to the primitive version.

In future, the SoundEx algorithm for Hindi language proposed in our approach can be used to develop an application in this era of Digitalizing India. The application having Hindi datasets can help people of various states in India to access the data in Hindi language and consequently removing the barrier known as "language" for the people living in rural areas.

Acknowledgements Aayush Pipal and Vishakha Gautam would like to thank their parents and their entire families for their constant motivation and unending support.

References

1. Shah R, Singh DK (2014) Improvement of Soundex algorithm for Indian language based on phonetic matching. Int J Comput Sci Eng Appl (Ijcsea) 4(3)
2. Beider A, Morse SP Phonetic matching: an alternative to Soundex with fewer false hits
3. Chaware S, Srikanthrao (2012) Analysis of phonetic matching approaches for Indic languages. Int J Adv Res Comput Commun Eng 1(2)
4. Chaware S, Srikanthrao (2012) Evaluation of phonetic matching approaches for Hindi and Marathi: information retrieval. Int J Adv Eng Technol
5. Pinto D, Vilariño D, Alemán Y, Gómez H, Loya N, Jiménez-Salazar H The Soundex phonetic algorithm revisited for SMS text representation. In: Lecture notes in computer science book series, vol 7499. LNCS
6. Parmar VP, Kumbharana CK (2014) Study existing various phonetic algorithms and designing and development of a working model for the new developed algorithm and comparison by implementing it with existing algorithm(s). Int J Comput Appl (0975–8887) 98(19)
7. Chaudhary A, Wakchoure N, Gotarne N, Nath P, Dhakulkar B (2016) A comparative study on name matching algorithms. Int J Res Advent Technol 4(5)
8. Bhatti Z, Waqas A, Ismaili IA, Hakro DN, Soomro WJ (2014) Phonetic based Soundex & Shapeex algorithms for Sindhi spell checker system. Aensi-Aeb 8(4):1147–1155
9. Zobel J, Dart P (1996) Phonetic string matching: lessons from information retrieval. In: Proceedings of the 19th SIGIR on research and development in information retrieval, pp 166–172
10. Name and Address Matching Strategy, White Paper, December 2007
11. Beider A, Morse SP (2010) Phonetic matching: a better Soundex
12. Hettiarachchi GP, Attygalle D (2012) SPARCL: an improved approach for matching Sinhalese words and names in record clustering and linkage. IEEE, Colombo

Sentiment Analysis and Feature Extraction Using Rule-Based Model (RBM)

Raghavendra Kumar Dwivedi, Mukul Aggarwal, Surendra Kr. Keshari and Ashwini Kumar

Abstract In the current era, each and every individual get benefited with available information on Internet that is used in decision-making by perceiving others' attitude, opinions, sentiments, and emotions. As we know sentiment analysis is one of the fields of natural language processing (NLP) that evaluates opinion of users and their sentiments, based on these sentiments and opinions, an individual or group can review their product and services. In this paper, we compared our rule-based model (RBM) sentiment lexicon features precision, recall, and F1 score to other known and established sentiment lexicons on Cornell movie review dataset and found the result is better and more accurate.

Keywords Sentiment analysis · Opinion mining · NLP
Rule-based model (RBM) · Positive (Pos) and negative (Neg) reviews
Support vector machine (SVM)

1 Introduction

It has been observed that there is an open challenge in research work to classify and analyze data with the objective to assign predefined category labels based upon

R. K. Dwivedi (✉) · M. Aggarwal · S. Kr. Keshari
Department of Information Technology, KIET Group of Institutions, Ghaziabad, Uttar Pradesh, India
e-mail: raghavendra.dwivedi@gmail.com

M. Aggarwal
e-mail: mukul.digital@gmail.com

S. Kr. Keshari
e-mail: surendra.keshari@gmail.com

A. Kumar
School of Computing Science & Engineering, Galgotias University, Greater Noida, Uttar Pradesh, India
e-mail: ashwinipaul@gmail.com

© Springer Nature Singapore Pte Ltd. 2019
S. Bhattacharyya et al. (eds.), *International Conference on Innovative Computing and Communications*, Lecture Notes in Networks and Systems 56,
https://doi.org/10.1007/978-981-13-2354-6_7

learned models [1]. As we know that anyone can convey their opinion in simple or complex way since each and every individual is prejudiced with area they work, region they live, and various interpersonal attributes. Many people use slang, sarcasm, irony, and ambiguous pattern in their language [2]. Sentiment analysis and classification are performed to check the attitudes of a service or product, person using, and then we compare both the service and product recommended by one user to another, and fetch positive or negative review on same topic. Nowadays, many brands have been working on their services and products rating and enhance user's views that are recommended to their new set of customers. It also helps to enhance customer relationship with company, in retention and serve customers to get their best product or services. Ping and Lee emphasize on subjectivity summarization as sentiment education [3], while Thomas et al. work on practical approach to analyze opposition supports on floor debate using sentiment analysis [4].

1.1   Statement of the Problem

Sentiment analysis is a subset of natural language processing and data mining. Whenever users visit a website to purchase any product, first they look at previous reviews for the same category a product belongs to. Summary of these set of reviews sets the opinion of a buyer toward that product. However, such approaches to classify sentiment in documents have not been efficient. This is due to the fact that text classification involves automatically sorting a set of documents into categories from a predefined set. Sentiment analysis sometimes treated beyond relevant classification of texts to count finding opinions and classifying them as positive or negative, and favorable or not favorable. There is need for a classification tool or a system that can classify text sentiments with higher accuracy as simple text classification techniques are not well enough to recognize hidden parameters. Requirement of sentiment analysis is being increased due to the application of sentiment analysis in diverse areas such as market research, business intelligence, public relations, electronic governance, web search, and email filtering. Machine learning algorithm has been grateful tool to sort out commercial challenges in current market scenario [5, 6].

1.2   The Process of Sentiment Analysis and Classification

The preprocessing step involves extracting the reviews from a source dataset. Terms in each review are parsed by a part-of-speech (POS) tagger such as the Stanford POS tagger [7]. The tags are then used by the lexical resource such as SentiWordNet and LIWC to determine the sentiment score for each item. The term in each document, together with their sentiment scores, is stored as feature vectors to be used as input to text classifier. The classification step involves using a text classifier which selects minimal features as either positive or negative.

1.3 Factors that Affect Sentiment Analysis and Classification

To classify whole text documents as positive or negative is more challenging as compared to individual words in review process. Let us consider a review: "This film should be brilliant." Words like "good", "best", or "excellent" depict a positive sense while in reality these set of words do not give a guarantee to produce positive review. However, authors Pang and Lee prove that the presence of human-generated keywords in a phrase is less accurate than machine learning technique-generated keywords [8]. The main factors that affect how opinions are analyzed and classified include the domain of the datasets, the size of the datasets to be analyzed, the format of the datasets (labeled or unlabeled), and quality of the dataset. Sentiment analysis has domain-specific challenges like same word or sentence of one domain may have different meanings or different sentiments in other domain. A positive sentiment for a book review section "go read the book" is negative for movie review. Similarly, review such as "it is so easy to predict the next action" is a positive sentiment for a political review but has negative remark for a movie plot. Depending on the size of datasets, it is always best to use a combination of manual and automatic approaches as it has been realized from a number of experiments that even the worst results obtained from using both approaches are superior to either approaches. Quality control on web is a measured concern in the current era to find out whether it is original review or duplicate or just statement to harm any product/services. There is no quality check on reviews of services or product. However, few brands keep their secure moderator's group to crosscheck not only reviews but also user's behavior.

2 Related Work

There are various techniques and algorithms like machine learning, part-of-speech tagging, n-gram features, and document-level classification which are used to classify sentiments. Pang and Lee applied machine learning techniques to classify movie reviews according to sentiment [8]. Baccianella Stefano et al. emphasize over SentiWordNet 3.0 where they work on lexical resource for sentiment analysis and opinion mining [9]. Mullen and Collier have worked in the area of part-of-speech and sentiment orientation values using pointwise mutual information. Mullen Collier et al. propose support vector machine (SVM) in combination with other hybrid models to classify their datasets [10]. Ohana et al. use SentiWordNet database to count polarity score. Their work depicted counting positive and negative term scores to decide sentiment orientation. Authors also built a dataset of relevant features and design an algorithm to detect negation [11]. Dave et al. focused on a document-level opinion classifier to synthesize product reviews using POS tagging approach and statistical techniques. They checked the average of positive and negative reviews, and a review will be referred to others only when it will be positive [12]. White et al. worked on appraisal groups where they analyze set of words in partial, full positive, or full negative category [13]. This work showed that using large feature vectors in combination with feature reduction, linear SVMs can be trained to achieve high classification accuracy on data that present classification challenges for a human annotator. All the literature and its various areas with different parameters found fit to be discussed in the section that helps us to get significant improvement in our work. The dataset taken from Cornell found suitable for sentiment classification and feature extraction consisting around 2,000 text files with positive (Pos) reviews and negative (Neg) reviews where we can apply our approach rule-based model (RBM) and compared with other existing approaches in next section.

3 Results and Discussion

3.1 Datasets

The Cornell movie review dataset and the Convote dataset were used for our experiments. The movie Cornell's review dataset is a popular dataset of sentiment classification that consists of 2000 processed text files with positive (Pos) reviews and negative (Neg) reviews. Usually, file gets the name of category they belong to. File with positive review starts with suffix "POS", while negative files are recognized with "NEG". A document which has POS(10) and NEG(10) represents the number of files existing with both sentiments (Fig. 1).

There are three segments: the development set, training set, and test set. However, the training set was used since it has majority of the data including 1,200 positive and negative documents.

Fig. 1 Division of datasets

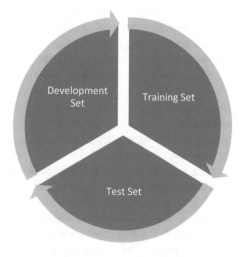

3.2 Result

There are following terms needed to be discussed for the analysis of result. Precision: Precision has significant role in classifier performance in any statistical analysis. It is a number of item labeled classes that are divided by sum of numbers labeled elements of a class including both correct and incorrect classifications. Recall: It is a value of true classification that comes after division of total classification of a class. Low Recall: Low recall depicts that how many known numbers are missing from the class. F1 score: F1 score is calculated through precision and recall and it shows the accuracy of the sentences [3]. Accuracy refers to how exactly the classifier classifies each document. Recall measures the completeness, or sensitivity, of a classifier. As per our assessment, our results are better in all three parameters: precision, recall, and F1 score. We compared our rule-based model (RBM) sentiment lexicon features (Precision, recall, and F1 score) to other known and established sentiment lexicons and found the analysis in Fig. 2:

1. Linguistic inquiry word count (LIWC),
2. General inquirer (GI),
3. Affective norms for English words (ANEW),
4. SentiWordNet (SWN),
5. SenticNet (SCN),
6. Word-sense disambiguation (WSD), and
7. Hu-Liu04 opinion lexicon.

Our rule-based model (RBM) is overwhelming on other lexicon features for precision, recall, and F1 score illustrated in Fig. 2. Rule-based model (RBM): After several research discussions, we reached on state to design a rule-based modeling tool to compute sentiment analysis. It is useful to generalize various domains and text styling on social media. It can work without training dataset and constructed through human curetted gold standard sentiment lexicon.

As we know handling streaming data is one of challenges to maintain performance of the system, our approach is well enough to handle these challenges. It is

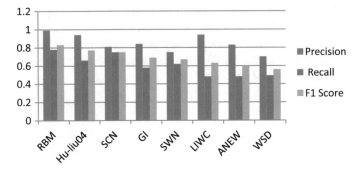

Fig. 2 Illustration of comparative approaches with rule-based model (RBM)

(a)   **(b)**

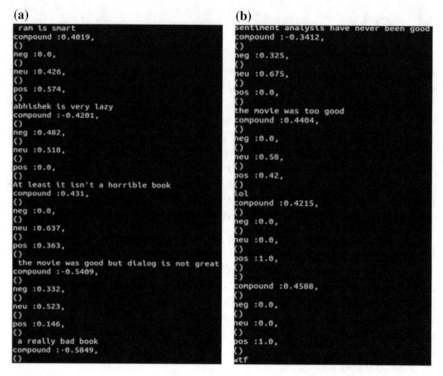

Fig. 3 **a** Sentences tagged with positive, negative, neutral, and compound. **b** Tricky sentences tagged with positive, negative, neutral, and compound

also secure enough from tradeoff of speed performance. Comprehensive sentiment lexicon, created by user, is rigorous process and has the possibilities of having error. Therefore, usually researchers depend on existing lexicons as primary resources.

(1) Rule-based model (RBM) does not require complex computation and other advanced system and specification. RBM is more accurate than other approaches.

(2) The lexicon and its rules can be easily accessed by the user. It is not hidden within a machine access. It is easily inspected, understood, extended, or modified.

(3) Since this approach does not require any training dataset, that is why, it is able to work in various domains. We focused on general sentiment lexicon that is related to grammar and syntax (Fig. 3).

4 Conclusion

As we know sentiment lexicons help to measure the precision, recall, and F1 score on datasets, Cornell's review dataset is a popular dataset consisting of enough processed text files with positive (Pos) reviews and negative (Neg) reviews. In our work, we found that rule-based modeling gives better result as compared to other seven sentiment lexicons discussed in the paper. We worked on five general rules that represent syntactical and grammatical reunion for lexical feature to get sentiment intensity. Based on these five rules, our approach rule-based model (RBM) yields the precision (0.99), recall (0.78), and F1 score (0.83) that overwhelm other existing approach's values. The results are quite remarkable to make rule-based model (RBM) highly regarded as sentiment analysis and feature extraction tool.

References

1. Pang B, Lee L (2008) Opinion mining and sentiment analysis. Found Trends Inf Retr 2(1–2): 1–135
2. Prasad S (2010) Micro-blogging sentiment analysis using Bayesian classification methods
3. Pang B, Lee L (2014) A sentimental education: sentiment analysis using subjectivity summarization based on minimum cuts. In: Proceedings of the Association for Computational Linguistics (ACL), pp 271–278
4. Thomas M, Pang B, Lee L (2006) Get out the vote: determining support or opposition from congressional floor-debate transcripts. In: Proceedings of EMNLP, pp 327–335
5. Liu B (2010) Sentiment analysis and subjectivity. In: Indurkhya N, Damerau FJ (eds) Handbook of natural language processing, 2nd edn
6. Yessenalina A, Yue Y, Cardie C (2010) Multi-level structured models for document-level sentiment classification. In: Conference on empirical methods in natural language processing, pp 1046–1056
7. Ho T, Cheung D, Liu H (2010) Advances in knowledge discovery and data mining. In: 9th Pacific-Asia conference on knowledge discovery and data mining, pp 301–311
8. Pang B, Lee L (2002) Thumps up? sentiment classification using machine learning techniques. In: Proceedings of the conference on empirical methods in natural language processing (EMNLP), pp 79–86
9. Stefano B, Esuli A, Sebastiani F (2010) SentiWordNet 3.0: an enhanced lexical resource for sentiment analysis and opinion mining. In: Seventh conference on international language resources and evaluation (LREC '10), pp 2200–2204
10. Mullen T, Collier N (2010) Sentiment analysis using support vector machines with diverse information sources. In: Proceedings of EMNLP-2004, Barcelona, Spain, July 2004. Association for Computational Linguistics, pp 412–418
11. Ohana B, Tierney B (2009) Sentiment classification of reviews using SentiWordNet. In: IT&T Conference
12. Dave K, Lawrence S, Pennock DM (2003) Mining the peanut gallery: opinion extraction and semantic classification of product reviews. In: Proceedings of WWW2003, Budapest Hungery, pp 519–528
13. Whitelaw C, Garg N, Argamon S (2010) Using appraisal groups for sentiment analysis. In: Proceedings of the 14th ACM conference on information and knowledge management, pp 625–631

The Role of Big Data Predictive Analytics Acceptance and Radio Frequency Identification Acceptance in Supply Chain Performance

Muhammad Nouman Shafique, Haji Rahman and Hussain Ahmad

Abstract In recent years, organizations are extracting knowledge from the huge volume of data to predict future trends. Specific applications have been developed for big data predictive analytics to utilize the current data in different industries. The efficiency of big data can be enhanced through the use of radio frequency identification (RFID) technique in supply chain management (SCM). The objective of this study is to establish and empirically investigate the relationship among big data predictive analytics (BDPA) acceptance, RFID acceptance, and supply chain performance (SCP). The population of this study is logistics industry in China. Results showed the positive direct effect between BDPA acceptance and SCP, and RFID acceptance has partially mediated. The implementation of this study will enhance supply chain performance in the logistics industry. This study also fills the literature gap because previous studies have not established the relationship between big data analytics acceptance and RFID acceptance in SCM.

Keywords Supply chain management (SCM)
Big data predictive analytics (BDPA) acceptance · Big data · Logistics
Radio frequency identification (RFID) acceptance
Supply chain performance (SCP)

M. N. Shafique (✉)
Dongbei University of Finance and Economics, Dalian, China
e-mail: Shafique.nouman@gmail.com

H. Rahman
University of Buner, Buner, Pakistan
e-mail: Haji616@yahoo.com

H. Ahmad
Federal Urdu University of Arts Science and Technology, Islamabad, Pakistan
e-mail: Hussainahmad251977@gmail.com

© Springer Nature Singapore Pte Ltd. 2019
S. Bhattacharyya et al. (eds.), *International Conference on Innovative Computing and Communications*, Lecture Notes in Networks and Systems 56,
https://doi.org/10.1007/978-981-13-2354-6_8

1   Introduction

In recent years, information communication technology became the essential part of life for everyone. The usage of smartphones, laptops, and other portable devices made the life more convenient, smooth, and systematic. In contrast, the usage of the Internet and electronic devices is collecting the variety and velocity of large data [1]. So, it is challenging situation for organizations to organize, utilize, and predict from big data to enhance their performance.

In the current scenario, distributive logistics companies in China are using their own customized logistic applications, and they are collecting data every day from whole channel. It is enormous in volume, and velocity of data is increasing regularly [1]. Big data is generated by their customers and employees. But, unfortunately, most of the times this big data is useless, because organizations are not using this big data to predict future trends.

In China, most of the logistic organizations are also using the barcode scanner for delivery and inventory purposes. Barcode scanner is the old technology, because it cannot track the products in real time and it has slow processing. In China, Taobao shopping required 4 days to deliver products. In addition, customers are unaware of the exact location of their products. They just got information for few central places, which were updated through barcode scanners. So customers need to know the precise location of their products; on the other hand, organizations need to track their inventory in real time. So they need advanced technologies to enhance their SCP.

This study has focused on utilization of big data through BDPA and RFID in SCM. Nevertheless, software's and soft techniques cannot work efficiently without hardware. So, BDPA needs to implement new hardware technology, RFID, to enhance logistics performance. This study has focused on the practical and operational gap in the logistics industry. Furthermore, this study fills the literature gap, because this study has established the relationship among BDPA acceptance, RFID acceptance, and SCP, which was not developed and tested in literature. So, this study opens the new horizons for academicians and management to the field of SCM. So, the research question is *"How the acceptance of technological resources will enhance SCP?"*

2   Literature Review

Big data is the massive volume of data; it has three main characteristics of variety, velocity, and size. The wise decision for organizations is to utilize and analyze big data to predict future. Big data predictive analytics is the systematic process to capture, store, organize, use, transfer, and visualize data to predict future trends [1]. So, organizations can improve their SCP through the visualization of big data.

BDPA in SCM is a powerful tool to enhance the buyer and supplier relationship to boost the performance and efficiency of SCM [2]. BDPA can be used to predict the future, and increase productivity and operational performance in SCM through the use of ongoing processes and existing customers. So, big data predictive analytics has the positive effect on supply chain performance [3].

RFID technology is a fast-growing technology used in logistics to increase the SCP. The adoption of RFID has been well elaborated in the literature [4]. RFID technology will reduce the operating cost, track inventory, reduce lead time, increase accuracy, efficiency, and effectiveness and strengthen buyer and seller relationship through the exchange of information, which will enhance SCP [4].

RFID acceptance can increase SCP because RFID technology can track the inventory, reduce operating costs, and enhance the efficiency of logistics, which will lead to SCP [5]. Moreover, the implementation of RFID technology in SCM has many benefits for organizations such as saving costs, decreasing time, maintain inventory level, track products, and lessen misplacing and theft of products to enhance supply chain performance [6].

Resource-based theory (RBT) provided the foundation for this study, because BDPA acceptance and RFID acceptance are organizational resources and capabilities, which will enhance SCP. RBT has suggested that organizations utilize their resources and skills to improve their performance through competitive advantage [7]. Because organizational resources are valuable (V), rare (R), imitable (I), and organized (O) according to VRIO model, it can enhance organizational performance. The relationship between RBT and organizational performance has been established in the literature [7, 8]. Literature review provides the direction of the following hypotheses:

H_1: BDPA acceptance has positive effect on SCP.
H_2: RFID acceptance has mediating effect between BDPA acceptance and SCP.

3 Methodology

The methodology is the set of all activities performed during research. It is based on population, sampling, data collection and data analysis techniques, and usage of software.

3.1 Sample and Data Collection Procedure

Logistic industry in China is the population for this study. Moreover, the sample was selected through simple random sampling method from 200 employees from logistic distribution companies from Dalian. The reason for the sampling is the nature of the study. Distribution logistics companies are directly associated with supply chain throughout China. Consequently, these companies are collecting data in huge volume every day. So, they can also deal with RFID technology.

Data has been collected through survey method based on the adopted questionnaire from September 2017 to November 2017. Totally, 200 questionnaires have been distributed among samples for data collection, but total 155 questionnaires are wholly filled for further analysis. The survey response rate is 77%.

3.2 Research Instrument

The instruments have been adopted from previous studies. BDPA has been measured through three adopted items [9]. Moreover, SCP has been measured through eleven adopted items [10]. Additionally, RFID technology acceptance is measured through four adopted items [11]. Likert scale, the five-point anchor, is used in this study.

3.3 Measurement Model

Reliability and validity are two main components of measurement modeling. In this study, reliability is measured through Crona Bech Alpha and composite reliability [12], whereas validity is measured through content validity, face validity, convergent validity, and discriminant validity [12].

Reliability is the consistency of items [13]. In this study, both individual and composite reliabilities are measured. The individual reliability is measured through Crona Bech Alpha (α) value; the minimum accepted value is 0.6 [14]. While the collective reliability is measured through composite reliability (C.R.), the threshold value of composite reliability is 0.6 [15]. Both Crona Bech Alpha (α) and composite reliability (C.R.) values are mentioned in Table 1, which shows data is reliable for further analysis.

Validity is measured by face validity, content validity, convergent validity, and discriminant validity. Face validity and content validity can be checked through the logic of the content and grammar. Convergent validity shows that how much items are theoretically correlated with each other. All these validities can be measured through factor loadings. If items have more than 0.6, factor loading, then they are valid [15]. The values of factor loadings are mentioned in Fig. 1, which shows items

Table 1 Mean, standard deviation, reliability, correlation, and square root of AVE

Constructs	Mean	S.D.	Reliability	Correlation matrix		
			α/C.R.	(1)	(2)	(3)
1. BDPA	1.68	0.662	0.807/0.886	**0.849**		
2. RFID	1.80	0.729	0.881/0.918	0.741**	**0.858**	
3. SCP	2.03	0.818	0.954/0.961	0.691**	0.751**	**0.831**

** Significant at 0.01, bold values in diagonal are the square root of AVE

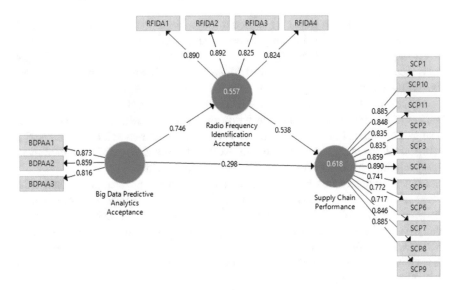

Fig. 1 Path analysis of SCP

are valid. On the other hand, discriminant validity shows how factors are different from each other. It can be measured through the square root of average variance extract (AVE). If the square root of AVE is greater than correlation values, then items have discriminant validity. The values are mentioned in Table 1, which shows items are discriminant valid [12, 15].

4 Results

The hypotheses have tested through structural modeling. The structural model can be assessed through path coefficient, explanatory factor (R^2), and their significant values in partial least square—structural equation modeling (PLS-SEM) [16]. The structural model has been tested through the use of Smart PLS 3. The significant values can be measured through bootstrapping method; minimum bootstrapping is 500. In this study, 5000 bootstrapping samples have been taken to more generalize the study. BDPA acceptance has the positive effect on SCP. This relationship has been accepted through statistical values ($R^2 = 0.298$, t statistics $= 2.7$, and p-value $= 0.02$) to prove this relationship. Besides, RFID acceptance has been mediating effect between BDPA acceptance and SCP. This indirect relationship has also tested statistically. The relationship between BDPA and RFID acceptance shows ($R^2 = 0.746$, t statistics $= 12.630$ and p-value $= 0.000$), while the relationship between RFID and SCP has been analyzed ($R^2 = 0.538$, t statistics $= 4.462$, and p-value $= 0.000$). So, the results have accepted the direct and indirect both relationships.

The mediation effect has also checked through the access of variance accounting for (VAF). It is the indirect method to use the mediation. The results from indirect effect and total effect have been noted. Furthermore, these results were divided to get VAF. The value of VAF is 57 which is found between 20 and 80. The results show that RFID acceptance has partially mediated between BDPA acceptance and SCP.

The results also show that software is useless without hardware, and hardware is useless without software. RFID is the hardware technology which can give the better results if organizations also concentrate on improving their soft skills of BDPA. So, if organizations are good enough in their soft and hardware capabilities, then there are maximum chances to develop their operational activities.

Logistic industry in China has the suitable environment to improve their SCP because every logistic industry is using the barcode and barcode scanner which is the previous type of RFID. So, they just need to improve their technology to track every packet in real time. Second, every organization has collected data from their customers, and they just need to analyze the data to see the customers' trends and make their packages according to customers' exact needs. So, they can use their big data efficiently to improve their SCP.

The overall model fitness has been analyzed through PLS-SEM model fit indexes. In PLS-SEM model, fit can be tested through the standardized root mean square residual (SRMR) and normed fit index (NFI). The minimum acceptable value for SRMR must be less than 0.08 [17, 18]; in this study, the value SRMR is 0.073, which is less than 0.08. So, it is accepted as model fit. On the other hand, the value of NFI is between 0 and 1. So, the maximum value shows more model fits. In this study, the value of NFI is 0.891, which is near to 1. So, it shows the goodness and overall model fit. Thus, the considered conceptual model is overall fit and shows the goodness of the model.

5 Conclusion and Discussion

This study has developed the relationship among BPDA acceptance, RFID acceptance, and SCP in logistic industry. In the previous studies, the relationship between RFID and SCP has been established and tested [11]. If organizations use RFID technology, then their SCP will be increased. The results from this study, consistent with previous studies, show that the use of RFID technology will enhance SCP in the logistics industry in China.

Big data is playing a significant role in the development of organizations and their performance. The use of BDPA will also increase the SCP. In previous studies, the positive effect of big data on supply chain performance has been established and tested [3]. The results of this study support the previous reviews because BPDA acceptance has the positive effect on SCP.

In this study, RFID acceptance has been taken as mediating variable in the relationship between BDPA acceptance and SCP. This mediating relationship has not been tested in previous studies. The results of this study show that if organizations use BDPA and RFID together, then their performance will be increased.

This study has the theoretical and practical contributions in the field of SCM. This study has integrated soft and hard resources of organizations to boost SCP. There is the literature gap in the relationship among BDPA acceptance, RFID acceptance, and SCP. So, this study has developed and tested this relationship, which will provide new directions for future researchers.

This study has focused only on logistics industry in China. However, for generalization, this study can be tested in other developing or underdeveloped countries. Moreover, this study can also be implemented in other industries like health, medical, and chemical industries for more generalization. This study is consisting of only big data and RFID technology to enhance SCP, while other factors like regulatory, packaging, customer relationship, and inventory management have been ignored in this study. In future studies, these factors can be considered to enhance SCP.

References

1. Duan L, Xiong Y (2015) Big data analytics and business analytics. J Manage Anal 2:1–21
2. Motamarri S, Motamarri S, Akter S, Akter S, Yanamandram V, Yanamandram V (2017) Does big data analytics influence frontline employees in services marketing? Bus Process Manage J 23:623–644
3. Gunasekaran A, Papadopoulos T, Dubey R, Wamba SF, Childe SJ, Hazen B, Akter S (2017) Big data and predictive analytics for supply chain and organizational performance. J Bus Res 70:308–317
4. White A, Johnson M, Wilson H (2008) RFID in the supply chain: lessons from European early adopters. Int J Phys Distrib Logist Manage 38:88–107
5. Pramatari K (2007) Collaborative supply chain practices and evolving technological approaches. Supply Chain Manage Int J 12:210–220
6. Veeramani D, Tang J, Gutierrez A (2008) A framework for assessing the value of RFID implementation by tier-one suppliers to major retailers. J Theor Appl Electron Commer Res 3
7. Barney J (1991) Firm resources and sustained competitive advantage. J Manage 17:99–120
8. Brandon-Jones E, Squire B, Autry CW, Petersen KJ (2014) A contingent resource-based perspective of supply chain resilience and robustness. J Supply Chain Manage 50:55–73
9. Hazen BT, Overstreet RE, Cegielski CG (2012) Supply chain innovation diffusion: going beyond adoption. Int J Logist Manage 23:119–134
10. Dwayne Whitten G, Green KW Jr, Zelbst PJ (2012) Triple-A supply chain performance. Int J Oper Prod Manage 32:28–48
11. Kros JF, Glenn Richey R, Chen H, Nadler SS (2011) Technology emergence between mandate and acceptance: an exploratory examination of RFID. Int J Phys Distrib Logist Manage 41:697–716
12. Chin WW (1998) The partial least squares approach to structural equation modeling. Modern Methods Bus Res 295:295–336
13. Henseler J, Ringle CM, Sinkovics RR (2009) The use of partial least squares path modeling in international marketing. In: New challenges to international marketing, pp 277–319. Emerald Group Publishing Limited
14. Nunnally JC (1967) Psychometric theory
15. Fornell C, Larcker DF (1981) Evaluating structural equation models with unobservable variables and measurement error. J Mark Res 39–50

16. Chin WW (2010) How to write up and report PLS analyses. In: Handbook of partial least squares, pp 655–690
17. Hu L-T, Bentler PM (1999) Cutoff criteria for fit indexes in covariance structure analysis: conventional criteria versus new alternatives. Struct Equ Model Multidiscip J 6:1–55
18. Hu L-T, Bentler PM (1998) Fit indices in covariance structure modeling: sensitivity to under-parameterized model misspecification. Psychol Methods 3:424

Peering Through the Fog: An Inter-fog Communication Approach for Computing Environment

Bhawna Suri, Shweta Taneja, Hemankur Bhardwaj, Prateek Gupta and Udit Ahuja

Abstract The advent of cloud computing has brought forth a wave of IoT devices and the number of devices connected to the cloud has been growing expeditiously. With this increase in demand, there is an imminent need for a more efficient solution for managing the IoT devices—Fog computing. Fog computing brings the concept of cloud computing closer to the edge of the network. These fog devices may be interconnected across, in addition to a hierarchical connection. In this paper, we discuss the advantage of installing this auxiliary connection, i.e., linking of adjacent fog nodes at the same level. We have proposed an architecture that would provide a low latency alternative to traditional fog network connection. An algorithm is defined to support the proposed architecture which is further stimulated by the case of a parking scenario. The experiment is conducted, and the results have proved that our proposed architecture shows lower latency as compared to the traditional fog architecture.

Keywords Cloud computing · Fog computing · Inter-fog communication · IoT

B. Suri · S. Taneja · H. Bhardwaj (✉) · P. Gupta · U. Ahuja
Department of Computer Science and Engineering, BPIT, GGSIPU, New Delhi, India
e-mail: hemankur@hotmail.com

B. Suri
e-mail: bhawnasuri@bpitindia.com

S. Taneja
e-mail: shweta.madhur21@gmail.com

P. Gupta
e-mail: umang.2497@gmail.com

U. Ahuja
e-mail: udahuja1996.ua@gmail.com

© Springer Nature Singapore Pte Ltd. 2019
S. Bhattacharyya et al. (eds.), *International Conference on Innovative Computing and Communications*, Lecture Notes in Networks and Systems 56,
https://doi.org/10.1007/978-981-13-2354-6_9

1 Introduction

Internet of Things (IoT) device has been dominating the Internet for the better part
of the century. Ever since the dawn of cloud computing, IoT devices have been
increasing rapidly. Estimates suggest that the number of such devices will reach 8.4
billion in 2017, which is 31% more than 2016 [1]. Cloud computing is a concept that
provides the users with a shared pool of resources hosted remotely over the Internet
to store and process data rather than deploying individual hardware and software
for its users [2]. Cloud services are broadly categorized into Software as a Service
(SaaS), Platform as a Service (PaaS), and Information as a Service (IaaS) [3]. There
is a need for a solution to manage the increasing amount of data while maintaining
integrity and security. Fog computing, also known as fogging or fog networking, is a
term coined by Cisco Systems, Inc. [4–6]. It is an extension of the cloud computing
paradigm to the edge of the network [7]. It can be used to take the vehicular cloud
further by providing a lower latency than cloud networks [8].

The devices used for fog communication are known as fog nodes [9]. Alternatively,
these fog nodes can be connected to an intermediate or aggregated node instead of
being directly connected to the cloud. Addition of this extra layer can be particularly
useful in large network systems. A simple fog network architecture consisting of an
aggregated node can be seen in Fig. 1.

In order to properly utilize the available resources, the data of different types are
handled differently. The fog node in the vicinity of the point of origin of data processes
the data which is the most time-sensitive and requires an immediate response. The
aggregated node can be used to manage fog devices connected to it and perform
extensive calculations when necessary. The data which is the least time-sensitive
such as archaic data which is stored in the data warehouse is further sent to the
cloud. Developers can write IoT applications which can direct this data to their

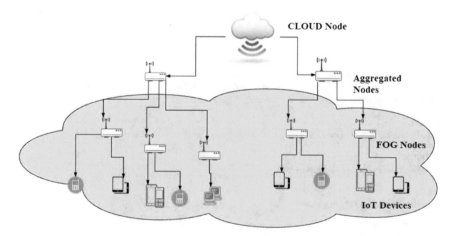

Fig. 1 Fog network with aggregated node

respective place. In order to reduce the response time, we have proposed an inter-fog network architecture in Sect. 3 that uses an auxiliary connection to connect adjacent fog nodes. To the best of our knowledge, no existing study focuses on the merits of using an auxiliary connection in a fog network.

This paper is organized as follows. Section 2 discusses work and studies related to our work. Section 3 shows our proposed work and architecture. Section 4 follows up with a case study that is used for simulating our proposed work and the algorithm. Section 5 contains the results of the simulation, and conclusion is drawn in Sect. 6.

2 Related Work

Cloud computing and its applications are growing by the minute and with the inflating amount of data, fog computing is becoming salient, now more than ever. There are many existing applications of fog computing. IOx by Cisco is one of them. Cisco IOx can be used to build, develop, and monitor applications designed for the fog networking paradigm. IOx is Cisco's implementation of fog computing. The applications on the fog node can be managed with the help of Cisco application hosting framework (CAF) [10]. The CAF can be used to host applications built using the IOx SDK.

Paradrop [11] is another implementation of fog computing that replaces traditional Wi-Fi routers and brings computing resources closer to the edge. Paradrop allows third-party developers to access these resources. This framework can reduce the dependency of the edge devices on communication with the cloud and improves performance and data privacy.

A study on solving the parking problem in an urban setting is done using cloud and fog environment [12]. The study highlights a fog computing-based approach to provide a reliable solution for finding a parking spot. In the study, the fog nodes delivered parking data to their respective RSUs (Roadside Units) which is like an aggregated node. These nodes communicate with the fog nodes to find the optimal parking spot. For parking scenarios, a prediction model can also be used [13]. This model can help find an optimal parking space. However, there was no direct communication between the fog nodes.

An approach toward minimizing the delay in IoT applications is seen in the model based on load sharing [14]. The paper illustrates an IoT-Fog-Cloud model based on an offloading approach to better reduce the delay in such a network. To achieve this, an analytical model has been presented and a framework is given. The framework is based on fog nodes collaborating with each other to handle the requests from the IoT layer. These fog nodes are connected to each other, and they offload the tasks after finding the best neighbor. In our proposed architecture, we suggest connecting the nodes adjacent to each other in a proper sequence as opposed to an ad hoc connection. This connection is further explained in Sects. 3 and 4.

This study reflects upon the merits of using the shortest path. Our work provides a low latency alternative to the existing paradigm. This study also focuses on the merits of using this auxiliary connection. Previous studies have focused on improving the efficiency of fog networks by changing factors such as the scheduling algorithm. Our proposed architecture can be combined with such factors to augment the results. With the help of this auxiliary connection, we aim to reduce the latency in fog environments. To relate and understand the effects, we further discuss a similar case study of a parking scenario.

3 Proposed Work

In the current scenario for any particular task, the communication between two fog devices takes places with the help of the cloud. Data that needs to be sent from one fog node to the other passes through the cloud. This leads to an increase in the number of hops required to send a packet back and forth between the fog nodes, which is directly proportional to the overall latency in the application. Therefore, the more are the hops, the higher is the latency. Thus, it reduces the performance of the network. In this paper, we propose and highlight the advantages of an auxiliary connection that links adjacent fog nodes, located at the same level, to each other. The proposed architecture can be seen in Fig. 2. In the figure, each fog node at level 1 has been connected to their adjacent fog device in addition to their aggregated node.

The architecture may or may not contain an aggregated node. The presence of the aggregated node does not change the basic idea, which is to share the data directly with the peers (fog nodes) and not through the cloud or an aggregated node. Further, to support the architecture given in Fig. 2, we have defined an algorithm shown in Fig. 3. The algorithm depicts the flow of a request processing which is being originated at a node called the front node.

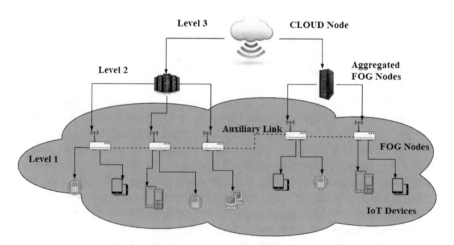

Fig. 2 Proposed inter-fog architecture

Algorithm 1

```
1. Initialize fog[]
            //list of traversed nodes - Initially empty
2. Input: Request for Service
            //originating from the front node
3. Forward the request to the nearest fog node F_i
            //F_i represents the current node
4. Add current node to fog[]
5. Check for result
6. If result is not found
      Forward the request to the nearest fog node F_j : F_j ∉
      fog[]
      go to step 5
7. If result is found
      Find Shortest path to the front node using shortest
      path algorithm and return the result
8. Output: Result on the front node
```

Fig. 3 Algorithm of the proposed architecture

Whenever a request is generated, the request is forwarded to the fog nodes one by one until the result is found. If there is no auxiliary link between adjacent fog nodes, then they communicate through the cloud which in turn increases the path length and thus the latency. By providing the auxiliary link between the fog nodes, we are reducing the dependency on the cloud. Not only does the auxiliary link reduce the path length but also provides a low latency alternative. When the result is found on some fog node, it is returned to the front node using the shortest path. This path may be via the auxiliary link or the cloud.

In Algorithm 1 (As shown in Fig. 3), fog[] is a list of traversed fog nodes, which is initially empty. The input for the system would be a request for service. The front node (as shown in Fig. 4) is the node from where the request is generated. Now, the request is passed to the nearest fog node F_i, which represents the current fog node. If the result is not found at the current node F_i, the request is then forwarded to the next nearest fog node F_j : $F_j \notin$ fog[] and F_i is inserted to fog[]. Otherwise, the result is returned to the front node using shortest path algorithm. The algorithm can be further understood with the help of the case study of a parking scenario given in Sect. 4.

4 Case Study

To analyze the effect on the latency of the proposed work, let us consider a simple multi-storey parking. In such a scenario, each floor would be equipped with a fog device that manages to take data from the sensors on that floor and perform necessary calculations. Traditionally, these devices would be connected only to the cloud server.

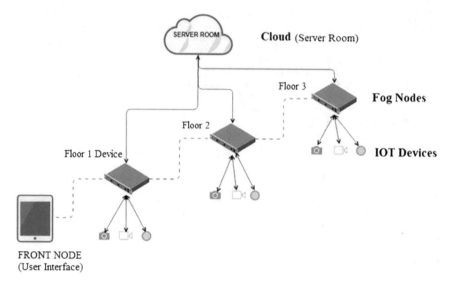

Fig. 4 Parking scenario

As per our requirement, we install an additional connection that joins two adjacent fog nodes. Figure 4 shows the discussed architecture.

In the given scenario, whenever a new user enters the parking building, a request is raised at the front node. Earlier, in the absence of the auxiliary link, this node would send the request to the main server which would then either simultaneously ask all the fog nodes or requests them one by one. Asking each fog node one by one is long and time-consuming. Forwarding the request to all the fog nodes does save time but is extravagant in the use of cloud resources, whereas with the help of an auxiliary connection between adjacent fog nodes, when the request is raised, it is sent to the first node only. The first node then checks for a parking space and returns the value to the front node if it is available. If not, the first fog node, situated on the ground floor, forwards the request to the second node which is on the first floor and so on. This process continues until a parking space is found or the entire lot has been searched.

For the purpose of comparison, we created graph data structure for both the architectures. The first one links all the fog nodes to the cloud, while the second graph includes an additional connection between all adjacent fog nodes, which is our proposed auxiliary connection. To prove our work, we have implemented our algorithm in Java and used A* heuristic algorithm approach for finding the shortest path. The request is forwarded by hops from one node to another, while the result is returned to the front node.

Parking found at	Traditional (ms)	Inter-fog (ms)
F = 0	3	3
F = 1	15	8
F = 2	21	13
F = 3	27	16.5
F = 4	33	19
F = 5	39	21.5
F = 10	69	34
F = 15	99	46.5
F = 20	129	59

Table 1 Comparison of latency

5 Simulation Results and Performance Comparison

For the simulation, we implemented our proposed algorithm on Java 1.8.0_144 with the help of the GraphStream [15] library. We implemented our algorithm discussed above on the parking scenario and found the results for both the traditional method (without Inter-fog communication) and our proposed architecture (with Inter-fog communication). To implement our algorithm on the case in hand, we made the following assumptions.

The results obtained by simulating the traditional method have been compared to the results obtained from our proposed architecture in Table 1.

The latency between the user interface node and the first fog node (the node on the ground floor) is reasonably fixed at 3 ms. We have assumed, to the best of our knowledge, that the round-trip time (RTT) for data between two fog nodes can be 5 ms and that between a fog node and the cloud can be greater than 5 ms. Therefore, for our simulation, we consider the RTT for adjacent fog nodes to be 5 ms and that for the cloud to be 6 ms.

Figure 5 shows a graphical representation of the results found on simulating both the cases, where red line indicates the results found on simulating our proposed algorithm while the blue line shows the results of the traditional fog architecture.

From the graph shown in Fig. 5, we can clearly see that the latency is considerably lesser in our proposed inter-fog method as compared to the traditional fog architecture.

Fig. 5 Graphical results showing a comparison of latency

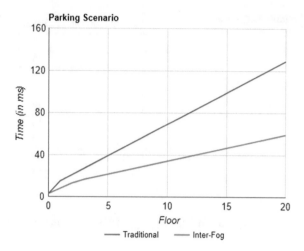

6 Conclusion and Future Scope

In our proposed architecture, an auxiliary connection has been made that links two fog nodes at the same level, adjacent to each other, directly in the fog computing environment. From the simulation results, as shown in Fig. 5, it is evident that using an auxiliary connection helps to reduce the latency in a fog computing environment. The time taken by the traditional method also increases at a much faster rate because the time delay due to the hopping between fog node and cloud is more than that of adjacent fog nodes. Therefore, the proposed architecture is more suitable for environments where the number of nodes is higher. In future, the proposed architecture can be extended to other real-time applications such as traffic monitoring, etc. There is a need to study and analyze the trade-off between the cost and the benefits reaped using an auxiliary connection.

References

1. Khan I, Sawant SD (2016) A review on integration of cloud computing and internet of things. Int J Adv Res Comput Commun Eng 5(4):1046–1050
2. Mell PM, Grance T (2011) The NIST definition of cloud computing. Technical report National Institute of Standards and Technology, U.S. Department of Commerce, NIST Special Publication 800-145 (2011). https://doi.org/10.6028/nist.sp.800-145
3. Luo J-Z, Jin J-H, Song A-B, Dong F (2011) Cloud computing: architecture and key technologies. J China Inst Commun 32(7):3–21
4. Stojmenovic I, Wen S (2014) The fog computing paradigm: scenarios and security issues. In: Proceedings of the federated conference on computer science and information systems (29 Sept 2014). https://doi.org/10.15439/2014f503
5. Sarkar S, Chatterjee S, Misra S (2015) Assessment of the suitability of fog computing in the context of internet of things. IEEE Trans Cloud Comput 1–1. https://doi.org/10.1109/tcc.2015. 2485206

6. Rahmani A, Liljeberg P, Preden J-S, Jantsch A (2017) Fog computing fundamentals in the internet-of-things. In: Fog computing in the internet of things, pp 3–13. https://doi.org/10. 1007/978-3-319-57639-8_1
7. Hong K, Lillethun D, Ramachandran U, Hong K, Lillethun D, Ramachandran U et al (2013) Mobile fog: a programming model for large-scale applications on the internet of things. In: Proceedings of the second ACM SIGCOMM workshop on mobile cloud computing, Hong Kong, China, pp 15–20 (Aug 2013)
8. Gerla M, Lee EK, Pau G, Lee U (2014) Internet of vehicles: from intelligent grid to autonomous cars and vehicular clouds. In: 2014 IEEE world forum on internet of things (WF-IoT) (March 2014). https://doi.org/10.1109/wf-iot.2014.6803166
9. Yi S, Hao Z, Qin Z, Li Q (2015) Fog computing: platform and applications. In: Third IEEE workshop in hot topics in web systems and technologies (HotWeb), pp 73–78. IEEE
10. Hoque S, de Brito MS, Willner A, Keil O, Magedanz T (2017) Towards container orchestration in fog computing infrastructures. In: Proceedings of IEEE 41st annual computer software and applications conference, (COMPSAC) (July 2017). https://doi.org/10.1109/compsac.2017.248
11. Liu P, Willis D, Banerjee S (2016) ParaDrop: enabling lightweight multi-tenancy at the network's extreme edge. In: Proceedings of IEEE/ACM symposium on edge computing (SEC) (Oct 2016). https://doi.org/10.1109/sec.2016.39
12. Kim OTT, Tri ND, Nguyen VD, Tran NH, Hong CS (2015) A shared parking model in vehicular network using fog and cloud environment. In: Proceedings of 17th Asia-Pacific network operations and management symposium (APNOMS) (Aug 2015). https://doi.org/10.1109/apnoms. 2015.7275447
13. Hsieh M-Y, Lai Y, Lin HY, Li KC (2017) A model for predicting vehicle parking in fog networks. In: International conference on frontier computing, pp 239–249 (27 Sept 2017). https://doi.org/10.1007/978-981-10-3187-8_25
14. Yousefpour A, Ishigaki G, Jue JP (2017) Fog computing: towards minimizing delay in the internet of things. In: Proceedings of IEEE international conference on edge computing (EDGE) (June 2017). https://doi.org/10.1109/ieee.edge.2017.12
15. GraphSteam—A Dynamic Graph Library. http://graphstream-project.org/doc/

Leaf Identification Using HOG, KNN, and Neural Networks

Prerna Sharma, Aastha Aggarwal, Apoorva Gupta and Akshit Garg

Abstract The main objective of this paper is to identify the leaves using the concepts of image processing. A dataset comprising 1900 images of 18 leaf species has been used to train our machine. Three major steps—image preprocessing, feature extraction (using Histogram of Oriented Gradients (HOG)) and classification—have been performed. The initial step includes grayscale conversion and represents the input image as a zero-one matrix. In the next step, 900 features have been extracted using HOG. The last step comprises classification of two supervised learning methodologies—K-nearest neighbors and backward propagation algorithm using artificial neural networks. Performance of the two methods has been compared, and artificial neural networks have proved to be a better choice with an approximate accuracy of 97%. The implementation has been carried out using MATLAB and its toolboxes.

Keywords Leaf identification · Histogram of oriented gradients (HOG)
K-nearest neighbors (KNN) · Support vector machine (SVM)
Artificial neural networks (ANN) · Backward propagation

P. Sharma (✉) · A. Aggarwal · A. Gupta · A. Garg
Department of Computer Science, MAIT, New Delhi, India
e-mail: prernasharma@mait.ac.in

A. Aggarwal
e-mail: aastha.aggarwal2@gmail.com

A. Gupta
e-mail: apoorvagupta96@gmail.com

A. Garg
e-mail: gargaksh97@gmail.com

© Springer Nature Singapore Pte Ltd. 2019
S. Bhattacharyya et al. (eds.), *International Conference on Innovative Computing and Communications*, Lecture Notes in Networks and Systems 56,
https://doi.org/10.1007/978-981-13-2354-6_10

Fig. 1 Dataset of different leaf species

1 Introduction

Leaf identification has several utilities. It can help in the creation and set up of a plant directory to identify edible and medicinal leaves, rare or exotic species, invasive plants, etc. It can greatly assist farmers, foresters, botanists, and researchers who need identification tools to study the distribution of plants and their evolution. Amateurs can also learn and contribute to such a humongous and informative database.

To train our system to recognize a leaf, we forged a training dataset based on databases from "Department of Horticulture, Auburn University, Alabama, USA", "Agricultural Research Service, United States Department of Agriculture", "Integrated Taxonomic Information System (ITIS)", and "Harvard University Herbaria & Libraries". It contains 18 leaf species, with a total of 1900 images. The leaf species are shown in Fig. 1.

The subsequent sections of the paper have been categorized as follows: Sect. 2 examines the previous research methodologies for leaf recognition. Section 3 illustrates our implemented algorithms and techniques. Section 4 elucidates the results attained. The final segment, Sect. 5, discusses the results and future scope of our work.

2 Literature Review

According to Bhanu and Pen [1], image segmentation is performed to extract contextually meaningful content from an image. Their object recognition system was based on a model with image segmentation, feature extraction, and model matching as the main components [2, 3]. They assumed that models of objects that are necessary for image recognition are known but the number of those objects and their locations are unknown. They used reinforcement learning to optimize a set of segmentation parameters adaptively, based on a real-valued confidence factor. It guaranteed convergence and performed efficient hill climbing in a statistical sense with less computational requirements. To cater to different types of images, they used Phoenix algorithm developed by CMU for color image segmentation.

Dalal and Triggs [4] were the first to propose another methodology for feature extraction [5, 6], Histogram of Oriented Gradients (HOG), for human detection. It was later used by Xia et al. [7] for leaf recognition. HOG works by dividing the leaf images into cells. A gradient histogram for each cell is calculated and normalized. Feature extraction is done from the normalized histograms.

As illustrated by Cunningham [8], K-nearest neighbor is a supervised learning algorithm, widely preferred for image classification. The basic assumption of this algorithm is that nearby objects belong to the same class. More than one neighbor is evaluated before assigning an object a class [9]. It is a lazy learner algorithm since classification is only based on training samples rather than learning from past experiences.

ArunPriya et al. [10] discussed Support Vector Machine (SVM) algorithm, another supervised learning paradigm used for classifying images due to its high accuracy, efficient output, simplicity, and computational complexity which is independent of the input space. The input data is mapped nonlinearly onto a high-dimensional space which is then divided linearly by a hyperplane [11, 12]. Classification is done by maximizing the margin between the two classes divided by the hyperplane. The optimal hyperplane that has a maximum distance from either class is chosen, and the classes on either side of the plane are used for classification. SVM supports multiclass classification using one-versus-all or one-versus-one approach. However, the algorithm requires longer training time, and understanding the weighted functions is complex.

As Boran et al. discussed in [13], neural network is a supervised learning technique which tries to mimic the human brain's decision-making process by learning patterns in the form of mathematical equations. They adapt and apply them as and when situations arise. The development of experiences in neural networks takes place by reinforcement of weights attached to the connecting neurons [14, 15]. Backpropagation is one of the ways in which neural networks can be implemented. It is a robust tool for pattern matching and associative memory [16, 17].

3 Proposed Work

The software implementation of our work has been done using MATLAB, developed by MathWorks, Inc. We identify a leaf species of our input image in three phases—image preprocessing, feature extraction, and classification.

3.1 Image Preprocessing

We first take a picture of a leaf and then convert it to a greyscale image. It is stored after resizing it to the dimension of 50 * 50 pixels in order to minimize the training period.

Equation (1) is used to convert RGB equivalent to a greyscale equivalent:

$$C_{grey} = 0.2989 * C_{red} + 0.8570 * C_{green} + 0.1140 * C_{blue} \tag{1}$$

The greyscale equivalent is then converted to a zero-one matrix according to the species to which it belongs.

3.2 Feature Extraction

Leaves are distinguished on the basis of features unique to them. We have utilized **Histogram of Oriented Gradients (HOG)** algorithm for this purpose. The chief idea behind using HOG is that the light intensity of locally distributed gradients is well suited to characterize an object's outline and form. The image can be segregated into cells which represent evenly juxtaposed and separated regions. These regions can be classified into blocks which are then normalized in order to remain consistent against illumination or photometric effects. Extracted features over blocks are computed for normalization. These blocks can be called Histogram of Oriented Gradient (HOG) descriptors.

We computed histograms from HOG descriptors by carrying out matrix assignment to the angles of each edge gradient. Angle $= I_X / I_y$ and magnitude $= \sqrt{I_x^2 + I_y^2}$ were determined, where I_X and I_Y stand for gradients in the x- and y-axes, respectively. We initialized a Gaussian filter (Gaussian filter with sigma $= 0.5 *$ block width) with the intent to wipe out extraneous and noisy data. In the next step, we removed the redundant pixels and performed the task of binning via bilinear interpolation. Binning puts pixels into "bins" by evaluating their neighboring pixels. Every bin contains similar kind of pixels. Bilinear interpolation was used for extending variables in the 2D space (x and y directions). For normalization across blocks, we applied L1 norm which signifies least absolute error. This function minimizes the sum of absolute difference between the target and actual values over blocks. After the concatenation of blocks, we obtained the feature vectors (representative of leaf features). Normalization of feature vectors was performed via L2 norm. L2 norm function was employed to diminish the effect of large errors by squaring them.

In our work, a total of 900 leaf features have been taken into consideration. The primary ones are listed below:

1. Length—The interval between the beginning and end of a leaf vein.
2. Width—The horizontal expanse from one edge of a leaf to another.
3. Aspect ratio—It indicates the ratio of length versus width of the leaf.
4. Perimeter—It represents the number of pixels obtained by summing up the number of pixels on the leaf's image.
5. Form factor—It defines the relation between a leaf and a circle, illustrated in Eq. (2).

$$\text{Form Factor} = 4 * pi * Area/Perimeter^2 \tag{2}$$

6. Circularity—It characterizes the disparity between a circle and a leaf. Equation (3) depicts circularity.

$$\text{Circularity} = 4 * Area/Perimeter \tag{3}$$

7. Compactness—It is calculated by Eq. (4).

$$\text{Compactness} = Area/Perimeter^2 \tag{4}$$

3.3 Classification

Classification algorithms assign a class to the input leaf image from a list of predetermined classes of leaf species based on the homogeneity of leaf features. We have used K-nearest neighbors and neural network classification algorithms and compared their performance.

3.3.1 K-Nearest Neighbors (KNN)

K-nearest neighbors is a supervised learning archetype involving a two-step estimation of nearest neighbors and classified using those neighbors. In the context of leaf classification, during the course of the training phase, the machine is made to learn about predesigned classes and leaf images belonging to those classes. These can also be labeled as trained samples. We have taken 18 species of plant leaves as the training samples.

In the next phase, the testing phase, the distance between a test sample and the trained samples is computed. Any distance measure must adhere to the properties of non-negativity, identity, symmetry, and triangle inequality. In our work, we have used Euclidean distance because it can take into account small deviations and still produce precise results, as given by Eq. (5).

$$\text{Euclidean distance} - \sqrt{p1^2 - p2^2} \tag{5}$$

The distance metric finds out similarities between the input test sample and the trained classes or K-nearest neighbors. The test sample is designated to the class it carries resemblance with. K is taken to be an integer value and if calculation occurs in decimal then it is rounded off for simplification. We received an accuracy of 92.3% via its usage. High computational cost and lazy learner characteristic render K-nearest neighbor as a less preferred choice. Thus, we have deployed neural networks to enhance classification capabilities.

3.3.2 Artificial Neural Network (ANN)

An artificial neural network contains processing elements that work similar to a neuron in the human nervous system. A neural network is a layered network of processing components where multiple layers work internally and feed input to the subsequent layers. Each processing component in a neural network is assigned an internal parameter, "weight", and the output behavior of the network is dependent on these weights. These are re-adjusted till we get the desired output. This process is defined as the training period. During this phase, the input data is individually weighed and summed as a result of the transfer or activation function. The output of this function is the output of the neural network.

There are several ways in which neural networks can be implemented. We have used backward propagation learning algorithm for classification of leaves. It is used as a training method of neural networks to converge to a set of weights. The convergence of these weights is done to reduce the mean square error in the output of the network. Error for each layer is estimated and propagated in a backward manner.

In the training phase, we utilized a network training function, "trainscg", available in MATLAB. It is suitable for computing image gradients in low memory situations. We fed 900 features of leaf species into the neural network. In order to create a pattern recognition network, we took the number of hidden layers to be equal to 10. This figure gave us optimum results in accordance with our dataset. There was a trade-off between efficiency and confusion in the case of less and more than 10 layers, respectively, as shown in Fig. 2. We then divided our training database into a ratio of 14:3:3 for training, validation, and testing. In subsequent steps, we trained and tested our network. The leaf species was determined via 18 output neurons.

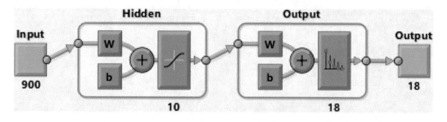

Fig. 2 Neural networks with 900 inputs, 10 hidden layers, and 18 output neurons

Table 1 Training, testing, validation, overall accuracies, and performance in terms of epoch and time for some leaf species of our dataset as obtained by implementation of neural networks

Specie name	Training (%)	Testing (%)	Validation (%)	Overall (%)	Epoch	Time
Chinese Cinnamon	98.8	89.7	91.1	96.2	80	00:00:03
Chinese Redbud	99	89.1	90.4	96.2	81	00:00:03
Deodar	99.4	92.9	90.4	97.1	98	00:00:04
Japan Arrowwood	99.7	92.9	90.4	97.3	86	00:00:03
Japanese Cheese-wood	99.9	90.4	91	97.1	94	00:00:03
Japanese maple	99.6	91.7	91.7	97.2	84	00:00:03
Maidenhair Tree	99	88.5	89.7	96.1	85	00:00:03
Nanmu	100	91	92.3	97.5	84	00:00:02
Pubescent Bamboo	99.9	90.4	89.1	96.8	92	00:00:03
True indigo	99.4	94.9	88.5	97.1	105	00:00:04

4 Results

We first extracted 900 features using histogram of oriented gradients. Next, we classified them using KNN with an output accuracy of 92.3%. To enhance computational efficiency, we utilized backpropagation method in neural networks, which produced an accuracy of 97% (approx.). Table 1 summarizes the training, testing, validation, and overall accuracies for some leaf species of our dataset as obtained upon the application of neural networks. Figure 3 depicts the confusion matrices of the training, testing, validation, and overall phases obtained during the implementation of neural networks.

5 Conclusion and Future Work

This work carries out leaf identification on a dataset of 18 leaf species, comprising 1900 images. In the premier step, features are extracted using HOG. Leaves are then classified into their respective species using KNN and ANN. The accuracy

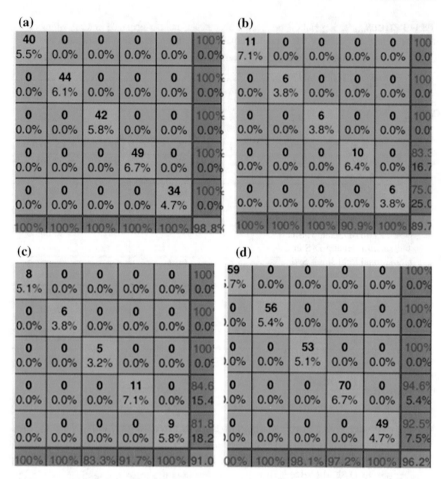

Fig. 3 Confusion matrices. **a** Training phase. **b** Testing phase. **c** Validation phase. **d** Overall obtained by implementation of neural networks

Table 2 Performance comparison of KNN and ANN accuracies

Algorithm implemented	KNN	ANN
Accuracy (%)	92.3	97 (approx.)

achieved by ANN is very high, and a comparison chart of the performance of the two classification techniques has been enlisted in Table 2. Our future work is aimed at maintaining the performance of our leaf recognition system with a wider and extensive database. It is also directed toward identifying different leaf diseases.

References

1. Bhanu B, Peng J (2000) Adaptive integrated image segmentation and object recognition. IEEE Trans Syst Man Cybern—Part C: Appl Rev 30(4)
2. Singh V, Varsha, Misra AK (2015) Detection of unhealthy region of plant leaves using image processing and genetic algorithm. In: 2015 international conference on advances in computer engineering and applications (ICACEA) IMS Engineering College, Ghaziabad, India
3. Barbedo JGA (2013) Digital image processing techniques for detecting, quantifying and classifying plant diseases. Springerplus 2(1), 660. Published online 7 Dec 2013
4. Dalal N, Triggs B (2005) Histograms of oriented gradients for human detection. In: CVPR 2005
5. Gopal A, Reddy SP, Gayatri V (2012) Classification of selected medicinal plants leaf using image processing. ©2012 IEEE
6. Vijayashree T, Gopal A (2015) Authentication of leaf image using image processing technique. ARPN J Eng Appl Sci 10(9)
7. Xia Q, Zhu H-D, Gan Y, Shang L (2014) Plant leaf recognition using histograms of oriented gradients. In: Huang D-S et al (eds) ICIC 2014, LNAI 8589, pp 369–374, 2014. © Springer International Publishing Switzerland 2014
8. Cunningham P, Delany SJ (2007) K-nearest neighbour classifiers. Technical report UCD-CSI-2007-4 March 27, 2007
9. Chaki J, Parekh R (2011) Plant leaf recognition using shape-based features and neural network classifiers (IJACSA). Int J Adv Comput Sci Appl 2(10)
10. ArunPriya C, Balasaravanan T, Thanamani AS (2012) An efficient leaf recognition algorithm for plant classification using support vector machine. In: Proceedings of the international conference on pattern recognition, informatics and medical engineering, 21–23 Mar 2012
11. Ankalaki S, Majumdar J (2015) Leaf identification based on back propagation neural network and support vector machine. ©2015 IEEE
12. Salman A, Semwal A, Bhatt U, Thakkar VM (2017) Leaf classification and identification using canny edge detector and SVM classifier. In: International conference on inventive systems and control (ICISC-2017)
13. Şekeroğlu B, İnan Y (2016) Leaves recognition system using a neural network. In: 12th International conference on application of fuzzy systems and soft computing, ICAFS 2016, 29–30 Aug 2016, Vienna, Austria
14. Singh S, Bhamrah M (2015) Leaf identification using feature extraction and neural network. IOSR J Electron Commun Eng (IOSR-JECE) 10(5), 134–140. e-ISSN: 2278-2834, p-ISSN: 2278-8735, Ver. I (Sep–Oct 2015)
15. Sumathi CS, Kumar AVS (2014) Neural network based plant identification using leaf characteristics fusion. Int J Comput Appl (0975–8887) 89(5)
16. Maqbool I, Qadri S, Khan DM, Fahad M (2015) Identification of mango leaves by using artificial intelligence. Int J Nat Eng Sci 9(3):45–53
17. Wable PB, Chilveri PG (2016) Neural network based leaf recognition. In: 2016 International conference on automatic control and dynamic optimization techniques (ICACDOT). International Institute of Information Technology (I^{2}IT), Pune
18. Ekshinge S, Sambhaji DB, Andore M (2014) Leaf recognition algorithm using neural network-based image processing. Asian J Eng Technol Innov 10:16

Meta-heuristic Techniques to Solve Resource-Constrained Project Scheduling Problem

Bidisha Roy and Asim Kumar Sen

Abstract Scheduling is a foremost vital activity broadly in most engineering fields; mostly Project Management and Operations Research. However, a more practical approach towards solving a scheduling problem would be to consider the applied constraints at hand, such as resources available at hand and other constraints. One such type of scheduling problem is the Resource-Constrained Project Scheduling Problem, abbreviated as RCPSP. The main objective of the RCPSP problem is to plan the project activities with optimal makespan keeping in view the fact that the availability of resources over the timespan of a project is limited. However, being a NP-Hard Combinatorial Optimization Problem exact methods have a problem with convergence as the problem size increases. In recent years, meta-heuristics have shown promising solutions to this problem. In this work, the usage of meta-heuristics to solve this problem is highlighted with possible future directions.

Keywords Resource-constrained project scheduling · Optimization
Meta-heuristics · Swarm intelligence

1 Introduction

The Resource-Constrained Project Scheduling Problem (RCPSP), RCPSP is the problem of generating the schedule for a project within the available limited resources while trying to resolve the resource conflicts and yet satisfying the scheduling objectives [1]. It has major applications in Scheduling industry as well as in Process and Project Management in industry like the cutting stock problem, Aircraft Maintenance,

B. Roy (✉)
St. Francis Institute of Technology, University of Mumbai, Mumbai, India
e-mail: bidisha.bhaumik.roy@gmail.com

A. K. Sen
Yadavrao Tasgaonkar College of Engineering and Management,
University of Mumbai, Mumbai, India
e-mail: asim_sen@linuxmail.org

© Springer Nature Singapore Pte Ltd. 2019
S. Bhattacharyya et al. (eds.), *International Conference on Innovative Computing and Communications*, Lecture Notes in Networks and Systems 56,
https://doi.org/10.1007/978-981-13-2354-6_11

Timetabling, etc. Hence, RCPSP has attracted researchers towards trying to find a solution which is near optimal. The early solutions to this problem were based on exact procedures like linear programming, Dynamic Programming and Branch and Bound [2]. However, like other scheduling problems, it is a Combinatorial Optimization Problem which is strongly NP-Hard [3]. Hence, these exact procedures failed to converge and provide a polynomial time solution to the problem. The research then turned towards using approximate techniques like heuristics and meta-heuristics to find an acceptable solution as the problem size increased. The research in this field will be discussed in further sections. Though they do not come with an assurance of an optimal solution, meta-heuristics have shown better results for a large category of NP-Hard problems.

The research using meta-heuristics started with the usage of Evolutionary Algorithms [4]. The performances of these algorithms exceeded the performance of standard heuristics in terms of standard deviation for large project instances. Inspired by this success, researchers moved towards Nature-Inspired meta-heuristics for solving the problem.

Paper Structure. The detailed definition of the RCPSP problem is given in Sect. 2. Section 3 discusses the various algorithms put forward towards solving the RCPSP problem in the Literature Review Section. Section 4 identifies the research challenges and future directions that have been observed after having studied the literature. Section 5 briefly summarizes the paper with final remarks.

2 Resource-Constrained Project Scheduling Problem

To represent the RCPSP for a project, the following information is needed [5]:

- $j = 1, 2, 3, \ldots, n$ activities each with processing time p_j
- Resources $k = 1, 2, 3, \ldots, r$ available for the project. R_k is the amount of resource k is available at any instant of the project. As the project is carried out, r_{jk} indicates the how much of resource *k* is utilized by activity *j*
- A set of precedence constraints $i \rightarrow j$ between a pair of activities i and j indicating that activity *i* needs to be completed before activity *j* can start.

Thus formally RCPSP would be defined by a tuple (V, p, E, R, B, b), where,

- $V = \{A_0, \ldots, A_{n+1}\}$ **activities** constituting the project
- $p \rightarrow$ duration of activities, $p_0, p_{n+1} = 0$
- $E \rightarrow$ Precedence relations represented by a set of ordered pairs such that $(A_i, A_j) \in E$ would indicate that activity A_i is predecessor of activity A_j
- $R = \{R_1, \ldots, R_q\} \rightarrow$ *Renewable resources* available for the project
- $B \rightarrow$ Availability of resources represented where every B_k denotes the availability of R_k
- $b \rightarrow$ *Demands* of activities for resources wherein the value of b_{ik} represents the amount of R_k utilized per unit time during the execution of A_i.

For the above information provided regarding a project, we need to construct an optimal schedule. A typical Schedule **S** is defined as a set of S_i representing the starting time of activity A_i. And the set C consisted of C_i denoting the completion time of activity A_i, wherein $C_i = S_i + p_i$.

S is a **feasible solution** if it is compatible with the

Precedence Constraints

$$S_j - S_i \geq p_i \quad \forall(A_i, A_j) \in E \tag{1}$$

Resource Constraints

$$\sum_{A_i \in A_t} b_{ik} \leq B_k, \quad \forall R_k \in R, \quad \forall t \geq 0 \tag{2}$$

The mathematical formulation for RCPSP would thus be:

*Min **Cmax** subject to* (1) *and* (2) *above.*

RCPSP can thus be defined as the problem of finding a non-premptive schedule S having minimal makespan S_{n+1} satisfying the Precedence as well as Resource constraints.

3 Literature Review

The problem of finding an optimal solution for RCPSP has been a field of active research for almost two decades now. Many techniques providing optimal solutions and their analysis have been presented. Also, there has been a wide research in the area of meta-heuristic nature-inspired algorithms which have proved effective in various optimization problems as well as constrained scheduling problems in other domains. Hence the literature review in this research is divided into two parts:

- Earlier solutions for RCPSP which is majorly divided into heuristic solutions, swarm intelligence based meta-heuristic solutions
- Use of Meta-heuristics to solve optimization problems in other domains.

3.1 Heuristic Techniques for RCPSP

Among the earliest heuristic solutions provided for RCPSP was this paper which compared heuristic techniques like Schedule Generation Schemes (SGS) and X-pass methods with meta-heuristic techniques like Tabu Search (TS), Simulated Annealing (SA) and Genetic Algorithms (GA) [4]. The SGS used both Serial and Parallel SGS for the generation of feasible schedule for a problem [6]. X-pass methods which are priority rule based heuristics used either Serial or Parallel SGS to generate an initial schedule and then applied priority-based heuristics to select the optimal solution. j30, j60, j120 datasets were used from PSPLIB [7]. All the techniques were evaluated on average deviation from optimal schedules as mentioned in the datasets and multi-variate Linear Regression Analysis. Serial and Parallel SGS were used to generate an initial schedule. It was found that meta-heuristic techniques performed better than heuristic techniques as the project size increased with Simulated Annealing and Genetic Algorithms having the best performance. It also put forward the idea of using SGS heuristics for generating initial schedules and then applying meta-heuristics over these schedules to generate faster optimal solutions.

3.2   Swarm Intelligence Meta-heuristics to Solve RCPSP

The earliest swarm intelligence (SI) technique used to find an optimal solution was the Ant Colony Optimization (ACO) technique [2]. Additional features like discarding of the elitist strategy in the usual ACO and using a Local Optimization Strategy were added. It was experimented on j120 dataset from PSPLIB [7]. The results of this research were compared with SA and GA. It was found that ACO offered a better standard deviation compared to the other two techniques. Based on ACO many variants and hybrid techniques were later developed to solve the RCPSP.

Particle Swarm Optimization (PSO) Algorithm was another algorithm used to solve the RCPSP. An implementation using PSO was proposed in [8] wherein the initial schedule was developed using Serial SGS upon which hyper-heuristic approach using PSO is applied. The proposed algorithm did not produce efficient results for lesser iterations. This was also the first instance of usage of hyper-heuristics for solving the RCPSP. Another adaptive model was put forward by Kumar and Vidyarthi in [9], which improved upon the previous research by using an operator called a valid particle generator (VPG) which optimally adjusted the particle velocity in the swarm. This too was tested on the j30, j60, and j120 instances from the PSPLIB [7]. Due to the velocity adjustment, this algorithm gave better results on standard deviation as compared to the standard PSO variant.

Ziarati et al. in [10] had used the concept of Bee Algorithms to solve RCPSP using the intelligent behavior of honey bees. Three variants of the Artificial Bee Colony (ABC) Algorithm were presented. To resolve the infeasible solutions, a constraint handling feature was introduced. All the three algorithms were tested on schedules developed over Serial SGS over all benchmark datasets. The results showed that all the three had higher performance rates for a lower number of activities, i.e. datasets for smaller projects. However, when compared with other swarm intelligence techniques mentioned in the literature, the performance of the algorithms decreased for larger datasets.

Many hybrid approaches based on few of these SI techniques were proposed with varying results.

3.3 Use of SI Meta-heuristics to Solve Optimization Problems in Other Domains

Use of meta-heuristic to solve other optimization problems including other applications of scheduling has been a major research focus in recent years too. This section discusses some major work in this front.

Marichelvam et al. [11] solved another NP-Hard problem, Hybrid Flow Shop Scheduling using Discrete Firefly Algorithm (DFA). Firefly Algorithm, as the name suggests, is a SI algorithm that emulates the social behavior of fireflies. In this paper, it was adapted to a discrete algorithm to support discrete scheduling problems. The system was tested and compared for both industrial data and random instances against GA, SA, and ACO solutions for the same problem. It was found that the DFA showed better results for both the objectives of makespan criterion as well as mean flow time criterion.

A recent algorithm called TLBO (Teaching Learning Based Optimization) has been proposed which is inspired from the teaching learning process of human beings. Research mentioned in [12] has mentioned the use of TLBO and its variations for various constrained combinatorial optimization problems.

Using Nature-Inspired Algorithms for solving optimization problems has been a growing research area. However, there has not been active research in the use of these algorithms for RCPSP.

4 Research Challenges and Future Directions

The RCPSP is an important problem in project management and manufacturing process. Also, being an NP-Hard Combinatorial Optimization Problem, it represents an interesting research area. Standard heuristics being class specific, hence did not have the capacity to solve a broader category of problems. Hence, the focus then shifted to meta-/hyper-heuristics. Over a period of time many solutions have been proposed and implemented. However, there still exists a need for

- Exploring the usage of the latest meta-heuristic and nature-inspired algorithms to check for alternate and better solutions. The latest constrained and unconstrained meta-heuristic optimization algorithms, like [11, 12], have not been used to solve the RCPSP problem although researchers have proposed the usage of these algorithms on different kinds of combinatorial optimization problems.
- Non-availability of standardized frameworks to solve the problem. The current trend is to just apply a meta-/hyper-heuristic algorithm over heuristic SGS. Designing an appropriate framework to solve the problem could be a potential area of research.
- Most of the literature uses average standard deviation from the critical path as a comparison and evaluation parameter. Study to find other evaluation parameters can also be an effective area of research.

- There is very little research done using hyper-heuristics to solve RCPSP, PSO being one of them [8]. Hence, this area can be also explored as an area of research.
- Time durations in the PSPLIB [7] datasets are deterministic. Modeling them as probabilistic durations and then obtaining good solutions could also be an area future study.

5 Conclusion

RCPSP is a computationally hard and crucial problem that occurs frequently in high-scale project management. Hence, the potential impact of an inefficient schedule can be critical to the business utilities. In this paper, we have highlighted and classified the major solution approaches that have yielded optimal results over major benchmark instances. We have also discussed solutions used in other domains and computationally hard problems which could be tested on the RCPSP problem. Some future research directions were also highlighted which mainly focused on the usage of advanced meta-heuristics, framework designing and exploring more evaluation parameters. Meta-heuristics have been found to be best suited to find solutions close to optimal ones for larger problem instances of NP-Hard problems. Since most of the meta-heuristic algorithms work on a common framework of recursively generating new solution from a currently acceptable solution, research can also be done in the direction of finding a common framework. Hopefully, the challenges and future directions highlighted in this paper would motivate future researchers to bring about fresh dimensions to this problem domain.

References

1. Artigues C (2008) The resource-constrained project scheduling problem. http://www.iste.co. uk/data/doc_dtalmanhopmh.pdf
2. Merkle D, Schmeck H, Middendorf M (2002) Ant colony optimization for resource-constrained project scheduling. IEEE Trans Evol Comput 6(4)
3. Blazewicz J, Lenstra JK, Rinnooy Kan AHG (1983) Scheduling subject to resource constraints: classification and complexity. Discret Appl Math 5(1):11–24
4. Hartmann S, Kolisch R (2000) Experimental evaluation of state-of-the-art heuristics for the resource-constrained project scheduling problem. Eur J Oper Res 127(2):394–407
5. Vanhoucke M (2012) Project management with dynamic scheduling. Springer, Berlin. https://doi.org/10.1007/978-3-642-40438-2
6. Kolisch R (1996) Serial and parallel resource-constrained project scheduling methods revisited: theory and computation. Eur J Oper Res 90(2):320–333
7. Sprecher A, Kolisch R (1997) PSPLIB-a project scheduling problem library: OR software-ORSEP operations research software exchange program. Eur J Oper Res 96(1):205–216
8. Koulinas G, Anagnostopoulos K, Kotsikas L (2014) A particle swarm optimization based hyper-heuristic algorithm for the classic resource constrained project scheduling problem. Inf Sci 277:680–693

9. Kumar N, Vidyarthi DP (2016) A model for resource-constrained project scheduling using adaptive PSO. Softw Comput 20(4):1565–1580
10. Ziarati K, Zeighami V, Akbari R (2011) On the performance of bee algorithms for resource-constrained project scheduling problem. Appl Softw Comput 11(4):3720–3733
11. Marichelvam MK, Prabaharan T, Yang XS (2014) A discrete firefly algorithm for the multi-objective hybrid flowshop scheduling problems. IEEE Trans Evol Comput 18(2):301–305
12. Wang Z, Lu R, Chen D, Zou F (2016) An experience information teaching–learning-based optimization for global optimization. IEEE Trans Syst Man Cybern Syst 46(9):1202–1214

A Novel Approach to Find the Saturation Point of n-Gram Encoding Method for Protein Sequence Classification Involving Data Mining

Suprativ Saha and Tanmay Bhattacharya

Abstract In the field of biological data mining, protein sequence classification is one of the most popular research area. To classify the protein sequence, features must be extracted from the input data. The various researchers used n-gram encoding method to extract feature value. Generally, to reduce the computational time, the value of n of n-gram encoding method is considered as 2, but accuracy level of classification degrades. So, it is an important research, to find the optimum value of n for n-gram encoding method, where computational time and accuracy level of classification both are acceptable. In this work, an experimental attempt has been made to fixed up the limit of scaling of n-gram encoding method from 2-gram to 5-gram. Standard deviation method has been used for this purpose.

Keywords Data mining · Neural network model · Rough set · String kernel
Protein hashing · Support vector machine · n-gram encoding method

1 Introduction

The exponential growth of data can be successfully handled by data mining process, instead of traditional data analysis system. Data mining is a technique to extract pattern from a large amount of data set like biological data. This approach involves the combination of machine learning, statistics and database system. Data mining achieves significant success in the area of classification, clustering, associativity, sequence analysis and regression of data. Classification of the biological data like protein sequence is also carried out by this process. The process of feature extraction

S. Saha (✉)
Department of Computer Science and Engineering, Brainware University,
Barasat, Kolkata, India
e-mail: reach2suprativ@yahoo.co.in

T. Bhattacharya
Department of Information Technology, Techno India, Salt Lake, Kolkata, India
e-mail: dr.tb1029@gmail.com

© Springer Nature Singapore Pte Ltd. 2019
S. Bhattacharyya et al. (eds.), *International Conference on Innovative Computing and Communications*, Lecture Notes in Networks and Systems 56,
https://doi.org/10.1007/978-981-13-2354-6_12

is applied to identify unique characteristic of the data set. This paper is dealing with the limitations of feature extraction procedure, which is used to classify the protein sequences. Here, we have given a brief literature survey on different types of protein sequence classification techniques involving data mining, proposed by various researchers in Sect. 2. Section 3 describes a review of the n-gram encoding method followed by the outcome analysis and experimental proof of the boundary of n-gram encoding method in Sect. 4, and finally the conclusion is in Sect. 5.

2 Literature Review

To classify the unknown protein sequence in its proper class, subclass or family, various researchers proposed different types of classification techniques involving data mining. The major steps of classification are to extract important features from the input data (like protein sequence), a knowledge value is created, match the feature values which are extracted from the input and finally, the unknown protein sequence is classified.

The classifier based on neural network model [1] was proposed by the different researchers with 90% accuracy level. This classifier used 2-gram encoding method and 6-letter exchange group method as global similarity and *len, mut, occur* as local similarity for extracting feature value. Zainuddin and Kumar [2] were enlarged 2-gram encoding method up to n-gram encoding method. The feature values, which were extracted from n-gram encoding method, applied on Self-Organized Map (SOM)-based probabilistic neural network model. To improve the accuracy of the previous classification, back-propagation technique was proposed by Nageswara Rao et al. [3]. Another most popular classification model, i.e., Fuzzy Adaptive Resonance Theory MAP (ARTMAP) model was proposed by Mohamed et al. [4] where, molecular weight, the isoelectric point, the hydropathy composition, the hydropathy distribution, and the hydropathy transmission of the protein sequence were used as feature value. This classification model is claimed 93% accuracy. Mansoori et al. [5] were applied the rank-based algorithm upon fuzzy ARTMAP to reduce the computational time.

A faster, approximately 97% accurate and efficient classification model involving rough set theory was designed by Cai et al. [6] to classify the protein sequence data. This approach was based on structural and functional properties of the protein. A new approach involving the combination of neural network system, fuzzy ARTMAP model and Rough set classifier was elaborated by Saha and Chaki [7, 8] with 91% accuracy level. Here, the Knuth–Morris–Pratt (KMP) string matching algorithm was used instead of window sliding in the phase of the neural network model.

String kernel-based model was proposed by Spalding and Hoyle [9] with 87.5% accuracy, based on a Local pair-wise method like *Basic Local Alignment Search Tool (BLAST), FASTA*, etc., and K-mar Composition to detect both functional and structural similarities between protein sequences. String Weighting Scheme based classifier was also developed by Zaki et al. [10] using Hidden Markov Model and the mix of this portrayal with a classifier equipped for learning in extremely inadequate high-dimensional spaces. Ali and Shawky [11] were invented a new classifier involving Fast Fourier Transform (FFT) method to classify the unknown protein sequence with 91% accuracy. Tree-Based Classifier [12] also used to do the same based on the application of TreeNN rules over neighborhood classifier and Receiver Operating Characteristic (ROC) method with 93% accuracy value. Hidden Markov-Based Classifier [13] is also one of the successful classifiers by executing three phases of extraction: *Training, Decoding,* and *Evolution*. It reaches 94% accuracy level.

Rahman et al. [14] was proposed a new classifier using Structural Analysis of Protein involving six major techniques like *Structure Comparison, Sequence Comparison, Cluster Index, Connectivity, Taxonomic, and Interactivity* and *Taxonomic* followed by Hash table implementation using selection and linear search with 98% accuracy level. On the other hand, Caragea et al. [15] invented a classifier using Feature Hashing involving K-gram encoding methods followed by Rank-based algorithm with 82.83% accuracy level. Zhao et al. [16] were also proposed a classifier using combination of Support vector machine and Genetic algorithm (GA/SVM) System, based on SVM hybrid framework. The accuracy level of this method was calculated as 99.24%.

3 Feature Extraction Using n-Gram Encoding Method

In respect to the one-dimensional structure, a protein sequence holds the 20 different amino acids, identified by 20-letter alphabets like A, C, D, E, F, G, H, I, K, M, N, P, Q, R, S, T, V, W, and Y. Different types of features are extracted from the protein sequence, which is applied to soft computing methodology like neural network model for classifying unknown protein sequence. n-gram encoding, and 6-letter exchange group method are most common approaches to calculate the global similarity of the protein sequences. The n-gram encoding method provides occurrence of patterns of n consecutive amino acids from a protein sequence data. The value of n may vary from 2 to $n - 1$. Lets consider the value of n equal to 2 which converts the n-gram encoding method to 2-gram. As an example of protein sequence, *ACDEFACD*, where 2-gram encoding method is applied and originates the following results of occurrence: AC occurs in 2 times, CD is also 2 times, DE occurs in 1 time, EF and FA are in 1 time, respectively. Now, this result of occurrences are used to calculate feature values like mean value and value of standard deviation, with the help of following formula [1, 2]. Finally those features are applied to neural network based classifier to classify unknown protein sequences.

$$M = \frac{\sum_{j=1}^{t} y_j}{t}, \quad SD = \sqrt{\frac{\sum_{j=1}^{t}(y_j - M)^2}{t - 1}}$$

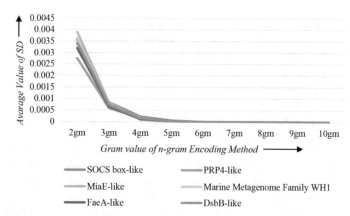

Fig. 1 Graphical representation of average value of standard deviation of 2-gram to 10-gram encoding method in respect to the six different protein classes on basis of n-gram encoding method

where M denotes mean and SD standard deviation. t is the number of distinct patterns extracted from the sequence and $y = ac/(len(str) - (g - 1))$. ac denoted the number of occurrence of the distinct pattern, $len(str)$ indicates the length of the sequence str and g means gram value.

4 Problem Analysis and Experimental Result

Feature extraction involving n-gram encoding method is a very efficient technique for neural network based classifier. Now, the problem of the research is, to decide the proper acceptable upper limit of n in case of n-gram encoding method. In general, the lower limit is fixed to 2 (2-*gram*) [1, 2] but the maximum upper limit may be $n - 1$. The problem is that, if the upper limit of n-gram encoding method is $n - 1$, takes longer time to compute. It is an important area of research to reduce the required time of analysis means bound the upper limit of n-gram encoding method and side by side increases the accuracy level of classification.

Table 1 shows a portion of the actual experimental result, where some value of standard deviation from 2-gram to 10-gram encoding method corresponding six different classes are presented.

Analyzing the experimental data, it can be proved that value of standard deviation, from 6-gram to 10-gram encoding method are bounded to zero. In this case, those data do not require to consider as a part of feature value. Table 2 is presented the summary of the total experiment up to 497 known protein sequences of six different classes. This table is also strongly fixed the executive upper limit of the n-gram encoding method to the 5-gram.

Now, considering the following chart (Fig. 1) of the average value of standard deviation extracted from the previous experimental result, up to 10-gram encoding method, applied on the protein sequence data of the six different classes. The mean value and the corresponding standard deviation up to 10-gram encoding method have been extracted from around 497 protein sequences of 6 different classes. From this result, it has been experimentally proved that the standard deviation of 2-gram encoding method provides a better distinguishable measurement of classification (Fig. 2). Similarly, 3-gram, 4-gram, and 5-gram are also providing some distinguishable measurement of classification, which is also being considered in some cases. It also proved that after 5-gram encoding method, the values of the standard deviation are zero. In this case, any distinguishable measurement of classification is not shown here. Thus, this experiment can fix up the boundary value of n-gram encoding method up to 5-gram.

Figure 1 is plotted on the average value of standard deviation of the 2-gm to the 10-gram encoding method, which is clearly shown that after 5-gram encoding method the standard deviation are leveled up, i.e., all are belongs to the zero region. In this case, all lines of 6-gram to 10-gram are plotted one above another which cannot be classified. In this scenario, it was strongly recommended that the value of standard deviation above 5-gram encoding method is totally useless in respect of classification.

Table 1 A sample of actual experimental result involving six different classes

Class name	Standard Deviation (SD)								
	2gm	3gm	4gm	5gm	6gm	7gm	8gm	9gm	10gm
SOCS box-like	0.00428	0.00129	0.00022	0	0	0	0	0	0
SOCS box-like	0.00341	0.00059	0	0	0	0	0	0	0
SOCS box-like	0.00411	0.00128	0.00051	0.00024	0	0	0	0	0
PRP4-like	0.00305	0.00062	0.000146	0	0	0	0	0	0
PRP4-like	0.00265	0.00057	0.000207	0.00012	0	0	0	0	0
PRP4-like	0.00298	0.00055	0	0	0	0	0	0	0
MiaE-like	0.00347	0.0006	0	0	0	0	0	0	0
MiaE-like	0.00368	0.00072	0.00024	0	0	0	0	0	0
MMF WH1[a]	0.00289	0.00058	0.00022	0.00011	0	0	0	0	0
MMF WH1[a]	0.00246	0.0004	0	0	0	0	0	0	0
MMF WH1[a]	0.00324	0.00062	0.000287	0	0	0	0	0	0
FaeA-like	0.00245	0.00055	0.000097	0	0	0	0	0	0
FaeA-like	0.00376	0.00073	0	0	0	0	0	0	0
FaeA-like	0.00335	0.00067	0.000288	0.00021	0	0	0	0	0
DsbB-like	0.00601	0.00066	0	0	0	0	0	0	0
DsbB-like	0.00822	0.00243	0.00082	0.00048	0	0	0	0	0

[a]MMF WH1 means Marine Metagenome Family WH1

Table 2 Summary of the actual experimental results involving six different protein classes

Class name	No of nonzero value of Standard Deviation									
	Total sequence	2gm	3gm	4gm	5gm	6gm	7gm	8gm	9gm	10gm
DsbB-like	63	63	59	42	21	0	0	0	0	0
FaeA-like	78	78	76	48	32	0	0	0	0	0
MMF WH1[a]	84	84	84	44	31	0	0	0	0	0
MiaE-like	83	83	80	49	38	0	0	0	0	0
PRP4-like	81	81	77	59	28	0	0	0	0	0
SOCS box-like	108	108	101	66	29	0	0	0	0	0
Total	497	497	477	308	179	0	0	0	0	0

[a]MMF WH1 means Marine Metagenome Family WH1

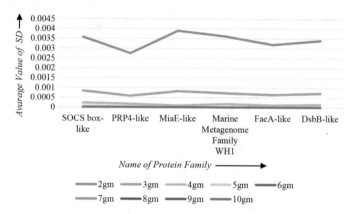

Fig. 2 Graphical representation of average value of standard deviation of 2-gram to 10-gram encoding method in respect to the six different protein classes on basis of protein classes

Figure 2 is also presents the average value of standard deviation from 2-gram to 10-gram encoding method in respect to the 6 different protein classes. It has been observed that in 2-gram encoding method, the standard deviation value provides more appropriate distinguishable measurement of classification. With the increment of gram value, the distinguishable measurement of classification value goes to low. Finally, after the 5-gram encoding method, this value is reached to zero which cannot be considered in terms of classification.

5 Conclusion

This article includes a brief review of different classification techniques involving data mining for classifying the unknown protein sequences. It has been perceived that feature extracted from the protein sequence, has been applied to any optimized technique involving soft computing is the prime idea of classification. From the classification rules, it is shown that n-gram encoding method applied to the neural network model, is the key techniques, which is used by most of the authors. Some of them also use only 2-gram encoding method to reduce the execution time but on other hand, it also decreases the accuracy level. This paper is concluded with the saturation point of n-gram to 5-gram with the details experimental proof. This is also shown that after 5-gram encoding method, i.e., from 6-gram to n-gram all the value of standard deviation is bounded to zero, which has not considered in the case of protein classification. This approach may increase the computational time in respect to the 2-gram encoding method but accuracy level has been reached at the top. It has been observed that by fixing the upper bound of n there is a significant improvement in the time of execution without hampering the accuracy level of classification.

References

1. Wang JTL, Ma QH, Shasha D, Wu CH (2000) Application of neural networks to biological data mining: a case study in protein sequence classification. In: KDD, Boston, pp 305–309
2. Zainuddin Z, Kumar M (2008) Radial basic function neural networks in protein sequence classification. Malays J Math Sci. 195–204
3. Nageswara Rao PV, Uma Devi T, Kaladhar D, Sridhar Gr, Rao AA (2009) A probabilistic neural network approach for protein superfamily classification. J Theor Appl Inf Technol
4. Mohamed S, Rubin D, Marwala T (2006) Multi-class protein sequence classification using Fuzzy ARTMAP. In: IEEE conference, pp 1676–1680
5. Mansoori EG, Zolghadri MJ, Katebi SD, Mohabatkar H, Boostani R, Sadreddini MH (2008) Generating fuzzy rules for protein classification. Iran J Fuzzy Syst 5(2):21–33
6. Cai CZ, Han LY, Ji ZL, Chen X, Chen YZ (2003) SVM-prot: web-based support vector machine software for functional classification of a protein from its primary sequence. Nucleic Acid Res 31:3692–3697
7. Saha S, Chaki R (2012) Application of data mining in protein sequence classification. IJDMS 4(5)
8. Saha S, Chaki R (2012) A brief review of data mining application involving protein sequence classification. In: ACITY 2012. AISC, vol 177. Springer, India, pp 469–477
9. Spalding JD, Hoyle DC (2005) Accuracy of string kernels for protein sequence classification. In: ICAPR 2005. LNCS, vol 3686. Springer
10. Zaki NM, Deri S, Illias RM (2005) Protein sequences classification based on string weighting scheme. Int J Comput Internet Manag 13(1):50–60
11. Ali AF, Shawky DM (2010) A novel approach for protein classification using fourier transform. Int J Eng Appl Sci 6:4
12. Boujenfa K, Essoussi N, Limam M (2011) Tree-kNN: a tree-based algorithm for protein sequence classification. IJCSE 3:961–968. ISSN 0975-3397
13. Desai P (2005) Sequence classification using hidden Markov models, electronic thesis or dissertation. https://etd.ohiolink.edu/

14. Rahman MM, Alam AU, Abdullah-Al-Mamun, Mursalin TE (2010) A more appropriate protein classification using data mining. JATIT 33–43
15. Caragea C, Silvescu A, Mitra P (2012) Protein sequence classification using feature hashing. Proteome Sci 10(Suppl 1):S14. https://doi.org/10.1186/1477-5956-10-S1-S14
16. Zhao X-M, Huang D-S, Cheung Y-M, Wang H-Q, Xin H (2004) A novel hybrid GA/SVM system for protein sequences classification. In: IDEAL 2004. LNCS, vol 3177. Springer, pp 11–16

A Comparative Evaluation of QoS-Based Network Selection Between TOPSIS and VIKOR

E. M. Malathy and Vijayalakshmi Muthuswamy

Abstract Significant research to establish a seamless connection through a vertical handover process has been carried out in the past. Selecting a suitable access network to support roaming, give rise to wide challenges. A mobile terminal decides when to transfer the call connection through a handover process and the entire control process is carried out by the mobile devices. A mobile terminal based decision often creates high latency and thereby gives way for high call drop as the terminal or the mobile user is unaware of the network conditions. The aim of this paper is to design and implement network-controlled selection for handover process. TOPSIS and VIKOR are compared to selection score to utilize MADM scheme to perform handover in a wireless network to avoid unnecessary handover.

Keywords Network selection · TOPSIS · VIKOR · Wireless network
Handover process

1 Introduction

Wireless communication technology is an effective modern alternative to offer real-time multimedia services over Next-Generation Network (NGN). Next-generation network is an evolution of 4G wireless communication Technology. NGN is a high-performance technology for application like multimedia, full motion video, and wireless teleconferencing. NGN is envisioned with wide support in global wireless roaming and extends service portability across next-generation network [1]. This enables different services and large coverage area and interoperability for seamless connection with 50 Mbps or more bit rates. IP-based working principle of NGN facilitates

E. M. Malathy (✉)
SSN College of Engineering, Chennai, Tamil Nadu, India
e-mail: malathyem@ssn.edu.in

V. Muthuswamy
College of Engineering, Anna University, Chennai, Tamil Nadu, India
e-mail: vijim@annauniv.edu

© Springer Nature Singapore Pte Ltd. 2019
S. Bhattacharyya et al. (eds.), *International Conference on Innovative Computing and Communications*, Lecture Notes in Networks and Systems 56,
https://doi.org/10.1007/978-981-13-2354-6_13

selection of any networks that are interconnected. This increases demand on new and multimedia services and that are too less cost. The features of NGN are to define and deploy different network in heterogeneous network. The technology has capability to extend to develop and use service independent of other access network. NGN technology is effectively equipped with multiple interfaces to offer rich multimedia service and connect to different Radio Access Network (RAN) for global roaming. These communication access networks include CDMA, GSM, GRPS, Bluetooth-based Personal Area Network, WLAN, WIMAX, WIFI, LTE, and satellite network. These interconnected technologies offer innovative mobile services to satisfy user demand. The ultimate objective of this next-generation wireless network technology development is to make way for mobile user to access services anywhere, anytime. Each of these network offer specific bandwidth, different coverage area, latencies and operate at different frequency. Therefore, well organized effective vertical handovers are essential to enable mobile users to switch over between these heterogeneous networks.

Vertical handover decision (VHD) aids in seamless connection to selected network. This selection of network plays an important role in resource utilization of the network. Inaccurate decision leads to failure of handover. Also frequent decision of network selection leads increase in traffic condition of network. This causes call blocking. Moreover, if handover decision is initiated by mobile terminal, there is complexity in handling the decision optimal in network as the device do not have enough information about network condition. The consequences are therefore,

- Increases in system load: Mostly traditional handover decision measures the single parameter of evaluation to select the network for switch over. This leads to unnecessary handover. Increasing mobile user in a particular network creates heavy network traffic which in turn results in poor network utilization of resources [2]. Any mobile controlled handover is unaware of network condition and keep selecting the particular network, keep network load condition to higher level. This causes poor QoS to mobile user [3]. Therefore heterogeneous network should incorporate resource utilization method to avoid performance degradation.
- Increases in packet losses: Always Best Connected (ABC) paradigm directs load into different interconnected network. This eventually increases the call blocking and call dropping probability handover (Narayanan et al. 2014). Access to right network at right time is an encouragement to mobile user. This increases the diversity in network. Hence there is a need for handover decision strategy to avoid call dropping.
- Increase in handover failure: The simplest schemes in literature [4] are all based on single parameters such as RSS, bandwidth, and cost as sufficient criteria to proceed with vertical handover decision-making. But sometimes, signal interferences and

unfavorable conditions may initiate unnecessary vertical handovers. This becomes the initial cause to affect the overall performance of the network. Although high throughputs are achieved high handoff delay is present with arrival of more number of mobile nodes. Therefore, an efficient handover to minimize blocking probability is still an open challenge. This paper proposes an MADM based network selection process in heterogeneous network by utilizing TOPSIS and VIKOR.

1.1 TOSIS Network Selection Process

Multi-Attribute Decision-Making (MADM) network score ranking algorithms are effectively utilized to demonstrate and perform the selection suitable best candidate network for handover among the other available wireless networks. The approach includes multi-attribute method such as Techniques for Order Preference by Similarity to Ideal Solution (TOPSIS) and VIKOR (VIseKriterijumsa Optimizacija I Kompromisno Resenje). Any designed handover scheme with QoS support targets reduced overhead in connection and should avoid additional design complexity for various traffic characteristics. The proposed work considers conversational traffic class, streaming traffic class, iterative traffic class and background traffic class. To enable any application service with suitable QoS requirement, the main key components for best user experience and suitable choice to switch with appropriate network capability is based on the traffic characteristics [5]. They are classified first as conventional for VoIP and video conference based application, second traffic includes streaming for audio- and video-based application, third includes interactive traffic for web browsing and gaming messages and finally the background traffic class where file transfer and e-mail type applications are included. In such scenario QoS factors have direct impact by specific policies that are adopted dynamically and therefore the network selection decision to facilitate handover by Multiple Attribute Decision-Making (MADM) method brings optimal policy. The various QoS parameters considered for computation includes RTT, network reliability factor, traffic handling priority value and signal strength with traffic classes for specific application. Right choice of network parameters selection reduces signaling overheads during handover process. The proposed work computes signal strength and network reliability to have valid handover computations. Traffic handling priority value enables network congestion check and therefore, offer load balance in network conditions as suggested in [6]. The cost factor computation gives the best choice of the network selection along with round trip time factor to have reduced network connection failure.

Step 1 : Decision matrix construction

$r_{ij} = x_{ij}/ (\Sigma x_{ij}^2)$

for $i = 1, ..., m; \; j = 1, ..., n$ (i)

Step 2: Weighted decision matrix formation

$V_{ij} = W_j r_{ij}$

for $i = 1, ..., m; j = 1, ..., n$

(ii)

Step3: Determine Positive and Negative ideal solutions

$A^* = \{ v_1^*, ..., v_n^* \}$ where $v_j^* = \{$ max (v_{ij}) (iii)

$A' = \{ v_1', ..., v_n' \}$, Where $v' = \{$ min (v_{ij}) (iv)

Step 4 : Separation measures calculation

$S_i^* = [\Sigma (v_j^* - v_{ij})^2]^{1/2}$ $i = 1, ... m$

(v)

$S_i' = [\Sigma (v_j' - v_{ij})^2]^{1/2}$ $i = 1, ...,$

m (vi)

Step 5: Calculation of the ideal solution

$C_i^* = S_i' / (S_i^* + S_i')$ $0 < C_i^* < 1$

(Vii)

1.2 VIKOR Network Selection Process

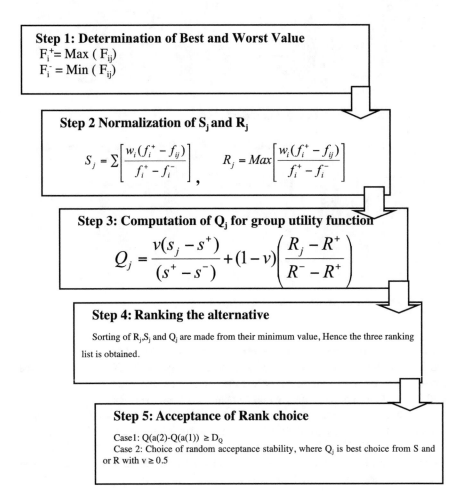

Step 1: Determination of Best and Worst Value
$F_i^+ = Max (F_{ij})$
$F_i^- = Min (F_{ij})$

Step 2 Normalization of S_j and R_j

$$S_j = \Sigma \left[\frac{w_i(f_i^+ - f_{ij})}{f_i^+ - f_i^-} \right] , \qquad R_j = Max \left[\frac{w_i(f_i^+ - f_{ij})}{f_i^+ - f_i^-} \right]$$

Step 3: Computation of Q_j for group utility function

$$Q_j = \frac{v(s_j - s^+)}{(s^+ - s^-)} + (1 - v)\left(\frac{R_j - R^+}{R^- - R^+} \right)$$

Step 4: Ranking the alternative

Sorting of R_j, S_j and Q_j are made from their minimum value, Hence the three ranking list is obtained.

Step 5: Acceptance of Rank choice

Case1: $Q(a(2)) - Q(a(1)) \geq D_Q$
Case 2: Choice of random acceptance stability, where Q_j is best choice from S and or R with $v \geq 0.5$

2 Simulations and Result Analysis

This initial stage gives the importance of parameter in determining the candidate network. The aim of the method targets on the value to be assigned on different traffic classes. The QoS traffic classes include Streaming, one-way transport class such as watching video via YouTube where the delay is not much important than throughput. The background traffic is also a one-way traffic that includes sending or receiving SMS or emails. The background traffic classes show less importance

114   E. M. Malathy and V. Muthuswamy

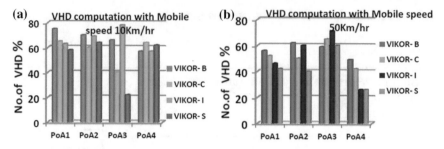

Fig. 1 a, b Network selection computation with VIKOR

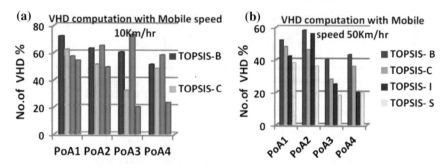

Fig. 2 a, b Network selections computation with TOPSIS

for delay and throughput. Conversational traffic class on other hand is two-way traffic where online video conference is an ideal example with delay is very sensitive parameter, where as throughput is of less important. Iterative traffic is also a two-way traffic class that includes chatting and truncations on web services.

The comparative evaluation is presented in Figs. 1 and 2 with two mobility scenarios. It is observed that TOPSIS executes network ranking in reduced number compared to VIKOR. VIKOR computes score in vague value for each traffic class which produced more number of network selection. Therefore TOSIS can be used to cooperate network selection process with quality support.

3 Conclusion

In this paper, network selection using multiple parameters is proposed for the wireless next-generation network. First, the work estimates the available network with the ranking schemes such as TOPSIS and VIKOR, the Multi-Attribute Decision-Making Method that yields candidate network for selection. QoS-based support is implemented with a mobility scenario with four access points. TOPSIS computes reduced network selection process thereby avoiding unnecessary handover in a wireless network.

References

1. Chou, SF, Yu, YJ & Pang, AC (2017) Mobile small cell deployment for service time maximization over next-generation cellular networks. IEEE Transactions on Vehicular Technology 66(6):5398–5408
2. Navaratnarajah S, Saeed A, Dianati M & Imran MA (2013) Energy efficiency in heterogeneous wireless access networks. IEEE wireless communications 20(5):37–43
3. Piamrat K, Ksentini A, Bonnin JM & Viho C (2011) Radio resource management in emerging heterogeneous wireless networks. Computer Communications 34(9):1066–1076
4. Miyim AM, Ismai LM, Nordin R (2014) vertical handover solutions over LTE-advanced wireless networks: an overview. Wirel Pers Commun 77(4):3051–3079
5. Malathy EM, Muthuswamy V (2015) Knapsack-TOPSIS technique for vertical handover in heterogeneous wireless network. PLoS ONE 10(8):e0134232
6. Navarro ES, Morales JDM, Rico UP (2012) Evaluation of vertical handoff decision algorithms based on MADM methods for heterogeneous wireless networks. J Appl Res Technol 10(4):534–548
7. Wu JS, Yang SF, Hwang BJ (2009) A terminal-controlled vertical handover decision scheme in IEEE 802.21-enabled heterogeneous wireless networks. Wiley Int J Commun Syst 22(7):819–834
8. Yan X, Ahmet Şekercioğlu Y, Narayanan S (2010) A survey of vertical handover decision algorithms in fourth generation heterogeneous wireless networks. Comput Netw 54(11):1848–1863
9. Nguyen-Duc T, Kamioka E (2016) An energy-efficient mobile-controlled vertical handover management for real IME services. J Comput Commun 4(1):59–75
10. Rathi R, Khanduja D, Sharma (2016) Efficacy of fuzzy MADM approach in six sigma analysis phase in automotive sector. J Ind Eng Int 12(3):377–387
11. Vasu K, Maheshwari S, Mahapatra S, Kumar CS (2013) Energy and QoS aware fuzzy-top vertical handover decision algorithm for heterogeneous wireless networks. IET Netw 2(3):103–114

Analysis and Implementation of the Bray–Curtis Distance-Based Similarity Measure for Retrieving Information from the Medical Repository

Bray–Curtis Distance Similarity-Based Information Retrieval Model

Narina Thakur, Deepti Mehrotra, Abhay Bansal and Manju Bala

Abstract Information retrieval involves similarity estimation of the documents in a repository. It is the measure of the closeness of documents which can be in general measured as a similarity/distance score for the user entered query. This score is used to rank and retrieve the documents from the repository based on user need. Distance-based similarity algorithms are generally of the order $O(n)$ rather than $O(n^2)$. A similarity measure finds its usage not only in estimating similarity score for document retrieval but also clustering and classification. Researchers in the past have suggested numerous similarity measures. This paper presents a new and efficient Information retrieval algorithm using Bray–Curtis Distance-based information retrieval from OHSUMED. Detailed analysis shows that the Bray–Curtis Distance-based similarity measure used for Information retrieval outperforms the other prevailing similarity methods.

Keywords Similarity · Bray–Curtis Distance · Precision · Recall
Information retrieval · Jaccard index · Cosine · Manhattan distance

N. Thakur (✉) · A. Bansal
Department of CS, Amity School of Engineering Technology,
Amity University Uttar Pradesh, Noida, Uttar Pradesh, India
e-mail: narinat@gmail.com

D. Mehrotra
Department of IT, Amity School of Engineering Technology,
Amity University Uttar Pradesh, Noida, Uttar Pradesh, India
e-mail: dmehrotra@amity.edu

A. Bansal
e-mail: abhaybansal@hotmail.com

M. Bala
Department of CS, IP College for Women, Delhi University,
New Delhi, Delhi, India
e-mail: manjugpm@gmail.com

© Springer Nature Singapore Pte Ltd. 2019
S. Bhattacharyya et al. (eds.), *International Conference on Innovative Computing and Communications*, Lecture Notes in Networks and Systems 56,
https://doi.org/10.1007/978-981-13-2354-6_14

1   Introduction

Current years have viewed rapid growth and development in structured and unstructured data measure; also the size of the data is expected to surge exponentially in the near future. This begets the need for an efficient Information Retrieval (IR) system and similarity measure for data retrieval from the repository. Apart from work on directly perceivable document clustering, classification, information retrieval, and the similarity measures have simultaneously gained importance in the field of web, ontology, and enterprise search. The IR from the medical data set or repository as the data demands of the users become more complicated. Also, the recall is much higher for medical data set rather than the precision. The prime objective of this paper is to exploit the Bray–Curtis Distance (BCD)-based similarity for retrieving information from OHSUMED Medical data set for efficient retrieval of documents. The accuracy of the IR system is reliant upon the description of the user query, and similarity measure, hence useful similarity measure is required. Giving the right similarity measure and a right set of keywords enables the user to get relevant result/documents in searching using keywords or user entered query/batch queries. Among the various existing similarity measures used in Information retrieval systems, Euclidean distance is unique as it is s standard distance, whereas there are many other non-Euclidean distances like Manhattan/L1, Jaccard index, and Cosine similarity which satisfies the basic metric is used as similarity measures in IR systems. Euclidean distance is simple and straightforward in its geometric interpretation, for this reason, it is much used in cluster analysis. The fundamental challenge of the Euclidean similarity distance is that the two documents seem as pretty similar to each different, despite the reality that they share no terms at all in common in IR. Additionally, Euclidean distance is not always normalized. Bray–Curtis Distance-based similarity measure is nearly as roburst as other distances measures in estimating and quantifying the differences between the documents. Bray–Curtis Distance-based similarity (BCD) is a dissimilarity measure rather than a similarity measure or a distance metric. BCD similarity measure for retrieving information from the OHSUMED medical dataset has been found very useful and efficient in retrieving information from a medical data set where batch queries are used rather than single query and documents. The sections are systematized as follows: Sections 2 and 3 elaborates IR and similarity measures. Section 4 presents Bray–Curtis Distance-based similarity measure for retrieving information from the OHSUMED dataset followed by the implementation of the proposed Bray–Curtis Distance-based similarity method. In Sect. 5, experimental results are discussed in Sect. 5 and Sect. 6 draws Conclusion.

2   Information Retrieval

Information Retrieval (IR) [1] process is to find a document of an unstructured nature that fulfills user need or an information need from within large collections or

Fig. 1 Information retrieval tasks

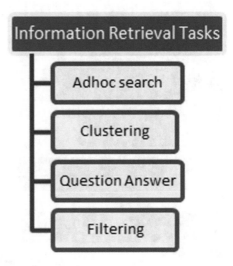

repository' IR can be defined as a process to partition the documents of the data set into matched or not matched documents in content for the given query.

2.1 Information Retrieval Tasks

Information Retrieval tasks are divided into four broad categories as discussed in Fig. 1. Adhoc Search [2–4] is finding the relevant document for the text query. Classification [5] is the process to identify the relevant label for the documents. The IR task of a Question [6] Answer system is to give a specific answer to a question and Filtering [7, 8] is the process to identify relevant user profile for the new document.

2.2 IR Process

IR process is to identify the set of documents that are a best possible match for the queries. The IR Process is as shown in Fig. 2 consists of indexing the preprocessed, stemmed documents. The preprocessing step involves the stopword removal followed by Stemming. Stemming is the process of removing or splitting down the terms/tokens to the base for example, looks, looked, looking terms are stemmed down to look, followed by similarity comparison and ranking of retrieved documents. The preprocessed QUERLs are tokenized to extract the keywords and are matched with the index objects and the matched documents are retrieved. Each retrieved document will be assigned a score using the IR model and similarity measure as discussed in Sect. 3. The similarity score is computed using the TF and the chosen similarity mea-

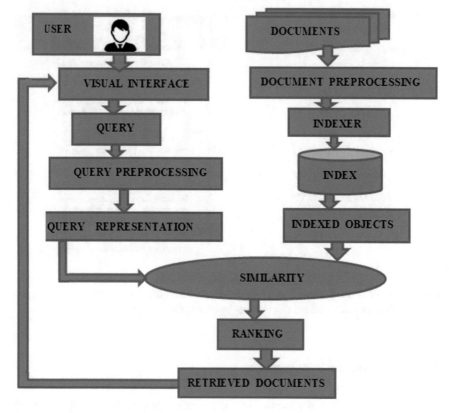

Fig. 2 Information retrieval process

sure. The documents are then reverse sorted by their score and further these sorted documents are retrieved [9]. A similarity score is calculated for the user entered query representation and the indexed objects/document. The documents are retrieved based on the sorted similarity scores of the documents.

3   Similarity Measure

The similarity measure used in the IR process is primarily inaccurate since user decision is required to identify that the IR system has retrieved the correct documents. Information retrieval and document classification, both the cases require the similarity or matching process; also after the user judgment or relevance, the documents are ranked. Similarity measure [10] compares the document with a document or the given query thereby partitioning all the documents of the data set into relevant or nonrelevant or matched or non-matched document labels. The similarity measure is broadly categorized into seven types as shown in Fig. 3. Set difference similarity

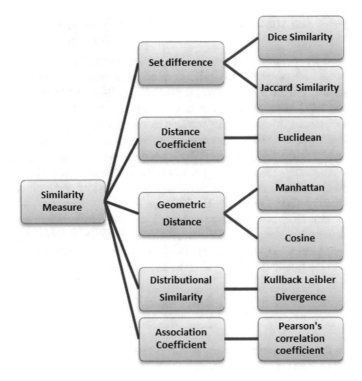

Fig. 3 Types of similarity measures

measure uses set comparison method to retrieve documents, Dice similarity coefficient is based on the F-score and it retains the sparsity property of cosine similarity measure with the discrimination of collinear vectors. Jaccard similarity coefficient is widely used in web clustering to calculate the similarity between the two groups. Jaccard similarity coefficient is the proportion of the size of the convergence to the extent of the association. The Distance Coefficient is the Euclidean similarity measure. Euclidean is a Geometric distance coefficient [11] widely used, well known and default similarity metric used in the k-NN Clustering. Manhattan distance and Cosine similarity are Geometric Distance measures. Manhattan distance or Taxicab metric is similar to Euclidean distance, but it is very sensitive to document length [12]. Cosine is the widely applied text/documents retrieval/Information retrieval method which is normalized inner product of vectors [13].

Cosine is a symmetric geometric similarity measure. This similarity measure is independent of document length. Distributional Similarity is widely used in Natural language processing. It improves the probability estimation for invisible/unseen co-occurrences [14]. Kullback Leibler Divergence (KLD) is anticipating the log variance or divergence among the probabilities of data in the original distribution with the approximating distribution. Pearson's correlation coefficient is an Association Coefficient. It is the fraction of variables covariance to the product of their standard deviations.

4   Bray–Curtis Distance-Based Similarity Measure

The proposed Bray–Curtis distance (BCD) [15] similarity measure calculates the matching score between the OHSUMED dataset and the QRELS. Better normalization and goodness to the fit to the ordinal proximity data are the foremost benefit of the proposed BCD based similarity measure. Bray–Curtis distance is the modified Manhattan distance [16] additionally referred to as Lance & Williams's metric. BCD is an asymmetrical, normalized measure often used in ecology ecological patterns, environmental science, and biology. BCD is not a distance metric as it does not fulfill triangle inequality. The formulae used to calculate the BCD score is as shown in Eqs. (1) and (2) below. It treats alike the variations among low and high values. The value of the Bray–Curtis similarity lies between 0 and 1, where zero means actual similarity objects BCD consider large and small variable values equally and does not treat them otherwise. The value of the Bray–Curtis similarity lies between 0 and 1, where 0 means exact similarity objects. The Bray–Curtis dissimilarity between two documents in the Euclidean space can be as shown in (1).

$$d_{ij} = \frac{\sum_{k=1}^{N} |X_{ik} - X_{jk}|}{\sum_{k=1}^{N} (X_{ik} + X_{jk})} \tag{1}$$

where $d_{i}j$ is the dissimilarity between document i and j, k is the keywords index and n is the keywords count in X document, i.e., TFIDF value for the documents and the query. Also,

$$d_{ij} = \frac{\sum_{k=1}^{N} |X_{ik} - X_{jk}|}{\sum_{k=1}^{N} X_{ik} + \sum_{k=1}^{N} X_{jk}} \tag{2}$$

The higher tf_idf difference values of terms in documents influence the BCD score and these terms are probable to distinguish amongst the documents. When both the documents are empty, the denominator becomes zero and the score is not defined. To address the above issue, work done by [17] proposed a zero-adjusted BCD coefficient which comprises a dummy variable being 1 for all documents; hence in the numerator, this variable subtracts to zero and in the denominator, it sums to 2 as in Eq. (3) below

$$d_{ij} = \frac{\sum_{k=1}^{N} |X_{ik} - X_{jk}|}{\sum_{k=1}^{N} 2 + (X_{ik} + X_{jk})} \tag{3}$$

To calculate the internal similarity between species arrangement and communications, the percentage BCD similarity were proposed for pollinator species and plants observational data taken through split investigations or collaboration surveys. The proposed BCD scoring model has been implemented in python using the whoosh package and TREC-9 OHSUMED datasets. The index is created first for the training set of 54,710 documents, followed by the similarity score computation by running the python script for BCD similarity model. The implementation of the scoring script and model uses Eq. (2) as discussed above for 63 QREL/Queries.

5 Experimental Results and Discussion

This section, evaluate the Bray–Curtis Distance-based similarity measure for retrieving Information from OHSUMED Medical dataset [16]. The evaluation uses the standard Trec_eval script (TREC conference evaluation script) in Ubuntu on the standard precision and recalls performance metrics and the results are discussed. The Precision, Mean average precision (MAP) and Recall of the proposed BCD similarity measure are compared with the state-of- the-art similarity measures TF_IDF, DFR, BM25 in Table 1.

It can be seen from Table 1 and Fig. 4 that considerably alike relationship exists between Interpolated precision and the recall for the Bm25 and DFR and the proposed BCD similarity model outperforms BM25 similarity model with Precision @ recall 0.20 to Precision @ recall 0.80 with peaks.

The mean average precision for the BCD model is 0.174 maximum followed by 0.1722 in BM25, 0.164 in DFR, and minimum 0.0636 in TF_IDF similarity model as seen from Table 1.

Table 1 Results of the BCD similarity with TF_IDF, DFR, BM25

runid	TF_IDF	DFR	BM25	BCD
map	0.0636	0.164	0.1722	0.174
gm_map	0.0288	0.0964	0.0972	0.0209
Rprec	0.1098	0.2222	0.2283	0.2048
bpref	0.4965	0.6072	0.6082	0.4161
recip_rank	0.3227	0.5625	0.5853	0.5088
iprec@recall_0.00	0.3662	0.619	0.6374	0.518
iprec@recall_0.10	0.1676	0.4115	0.4309	0.4271
iprec@recall_0.20	0.1184	0.319	0.3328	0.3365
iprec@recall_0.30	0.0923	0.2432	0.2505	0.2859
iprec@recall_0.40	0.0667	0.1906	0.197	0.2015
iprec@recall_0.50	0.047	0.1424	0.1424	0.1796
iprec@recall_0.60	0.0219	0.0728	0.0755	0.0791
iprec@recall_0.70	0.0122	0.0353	0.0362	0.034
iprec@recall_0.80	0.0035	0.0135	0.0154	0.0158
iprec@recall_0.90	0	0.0033	0.0052	0.0017
iprec@recall_1.00	0	0	0	0.0017

Fig. 4 Interpolated
Precision versus Recall
graph for the proposed BCD,
TF_IDF, DFR, BM25
similarity measures

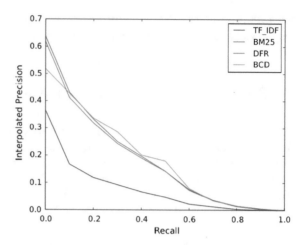

6 Conclusion

This paper proposes the Bray–Curtis similarity measure for retrieving information
from OHSUMED medical dataset. A similarity score is calculated for ranking and
retrieving documents for the user entered query. The proposed similarity model
makes use of the standard tf_idf score for documents and batched queries. A number
of the latest similarity measure used in the information retrieval system for calculating
similarity matching and retrieval were reviewed in terms of their functionalities and
the techniques proposed BCD similarity model is evaluated on precision and recall
measure. Experiments have been carried out on the OHSUMED dataset using the
proposed Bray–Curtis similarity measure. The comparative evaluation results of the
BCD similarity measure demonstration a significant improvement in performance
for BM25F, DFR and TF_IDF retrieval models. Later on, the research work can be
extended by hybridizing the BCD and another similarity measure to extemporize the
accuracy and precision in document ranking and retrieval.

References

1. Ounis I, Macdonald C, Lin J, Soboroff I (2011) Overview of the TREC-2011 microblog track.
   In: Proceedings of the 20th Text REtrieval conference (TREC 2011), vol 32
2. Zhai C, Lafferty J (2017) A study of smoothing methods for language models applied to ad
   hoc information retrieval. ACM SIGIR Forum 51(2):268–276
3. Voorhees EM, Harman DK (eds) (2005) TREC: experiment and evaluation in information
   retrieval, vol 1. MIT Press, Cambridge
4. Belkin NJ, Croft WB (1992) Information filtering and information retrieval: two sides of the
   same coin? Commun ACM 35(12):29–38
5. Niwa Y, Sakurai H (1999) Document retrieval-assisting method and system for the same and
   document retrieval service using the same with document frequency and term frequency. U.S.
   Patent 5,987,460, issued 16 Nov 1999

6. Salton G, Wong A, Yang C-S (1975) A vector space model for automatic indexing. Commun ACM 18(11):613–620
7. Lee MD, Navarro DJ, Nikkerud H (2005) An empirical evaluation of models of text document similarity. Proc Cogn Sci Soc 27(27)
8. Ilijoski B, Popeska Z (2017) A survey of text mining techniques, algorithms and applications, pp 141–144
9. Zhang Y, Callan J, Minka T (2002) Novelty and redundancy detection in adaptive filtering. In: Proceedings of the 25th annual international ACM SIGIR conference on research and development in information retrieval. ACM, pp 81–88
10. Alami N, El Adlouni Y, En-nahnahi N, Meknassi M (2017) Using statistical and semantic analysis for Arabic text summarization. In: International conference on information technology and communication systems. Springer, Cham, pp 35–50
11. Lee L (1999) Measures of distributional similarity. In: Proceedings of the 37th annual meeting of the association for computational linguistics on computational linguistics. Association for Computational Linguistics, pp 25–32
12. Blei DM, Kucukelbir A, McAuliffe JD (2017) Variational inference: a review for statisticians. J Am Stat Assoc (just-accepted)
13. Bray JR, Curtis JT (1957) An ordination of the upland forest communities of Southern Wisconsin. Ecol Monogr 27:325–349
14. Clarke KR, Somerfield PJ, Chapman MG (2006) On resemblance measures for ecological studies, including taxonomic dissimilarities and a zero-adjusted Bray-Curtis coefficient for denuded assemblages. J Exp Mar Biol Ecol 330:55–80
15. Field JG, Clarke KR, Warwick RM (1982) A practical strategy for analyzing multispecies distribution patterns. Mar Ecol Prog Ser 8:37–52
16. Chacoff NP, Resasco J, Vázquez DP (2018) Interaction frequency, network position, and the temporal persistence of interactions in a plant-pollinator network. Ecology 99(1):21–28
17. Sebastiani F (2002) Machine learning in automated text categorization. ACM Comput Surv (CSUR) 34(1):1–47

Information Retrieval on Green Mining Dataset Using Divergence from Randomness Models

Tanisha Gahlawat, Shubham Lekhwar, Parul Kalra and Deepti Mehrotra

Abstract Green computing is one of the emerging aspect of technology that focuses on developing computer software and hardware devices which helps to reduce the power usage and carbon consumption. Today, with proliferation of smartphones, tablets and other unwired gadgets, efficient use of energy is becoming a key consideration while designing any software or hardware component. Manufacturers and developers look forward to energy-efficient solutions across the stack, with more established results through research and innovations in application design, hardware/architecture, operating systems, and runtime systems. The aim of the research is to evaluate the recall and precision of the dataset using the various Divergence from Randomness (DFR) models. The corpus used in the research was a Green Computing dataset that consisted of about 300 questions along with their answers. The topic files and Query Relevance judgement (QREL) files were created for all the questions and the answers. The results indicate that OKAPI probabilistic model has the highest precision.

Keywords Green computing · Divergence from randomness
Information retrieval · Energy · Power · Consumption · QREL · Green mining
Usage · Topic

T. Gahlawat (✉) · S. Lekhwar · P. Kalra · D. Mehrotra
Amity School of Engineering and Technology, Amity University
Uttar Pradesh, Noida, Uttar Pradesh, India
e-mail: tanishagahlawat@gmail.com

S. Lekhwar
e-mail: shubhamlekhwar@gmail.com

P. Kalra
e-mail: parulkalra18@gmail.com

D. Mehrotra
e-mail: mehdeepti@gmail.com

© Springer Nature Singapore Pte Ltd. 2019
S. Bhattacharyya et al. (eds.), *International Conference on Innovative Computing and Communications*, Lecture Notes in Networks and Systems 56,
https://doi.org/10.1007/978-981-13-2354-6_15

1 Introduction

In this era, computers are used widely in every field to obtain the accuracy and speed of work, but the computer works on power and that is why increasing use of computers leads to increase of power consumption and greater heat generation. Green computing, one of the emerging aspect of technology focuses on developing computer software and hardware devices, which helps to reduce the power usage and less carbon consumption [1]. Today, with proliferation of smartphones, tablets, and other unwired gadgets, energy efficiency has become a key consideration while designing software and hardware components. Manufacturers and developers look forward to energy-efficient solutions, with more efficient results through research and innovations in application design, hardware [2, 3], operating systems [4], and runtime systems [5].

Nowadays, information retrieval systems and services are part of our many day-to-day activities which range from web application and database search, multimedia files, digital libraries, etc. Thus the process of information retrieval act as a back-bone for all databases and web. But as the use of web applications and services has increased it also increases the concerns about the amount of energy that they consume [6]. This concern puts light on a major issue that is sustainable development and more specifically environment sustainable development. Numerous measures are being suggested to reduce the impact of business, industries, etc., on the environment. A large number of researches on environmental sustainability are being done today. These researches focus on the suitable use of the technology in order to minimize the overall GHG emissions. The environmental impact of information and the communication innovation can be reduced with the use of green IT and distributed computing. In this manner, the servers and distributed computing setup are looking for innovation that is more proficient. In recent years, there is developing spotlight on application programming and more productive methodologies that are compliment to earlier equipment and software design arrangements [7].

Today, the energy demand in Information Technology is growing rapidly [8]. A basic measurement to enhance the energy effectiveness of programming frameworks is to see how the software engineers visualize the scenario. The requirements of engineers and the difficulties that come in their way would help the researchers remain concentrated on the present reality issues [9]. The research would help fill in as a reasonable guide for further advancements in energy-efficient software.

The subsequent paper has been organized as follows: Sect. 2 consists of the literature review. The research methodology containing the dataset description, transformation, and experiment procedure is described in Sect. 3. Section 4 consists of the results and discussion followed by the conclusion in Sect. 5.

2   Literature Review

Gustavo Pinto et al. in their paper [10] present an empirical study in order to understand the programmers views on the issues related to the energy consumption. The dataset that they used for the research was from StackOverflow, which consisted of about 300 questions along with their answers that had a count of about 500 along with the involvement of more than 810 users. Their research came up with the major questions that the programmers ask and the major bases for the energy consumption issues. Irene Manotas et al. [2] describe the way practitioners visualize the energy while designing, writing specifications, constructing, coding, testing, and maintaining the software. They conducted a survey with the involvement of about 460 programmers from various IT industries such as Google, IBM, and Microsoft, etc., along with in-depth interviews from Microsoft officials. The aim of the research was to develop strategies so that the energy consumption can be improved. Candy Pang et al. [11] conducted a survey revealing the knowledge level of the programmers present in the industry. They concluded that the programmers do not have accurate information about how actually the power consumption can be reduced and thus there is a need to train the practitioners. Abram Hindle in his paper [12] presents a methodology in order to examine the effect of change of software technologies. He investigates the effect of object-oriented programming on the energy consumption. Taking everything into account, they investigated the impact of programming change on power utilization on two activities, and also gave an underlying examination on the effect of programming measurements on power usage. He also developed a set of tests in his research [13] to run on various versions of a single product to record the power consumption by each of them. Then these results were compared to come to a conclusion. He analyzed the performance of about 520 builds of Firefox. Robert R. Harmon et al. in their research [14] studied about various techniques of sustainable Information Technology. They conferred the key focus areas and also recognized the principles that are required to be followed for sustainable IT design. Keke Gai et al. proposed a model by simulating a real-world instance and thus formed a solid base for justifying their evaluations [15]. Also, the paper gives various ways through which the problems related to the waste disposal can be solved. San Murugesan in his paper [5] discusses about various principles and practices for Green IT. Green IT is a necessity rather than an option. They have also mentioned several challenges of Green IT. Patrick Kurp in his article [16] about green computing discusses whether or not we are ready for personal energy meter. He proposes a system that is more productive, considerably less costly, and also that would immediately affect the world's power usage. It is constantly less expensive to move information than the energy.

3   Research Methodology

This section describes the green computing dataset and the research approach that
has been employed for the evaluation.

3.1   Data Collection and Transformation

The corpus used in the research was a Green computing dataset that consisted of about
300 questions along with their answers. Each question could have more than one
answer. The extracted corpus was unstructured and could not be directly loaded into
the tool that is, Terrier 4.1. It was a combination of JSON and XML format. Hence,
it had to be transformed into Terrier supported XML file along with the creation
of Query Relevance judgement (QREL) and topic files. The dataset transformation
included identifying the tags that would be required and then applying rules for final
transformation. The data has to be transformed into various files namely,

1. *Data files*—these are the XML files containing the data in the <text> field along
   with the unique document number, and also the title. Figure 1 depicts the transformed XML data files.
2. *Topic files*—the topic file containing the queries are formed and the retrieval is
   according to indexing of the files.
3. *QREL files*—these are the relevance judgment files according to which the evaluation is done.

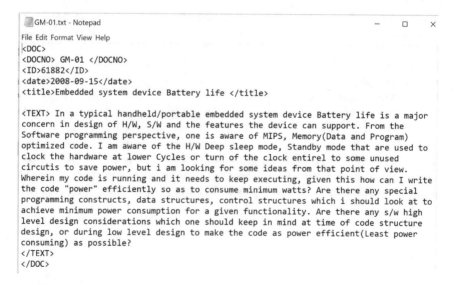

Fig. 1   The transformed dataset

3.2 Tool Used: Terrier IR Tool v4.2

Terrier is an open source, exhaustive, adaptable, effective, and straightforward stage for research and conducting experiments in content retrieval, which is promptly deployable on extensive scale accumulations of the corpus. Terrier first indexes the files and then retrieves them as per the users information need. It calculates the recall and precision of the queries and is capable of implementing various information retrieval models. There are number of IR tools like Lucene, Lemur, Sphinx, Xapian, Manatee etc. but we considered Terrier platform for our research, as it is most effective and efficient in retrieving the documents. It also implements "state of the art" indexing and retrieval functionalities [10]. Terrier works in three steps for retrieving and evaluating the desired documents namely; indexing the documents, running topic files and running QREL files. These steps can be described as:

Indexing in Terrier—In Terrier, we process the transformed data through indexing. The indexing is a four-step process that includes Collection, documents, term pipelines, and indexer. After the indexing is complete, the topic files are run on the indexed files. *Topic files in Terrier*—In Terrier supporting dataset format, which is used, as an information need that is "topic" to differentiate it from a "query" which is a data structure represented in the retrieval system. There are various statements of the topic in Terrier along with number of fields. Then these topic files are run on the corpus for the retrieval.

QREL files in Terrier—In order to test the relevance of topic files, relevance judgement files are made. They are often mistaken as the query files. However, the QREL files are the batch files that are created with the opinion of experts. After processing QREL file, we get an evaluation file according to the model used for evaluating it. Several models can be used in order to perform the mentioned task. The models can be weight models or field models. In this research, we have made the evaluations using the top four weight models (DFR models). After successful running of the QREL file, an evaluation file is obtained that clearly mentions the precision for different number of documents.

3.3 Procedure

The study began by loading the transformed dataset into the tool and then applying the weighing models that is, Divergence from Randomness (DFR) models. There are total 24 DFR models out of which the authors have considered the best four models for evaluating the green computing dataset. The recall and precision was calculated for each of the selected model, which was then compared to find out the model giving the highest results. Figure 2 shows the flow diagram of the process for evaluating the dataset.

The objective of the Load procedure is to load the data into files so that Terrier can work upon them. These files in Terrier are known as collection.spec file. The

Fig. 2 Flow diagram
depicting the methodology

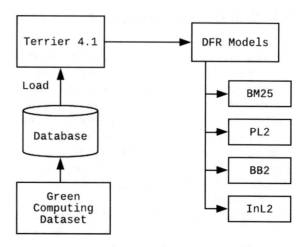

collection.spec file is created according to the terrier.properties file in Terrier. They
are created only after the successful setup and indexing. Collection.spec file can also
be created manually as and when required.

4   Results and Discussions

In this section the authors have summarized the outcomes and also provide discussions from the obtained results.

The evaluation brings out three important outcomes: one, it identified four main themes regarding green computing; two, it identified major causes for energy consumption problems and; three, it identified the model with the best recall and precision.

The evaluation files for the top four models that are BM25, PL2, BB2, and InL2 were compared and analyzed. The analysis identified the major themes, that are, *power consumption, code design, android devices and applications*, and *noise*. The major causes for energy consumption problems that were identified are, *CPU, wifi, battery, hardware, software, servers, sensors*, and *Bluetooth*.

Table 1 gives the number of queries and documents. The precision for more than p@10 had very similar values for almost all the four models. BM25 came out to be the best model with the highest Mean Average Precision (MAP). Table 2 compares the models followed by the graph that clearly explains Table 2 (Fig. 3).

Table 1 Document
specification

Number of queries	25
Retrieved	1662
Relevant	93
Relevant retrieved	90

Table 2 Precision @n for various models

S. No	Model name	MAP	R Pre-cision	P@1	P@2	P@3	P@4	P@5	P@10
1	BM25	0.8894	0.8	0.84	0.74	0.6667	0.57	0.504	0.312
2	PL2	0.8777	0.8	0.76	0.74	0.6933	0.57	0.488	0.312
3	BB2	0.8829	0.8	0.76	0.76	0.6533	0.55	0.496	0.304
4	InL2	0.8432	0.7387	0.80	0.72	0.68	0.57	0.504	0.312

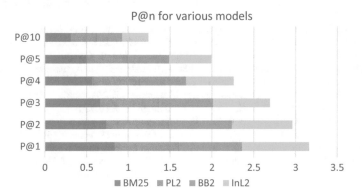

Fig. 3 Graph showing the values for P@n

5 Conclusion

In this research, we have examined the knowledge of users about green computing. The authors observed that the power consumption questions are more intriguing and furthermore are difficult to answer. The authors had to consider the weight models that are the DFR models for the retrieval. The recall and the precision had to be calculated for various DFR models. The top four models namely BM25, BB2CL, InL2 and Pl2 were considered. The recall and the precision for BM25 model was found to be the highest. Major themes were identified: *power consumption, code design, android devices and applications,* and *noise.* The major causes for energy consumption problems were also identified: *CPU, wifi, battery, hardware, software, servers, sensors,* and *Bluetooth.*

In future we will be analyzing the user based on the cognitive skills and find the skills that should be enhanced in order to bring out the maximum efficiency from the retrieval.

References

1. Kalra P, Mehrotra D, Wahid A (2018) Comparison and analysis of information retrieval DFR models. In: Smart computing and informatics. Springer, Singapore, pp 361–367
2. Farkas KI, Flinn J, Back G, Grunwald D, Anderson JM (2000) Quantifying the energy consumption of a pocket computer and a java virtual machine. In: SIGMETRICS
3. Wang S, Lo D, Jiang L (2013) An empirical study on developer interactions in stackoverflow. In: SAC
4. Treude C, Barzilay O, Storey M-A (2011) How do programmers ask and answer questions on the web? (NIER track). In: ICSE
5. Ribic H, Liu YD (2014) Energy-efficient work-stealing language runtimes. In: ASPLOS
6. Huang J, Liu Z, Duan Q, Atiquzzaman M, Jo M, Haas ZJ (2018) Green computing and communications for smart portable devices. Wirel Commun Mobile Comput
7. Vijaykrishnan N, Kandemir M, Kim S, Tomar S, Sivasubramaniam A, Irwin MJ (2001) Energy behavior of java applications from the memory perspective. In: JVM
8. AlMusbahi I, Anderkairi O, Nahhas RH, AlMuhammadi B, Hemalatha M (2017) Survey on green computing: vision and challenges. Int J Comput Appl 167(10)
9. Han G, Que W, Jia G, Zhang W (2018) Resource-utilization-aware energy efficient server consolidation algorithm for green computing in IIOT. J Netw Comput Appl 103:205–214
10. Amati G (2003) Probabilistic models for information retrieval based on divergence from randomness. PhD thesis, Department of Computing Science, University of Glasgow
11. Ge R, Feng X, Feng W, Cameron K (2007) CPU MISER: a performance-directed, run-time system for power-aware clusters. In: ICPP
12. Iyer A, Marculescu D (2002) Power efficiency of voltage scaling in multiple clock, multiple voltage cores. In: ICCAD
13. Merkel A, Bellosa F (2006) Balancing power consumption in multiprocessor systems. In: EuroSys
14. Morrison P, Murphy-Hill E (2013) Is programming knowledge related to age? An exploration of stack overflow. In: MSR
15. Rangan KK, Wei G-Y, Brooks D (2009) Thread motion: fine-grained power management for multi-core systems. In: ISCA
16. Tiwari V, Malik S, Wolfe A (1994) Power analysis of embedded software: a first step towards software power minimization. IEEE

Handling Big Data Using MapReduce Over Hybrid Cloud

Ankur Saxena, Ankur Chaurasia, Neeraj Kaushik and Nidhi Kaushik

Abstract Today is the world of digitalization and we cannot imagine digital world without Big Data over cloud infrastructures. The best financially savvy strategy to manage the expanding intricacy of huge information investigation is half and half cloud divide that leases brief off-start cloud assets to support the general limit amid top use of information. Half breed cloud framework make a domain for accessible for all intents and purposes boundless measure of computerized and informatics assets, which are overseen by outsiders and are gotten to by clients in secure way and as indicated by pay-examine way, with best Quality of Services. It empowers advanced figuring frameworks to be scaled all over as needs be to the measure of information to be handled. MapReduce is among the most famous models for advancement of Cloud applications. MapReduce is a famous procedure of Hadoop Big Data for dissecting expansive datasets and groups. It takes into account parallel handling of a lot of information over half breed distributed computing assets. Hadoop strategy of Big Data is an open-source usage of MapReduce and utilizations the Fair scheduler to relegate delineate lessen capacity to the different registering hubs. This paper we can fetch data using map and reduce function of Hadoop pig framework over hybrid cloud resources.

Keywords Cloud computing · Hybrid cloud · Big Data · Hadoop · MapReduce
Pig

A. Saxena · A. Chaurasia (✉) · N. Kaushik · N. Kaushik
Amity University Uttar Pradesh, Noida, Uttar Pradesh, India
e-mail: achaurasia@amity.edu

A. Saxena
e-mail: asaxena1@amity.edu

N. Kaushik
e-mail: nkaushik1@amity.edu

N. Kaushik
e-mail: nkaushik2@amity.edu

© Springer Nature Singapore Pte Ltd. 2019
S. Bhattacharyya et al. (eds.), *International Conference on Innovative Computing and Communications*, Lecture Notes in Networks and Systems 56,
https://doi.org/10.1007/978-981-13-2354-6_16

1   Introduction

In past recent days, we have been watching and breaking down that there is an expanded significance of gathering, putting away, and handling of gigantic measures of information in the fields of restorative science and advanced world. This pattern is seen in scholarly and corporate industry. New and propel procedures, for example, high-limit and fast stockpiling gadgets make snappier and less expensive the capacity and the entrance of such data. Numerous information, gathered from an assortment of sources, must be clubbed and investigated so connections between various components from various sources can be surmised from the information. The last objective of this sort of handling is the innovation of connections that can give upper hand to corporate world, or logical leaps forward to specialists and the scholarly community, there is an expanding significance in the limit of convenient investigation of immense measures of information with various zones.

In as of late, the best strategy for expanding handling limit of information was the arrangement of extensive scale Clusters. So the extension of the Cluster limit requests colossal interest being developed, support, and administration of the foundation, other than additional costs identified with controlling such frameworks.

To solve this issue of huge data and large scale clusters we can use MapReduce process of big data analytics techniques that facilitate near real-time analytics is iterative refinement [1], i.e. performing the same task over and over again to improve an intermediate result by reusing the initial data and the intermediate state updated by each successive job [2].

Hybrid Cloud: A hybrid cloud environment basically involves a combination of the Public and Private clouds. The goal is to create a well-managed, cost efficient and unified computing model by utilizing different deployment Cloud models [3] (Fig. 1).

An example of hybrid Cloud is a government department using private cloud to store critical information and using public cloud for non-critical information. Other examples of hybrid cloud include Amazon Web Services, Google Cloud, or Joyent Compute.

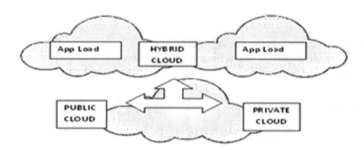

Fig. 1 Hybrid cloud infrastructure

A hybrid cloud combines the advantages of both private and public clouds. It offers the high security of private cloud and low cost of public cloud. A hybrid cloud environment is very beneficial to organizations as it is cost efficient, flexible and provides higher scalability. Basically a hybrid cloud provided best of both worlds.

In today digital world the private cloud is inside the firewall and as secure as any other system on a corporate world. Moreover its infrastructure, private cloud is managed by their own enterprises so it has a top level of control data, security and quality of services. Public cloud is provided by a third party and usually accessed through online, so that it has more advantages on operation and data sharing, and its resources can be considered higher range and extremely cheaper. Above all, hybrid cloud (a composition of two or more distinct cloud infrastructures) is the new wave of cloud computing.

MapReduce: MapReduce is the core part of the Hadoop and is involved in the processing or analysis of the hybrid data that is stored in the HDFS (Hadoop Distributed File System), MapReduce takes the input from the HDFS and then process or analyzes it [4]. MapReduce does not require any specialized hardware but can run on simple commodity hardware with enough memory [5]. MapReduce is divided into two phases. The map phase and the reducer phase. MapReduce is written in java in the earlier version and is written in java in the latest version also [6]. Whether it is any programming language (in the latest version of Hadoop) first before processing the data, the programming language (other than java) is decoded into java and then MapReduce process or analyzes the data [7]. MapReduce retrieves the data from HDFS for analyzing and after completing the task, the output or the result is stored in the HDFS. MapReduce also works on the master–slave architecture. The master in MapReduce is the Job tracker. Job tracker is the component that is in direct contact with the client [8]. The Task Tracker works as a slave node [9]. The job trackers do not do anything by itself, it assigns the task to the task tracker, the task tracker then processes the data and is divided into type phases [10]. One is the mapper phase and the other is the reducer phase [11], the input for the mapper phase comes from the HDFS where the file or the data is stored, then the output for the mapper is the input for the reducer phase, the final result is obtained from the reducer phase and is directly stored in the HDFS (Fig. 2).

2 Review of Literature

In recent days MapReduce with hybrid cloud infrastructure programming has been prominent used in academia and corporate world as well [12]. To implement MapReduce, an open source programming model is Hadoop framework [13]. This will helps improving and implementing basic features of Big Data [14]. Now mostly research target of MapReduce is focus on hybrid cloud infrastructure with public cloud resources. Matsunaga [15], Polo [16], Luo [17], and Fadika [18] proposed models that discuss the execution of MapReduce applications for multiple Clusters. Clusters resources are accessed using suitable techniques so that we can upgrade

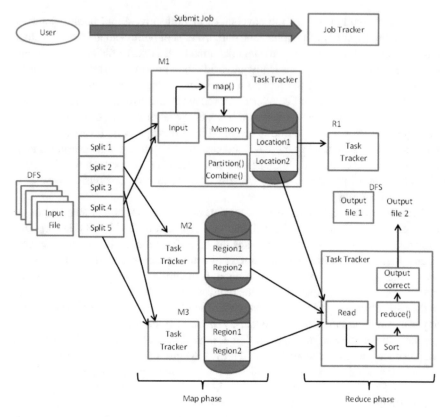

Fig. 2 MapReduce Hadoop framework

and speed up the execution of current application. Researcher final aim is to scale the calculations across the public cloud, where we can upgrade the scalability of available resources without any limit. This is very helpful to achieving soft deadlines defined by the user or client. Tsai [19] works for replication of executors for MapReduce tasks and scheduler. This is how, execution of MapReduce applications in Clouds takes place. Tian and Chen [20] and Verma [21] models for optimizing resource allocation for execution of MapReduce applications over public Clouds and Rizvandi [22] proposed a phenomenon of programmed setup of MapReduce design parameters with a specific end goal to improve execution of uses in a Cloud. Sehgal [23] proposed model for interoperable usage of MapReduce ready to execute applications on Clusters, Grids, and Clouds. The core inspiration of such a framework is to empower interoperation of utilizations that are emphatically attached to a given foundation. Dong [24] proposed an approach for meeting due dates of ongoing MapReduce applications running simultaneously to non-constant applications. Their approach organizes ongoing applications over non-continuous ones, however does not powerfully arrangement additional assets for meeting application due date as does our approach. Kc and Anyanwu [25] proposed an approach were a confirmation control instrument rejects demands for executing MapReduce applications when due dates can't be met.

3 Implementation

In this section we have fetch data from hybrid infrastructure with the help of Hadoop MapReduce Pig architecture in batch processing manner.

The map () procedure of MapReduce performs sorting and filtering data from Hybrid cloud infrastructure and reduce () procedure performs all the technical operation of data from Hybrid cloud infrastructure (Fig. 3).

To perform MapReduce framework we will perform two types of tasks that apply on data from Hybrid cloud infrastructure first is job tracker and second is task tracker.

In Fig. 4, we have shown the practical implementation of fetch data through hybrid cloud using MapReduce techniques over Cloudera.

In Figs. 5 and 6, customer presents a MapReduce application over pig advances to ace hub ready to oversee provisioning of assets and planning of MapReduce application with the assistance of pig. These undertakings are executed by nearby hubs that form the neighborhood Cluster (private Cloud). This nearby Cluster is progressively provisioned assets from an open Cloud to accelerate execution of various assignments.

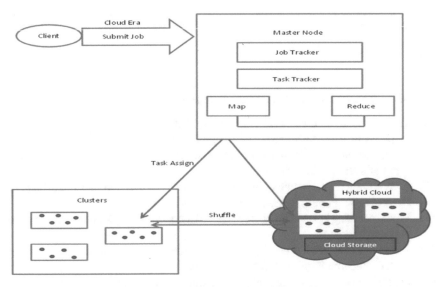

Fig. 3 MapReduce applications on hybrid Clouds

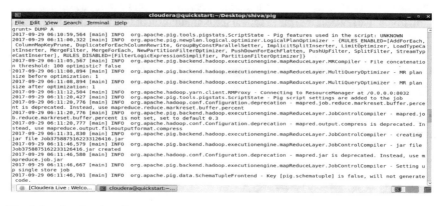

Fig. 4 Directory and file uploading on Pig

Fig. 5 Loading data into filesystem

In this graph we have settle that following key advances:

(1) The first stage comprises of neighborhood hubs enlisted with the ace hub and prepared to execute MapReduce assignments over Pig Framework. Datasets are accessible for neighborhood hubs and are likewise put away in the Cloud supplier's stockpiling administration.

(2) A MapReduce pig application is submitted to the ace hub and the scheduler starts to doling out Map assignments neighborhood hubs.

(3) Whenever Map errands finish, the scheduler appoints new Map undertakings to neighborhood hub.

(4) When the Map period of pig-based is finished, the scheduler starts allotting Reduce undertakings to neighborhood hub.

(5) Each pig-based Reduce errand gets the transitional information to be lessened from different hubs.

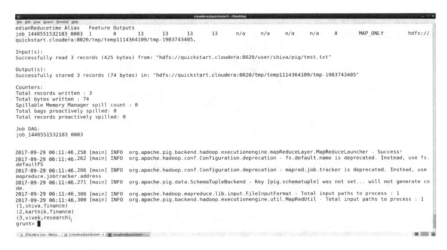

Fig. 6 Results of MapReduce function on Pig

(6) The consequence of the pig construct Reduce errand may stay in light of the hubs in foresight of another pig-based MapReduce application.

In this diagram our aim is to collect the raw data with the help of hybrid cloud and save the text file with the name "test.txt". The sample data file contains ID, Name, and department and many more fields. Our task is to read the content of this file from the HDFS and display all the columns of these records. To process this data is used Pig-based MapReduce. Create a directory at desktop for pig to operate on, then move the 'test.txt' file by using put command to transfer it to HDFS from local file system.

Finally, the utilization of Hybrid Cloud resources with Pig-based MapReduce technique minimizes the financial cost and less complexity to the client.

4 Result and Discussion

MapReduce pig-based method over half-and-half cloud foundation executed with the end goal of figuring a word check application. This procedure has various configurable documents and parameters that influence the insights of the application, which enables customer to watch the framework under various conditions. Guide process can total their own pledge tallies and just discharge one tally an incentive for every one of a kind word.

This conduct can be changed with another arrangement parameter, which directs the most extreme number for the check, implying that words that happened regularly would create different key-esteem sets. Lessen process total means each word produced by contrast Map process and total the relating esteems to create the last mean

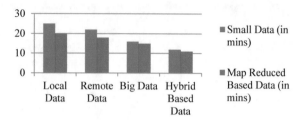

Fig. 7 Comparison between small data and MapReduce based data

word. In this part of the paper our aim to show the results of MapReduce Pig-based process over hybrid cloud infrastructure. This diagram shows the implementation of the Dump command of Pig, the MapReduce job for loading the data into the file system will be done. The result represents the successful query run in pig. In pig the data is stored in HDFS and load the data from HDFS as well, in MapReduce. The MapReduce shows the total number of records, size it occupy, Spill counters point out how many records that has been written in the local disks of the data nodes to avoid running out of memory.

Total bags proactively spilled indicate about the bags that have been spilled. Total records proactively spilled tell us about the records in those bags.

This chart collects the small data and MapReduce-based data in various categories like local data, remote data, big data, and hybrid data and show less execution time of MapReduce data in the comparison of small data.

5 Conclusion and Future Work

The MapReduce with Pig-based system proposed in this paper will help with the adaptability to plan MapReduce handling on information of various affectability levels to various mists prompting a more productive calendar. The last Results demonstrated that our paper, despite the fact that its lower multifaceted nature, conveys great outcomes. Our MapReduce pig-based system over cross-breed cloud framework could meet due dates of vital applications, which are characterized as far as finishing time of the Map portion, for expanding execution times of Map work and diminishing due dates (Fig. 7).

The proposed MapReduce Hadoop over Pig system in this model gives greater security and less multifaceted nature of the applications. Our arrangement to stretch out this procedure to advance the provisioning for more intricate and live situations, for example, different separate applications and composite MapReduce applications with Spark and hive system, where last application expends the yield of an underlying application. We will likewise create programming-based procedures for making up for variety execution of half and half Cloud. We will likewise grow such methods for lopsided hashing and dynamic rebalancing when a larger number of cans than Reduce assignments are available toward the finish of the Map stage.

References

1. Jain S, Saxena A (2016) Analysis of Hadoop and MapReduce tectonics through hive big data. IJCTA 9(14):3811–3911
2. Saxena A, Kaushik N, Kaushik N (2016) Implementing and analyzing big data techniques with spring framework in Java & J2EE. ICTCS ACM Digital Library
3. Saxena A, Kaushik N, Kaushik N, Dwivedi A (2016) Implementation of cloud computing and big data with Java based web application. In: 3rd International conference on computing for sustainable global development. BVICAM, pp 3043–3047
4. Chhawchharia A, Saxena A (2017) Execution of big data using map reduce technique and HQL. In: 4th International conference on computing for sustainable global development. BVICAM
5. Chand M, Shakya C, Saggu GS, Saha D, Shreshtha IK, Saxena A (2017) Analysis of big data using apache spark. In: 4th International conference on computing for sustainable global development. BVICAM
6. Sendre , Singh S, Anand L, Sharma V, Saxena A (2017) Decimation of duplicated images using Mapreduce in Bigdata. In: 4th International conference on computing for sustainable global development. BVICAM
7. Jain S, Saxena A (2017) Integration of spring in Hadoop for data processing. In: 4th International conference on computing for sustainable global development. BVICAM
8. Yesugade K, Bangre V, Sinha S, Kak S, Saxena A (2017) Analyzing human behaviour using data analytics in booking a type hotel. In: 4th International conference on computing for sustainable global development. BVICAM
9. Pathak N, Malik G, Verma A (2018) Security threats in cloud computing and their counter-measures. BVICAM, pp 708–712
10. Dean J, Ghemawat S (2008) MapReduce: simplified data processing on large clusters. Commun ACM 51(1):107–113
11. Shashi, Mishra A (2018) Review study on data security and privacy of cloud computing in service oriented architecture. BVICAM, pp 3838–3841
12. Mishra S, Misra A, Yadav S (2018) Fine tuning of MapReduce jobs using parallel K Map clustering. BVICAM, pp 552–557
13. Bhawna and G. Jagdev, "Analytic Implementation of Big Data in Sports Science Employing Apache Hadoop Framework and MapReduce Algorithm", BVICAM 2018, 320–325
14. Mishra S, Misra A, Yadav S (2018) Improving MapReduce performance using LATE scheduling in big data. BVICAM, pp 558–564
15. Matsunaga A, Tsugawa M, Fortes J (2008) CloudBLAST: combining MapReduce and virtualization on distributed resources for bioinformatics applications. In: eScience'08, Indianapolis, USA, Dec 2008, pp 222–229
16. Polo J, Carrera D, Bacerra Y, Beltran V, Torres J, Ayguadé E (2010) Performance management of accelerated MapReduce workloads in heterogeneous clusters. In: ICPP'10, San Diego, USA, Sep 2010, pp 653–662
17. Srinivasan M, Manjula KR (2018) A review on the implementation of MapReduce in classification algorithms. BVICAM, pp 3582–3584
18. Fadika Z, Dede E, Hartog J, Govindaraju M (2012) MARLA: MapReduce for heterogeneous clusters. In: 12th IEEE/ACM CCGrid'12, Ottawa, Canada, May 2012, pp 49–56
19. Tsai W-T, Zhong P, Elston J, Bai X, Chen Y (2011) Service replication with MapReduce in clouds. In: ISADS'11, Kobe, Japan, June 2011, pp 381–388
20. Tian F, Chen K (2011) Towards optimal resource provisioning for running MapReduce programs in public clouds. In: IEEE (CLOUD'11), Washington DC, USA, July 2011, pp 155–162
21. Bharti U, Bajaj D, Goel A, Gupta SC (2018) Experiences in setting up Hadoop environment and running MapReduce job for teaching big data analytics in universities. BVICAM, pp 4170–4179

22. Rizvandi NB, Taheri J, Zomaya AY, Moraveji R (2011) A study on using uncertain time series matching algorithms in map-reduce applications. The University of Sydney, Technical Report TR672
23. Sharma B, Kumar R, Hashmi A, Gunjan, Gupta P (2018) Map-reduce based parallel firefly algorithm for fast recommendations. BVICAM, pp 2429–2433
24. Dong X, Wang Y, Liao H (2011) Scheduling mixed real-time and nonreal-time applications in MapReduce environment. In: ICPADS'11, Tainan, Taiwan, Dec 2011, pp 9–16
25. Suhani, Jain N (2018) Proposed face recognition system using MapReduce. BVICAM, pp 4911–4914

Statistical Survey of Data Mining Techniques: A Walk-Through Approach Using MongoDB

Samridhi Seth and Rahul Johari

Abstract Big data is a term used for management of large, unstructured and complex data. It is used to organize data such that it is easy to read and understand. In today's world of digitization, it becomes important to track the growth and evolution of data mining techniques. This paper makes an effort in this direction as thorough and in-depth study has been carried out of several mining techniques. Also to show the effectiveness of data mining techniques, simulation has been carried out of the real-time dataset in MongoDB.

Keywords Machine learning · NoSQL · MongoDB · Data mining
Crowdsourcing · Internet of things

1 Introduction

Today, the amount of data is increasing exponentially. All big companies like Google, Facebook deals with massive data (Table 1). A big challenge to them is to maintain this data in a secure way.

Further, big data techniques, discussed below, are further contributing in the exponential increase of data. They have both merits and demerits. With digitization, the country has developed a lot but secured maintenance is a big issue. This data must be organized in an intelligent way and analysed wisely.

S. Seth (✉) · R. Johari
USIC&T, GGSIP University, New Delhi, India
e-mail: samridhiseth15@gmail.com

R. Johari
e-mail: rahul.johari.in@ieee.org

© Springer Nature Singapore Pte Ltd. 2019
S. Bhattacharyya et al. (eds.), *International Conference on Innovative Computing and Communications*, Lecture Notes in Networks and Systems 56,
https://doi.org/10.1007/978-981-13-2354-6_17

Table 1 Tabular representation of seven V's of big data [1]

S.no.	V's of big data	Description
1.	Volume	Volume means that today quantity of data generated is in large volume which needs to be maintained. For example, a railway site has large volume of details, about each and every station of India
2.	Variety	Variety indicates there is a variety of data to be maintained: text, mail, video, audio etc., and organizations collect data from a variety of sources
3.	Variability	Variability means that meaning of data is constantly changing
4.	Veracity	Veracity refers to reliability of the data
5.	Visualization	Visualization is how the complex data is displayed so that it is easy to understand. For example, some data is easy to represent in tabular form than in paragraphs
6.	Value	Value indicates worth of data, that is, it is of use for someone
7.	Velocity	Velocity is data must be consistent. Timing is an important factor with advanced technology. For example, in an online reservation system, it is important that money transactions take place at a faster pace

Nine more V's proposed as a part of this statistical survey are:

8.	Viscosity [2, 3]	Viscosity is the latency in the data relative to event being described, or how easily it can be changed into another scenario
9.	Virality [2]	Virality refers to the speed at which data spreads
10.	Validity [4]	Validity refers to correctness and accuracy of data
11.	Volatility [4]	Volatility refers to how long is the data a valid one, since, mostly the work is done on live data we also need to consider the relevance of data
12.	Vulnerability [4]	Vulnerability is susceptibility to security breach
13.	Verbality [3]	Verbality refers wide variety of unstructured data
14.	Verbosity [3]	Verbosity is to understand the semantics of redundancies between structured, unstructured and semi-structured data for efficient processing and reuse of data
15.	Versatility [3]	Versatility represents the importance of data in each sector, scenario and context
16.	Visibility [3]	Visibility is the access control of data

Various techniques of big data are discussed below:

1. **Machine Learning**: It involves automation of things. Big data needs to be organized in a way for automation. The categories of machine learning are:

   a. Supervised Learning.
   b. Unsupervised Learning.
   c. Reinforcement Learning.
   d. Deep Learning.

    e. Associate Rule Learning.

    f. Numeric Prediction.

2. **Data Mining**: Processing of large volumes of data using SQL Queries (used for small dataset) or Online Analytical Processing (used for large dataset) and retrieve important information from them, known as Knowledge Discovery in Database (KDD).

3. **NoSQL**: NoSQL gives a mechanism for maintaining (store and retrieve) data in large volumes, in means others than the tabular relations, which is the main advantage of NoSQL. No schema has to be maintained beforehand. Schema is maintained for each record individually. So, different types of data can be accommodated. NoSQL provides high-grade data consistency.

4. **Predictive Analytics**: Future behaviour is predicted from the collected data. The basic ideology used is that future plans are dependent on past actions. But another ideology opposes this, that is, future is unpredictable. So, it is said that improvements can be made based on past experiences.

5. **Crowdsourcing**: It is a practice to gather information on a project from large number of people, via internet, either paid or unpaid surveys to take advices, opinions of the people to know the needs of the user.

6. **Internet of Things (IoT)**: It is a vast network of devices, either living or nonliving, connected to Internet used for exchange of data. It is basically automation of each and everything.

The literature survey of different research papers undertaken as a part of the current research work is summarized below.

2 Literature Survey

In [5], the author(s) describe an efficient support system. Design support systems are important, but they should provide a solution in a timely manner. According to the author(s), existing systems are not that efficient as they do not use appropriate methods and tools for analysis. So, the author(s) have started a business process improvement methodology with the aim to improve the performance of existing decision support systems. Improvement methodology consists of five phases:

Phase 1—Define: discovers and defines the business process that is to be improved.

Phase 2—Configuration: to gather data of existing system and feed in later versions.

Phase 3—Execution: System configuration and execution of operational systems.

Phase 4—Control: Monitor and analysis of the outcome of decision support systems.

Phase 5—Diagnosis: Identify weaknesses of decision support systems.

In [6], the author(s) focuses on the unstructured data, that is, audio, videos, images and unstructured text, which constitutes a big part of big data. First, the author(s) describes big data and its seven V's. Big data is useless, unless, it is organized in some form and can be used in some decision-making. The process of extracting meaning is divided into two sub-processes: Data Management and Analytics.

Data management is the process of storing data and prepares it for retrieval for analysis. Data Management is a three-step process:

- Acquisition and recording.
- Extraction, cleaning and annotation.
- Integration, aggregation and representation.

Analytics are the techniques used for obtaining important information from the big data. Analytics is a two-step process:

- Modelling and analysis.
- Interpretation.

The paper also discusses big data analytic techniques and the tools available for big data analytics. Various big data techniques are:

- **Text Analytics** is to analyse textual data. The techniques are information extraction, text summarization, question answering and sentiment analysis.
- **Audio Analytics** (also known as Speech Analytics) is to acquire information from unstructured audio data. The technologies involved are transcript-based approach and phonetic-based approach.
- **Video Analytics** (also known as Video Content Analysis) is to acquire information from video streams. Video Analytics techniques are server-based and edge-based architecture.
- **Social Media Analytics** is to extract and analyse structured and unstructured data from social media and social media is a big hype these days.
- **Predictive Analytics** is to future prediction based on historical and current data.

In [7], the author(s) describes big data, its uses, process and characteristics, advantages and its features. Big data can be classified into five classes:

- *Data Sources*: Web and Social, Machine, Sensing, Transactions and IoT.
- *Content Format*: Structured, Semi-Structured and Unstructured.
- *Data Stores*: Document-oriented, Column-oriented, Graph based, Key value.
- *Data Staging*: Cleaning, Normalization and Transform.
- *Data Processing*: Batch and Real time.

Big data can be used in various fields like automotive industry, oil and gas industry, telecommunication sector, medical field, retail industry and health services. The author(s) also discusses various academic studies on big data.

In [8], the author(s) have discussed an application of big data, organization management and optimization of decision support systems. The author(s) have also described big data, its techniques and characteristics. Big data applications in different sectors and their benefits to the society have also been discussed.

In [9], the author(s) have proposed stages of decision-making process, which is highly dependent on big data. Decision-making is an important part for industry's success since raw unstructured data need to be organized in a proper format which can be used for the growth of industry. Quality of decision-making can be improvised by rule of 4F, that is, focus, fast, first and flexibility.

Proposed decision-making process has four stages, that is,

- Specifying authorized source of data to indicate the performance of enterprise.
- Developing and implementing scorecards based on assessment measures.
- Formulating and implementing business rules.
- Providing professional coaching.

Each company, either big or small has some amount of data stored. So, big data is an important part of each company. In [10], author(s) show the way how big data can be used for competitive advantage of the company using various analysis methods and well-developed tools. Collected data can be used for understanding the sentiments of customers, bringing transparency between workers of the organization, providing more accurate information and hence, improve decision-making.

Information processing is another issue in big data. Massive data is present, therefore, search engines should be an efficient one. Generally, keywords-based search, enterprise decision-making and passive service mode return nothing if no keywords are submitted and is very inefficient, enterprise decision-making is efficient for local data only. So, the author(s) in [11], provide a solution to make information processing (that is, search engine) automatic. Automation process includes two phases: domain knowledge construction and domain knowledge-based information service.

In [12], the author(s) have presented a big data assessment framework which helps organizations to analyse whether big data is actually involved in the respective project. This decision support system tool also prevents organizations from taking wrong decisions and to understand their project in a better way.

In [13], the author(s) have designed a big data analytics and decisions framework to incorporate big data analytics into decision-making. This framework can be applied in research and industry but has a little time and resource limitation. The framework based on Peffers et al's six-stage design science process includes identifying problems, defining objectives for solution, applying knowledge, framework development, evaluation and demonstration. Development includes identifying, storing, organizing and designing. Designing involves model planning, data analytics and analysing (defining appropriate actions). Evaluation is evaluating and deciding categories of data. Implementation is overall maintenance, that is, monitoring and feedback.

In [14], the author(s) have made three contributions in the field of big data. First, author(s) have contributed an archetype business process. Second, they have shown the role of big data in various fields as lack of clear vision in big data hampers growth of organization. Relationships between big data and dynamic capabilities have also been taken into consideration. The third contribution has been that it is generally thought that resource management is implemented via resource-based theory but now big data has taken over it. The author(s) argue that it should not be the case that organizations use big data once or twice, big data will be successful only if it could help organizations again and again and can be involved in each and every project.

In [15], the author(s) have taken into consideration the connection between IOT and big data. IOT is the new trend nowadays. The author(s) discusses researches in the fields of big data related to IOT. A new architecture on big IOT data analytics, its types, methods and technologies are also proposed. The paper also specifies the importance of big data in IOT and some open research challenges like privacy, big data mining, visualization and integration have also been highlighted. The paper concludes with prediction in the future, real-time analytics solution will be required.

In [16], the author(s) discuss the benefits and demerits of using the Big Data Technology. The paper describes the different phases of the Big Data Processing, different V's of Big Data along with the salient applications of the Big Data in the real-time applications.

3   Simulation and Analysis

For the implementation in MongoDB [17], the dataset used describes the rural and urban development in different states of India for the year 2010–11 to 2015–16. It specifies following points for each State and Union Territory in India for the period 2010–11 to 2015–16 [18]:

1. Number of households who had demanded employment.
2. Number of households who were provided employment.
3. Persondays in lakhs (Total).
4. Persondays in lakhs (SC).
5. Persondays in lakhs (ST).
6. Persondays in lakhs (Others).
7. Persondays in lakhs (Women).
8. Average person days per household.
9. Number of households who availed 100 days of employment.

It includes 36 tuples out of which few random are given below (Table 2).

Table 2 Rural and urban development in different states of India for the year 2010–11 to 2015–16 [18]

State/UT	Deml10–11	Deml11–12	Deml12–13	Deml13–14	Deml14–15	Deml5–16	Provl10–11	Provl11–12	Prmrl2–13	Provl3–14	Provl4–15
Assam	1803046	1354868	903548	1321079	1083011	1668379	1783230	1348958	903548	1261778	967179
Bihar	4344263	1743650	1529882	2378439	1473530	1923464	4330799	1716603	1529882	2059338	1034349
Goa	16415	11174	3782	5032	7250	5976	16315	11167	3782	5021	7225
Gujarat	1097470	836905	591576	642807	595720	642384	1093289	822039	591576	578674	513190
Haryana	238484	278556	213556	361761	362882	199585	238385	277834	213556	324919	217914
Kerala	1152800	1418062	1600827	1678824	1565148	1664788	1143552	1416444	1600827	1523863	1380236

State/UT	Prov15–16	tot10–11	tot11–12	tot12–13	tot13–14	tot14–15	tot15–16	SC10–11	SC11–12	SC12–13	SC13–14
Assam	1502410	468	353	178	298	211	486	50	20	10	20
Bihar	1438429	1482	657	499	862	352	671	670	162	120	251
Goa	5909	4	3	0	1	2	1	0	0	0	0
Gujarat	556810	492	313	181	230	182	225	72	24	17	18
Haryana	168724	88	109	70	118	62	48	42	54	36	57
Kerala	1505666	444	633	500	866	589	742	72	93	76	146

State/UT	SC14–15	SC15–16	ST10–11	ST11–12	ST12–13	ST13–14	ST14–15	ST15–16	Oth10–11	Oth11–12	Oth12–13	Oth13–14	Oth14–15	Oth15–16	wom10–11
Assam	13	25	127	80	34	48	32	91	290	253	133	231	166	370	125
Bihar	99	160	30	11	9	18	6	12	782	484	370	593	247	500	453
Goa	0	0	1	1	0	0	0	0	3	2	0	1	1	1	3
Gujarat	13	17	203	127	65	95	72	95	218	162	99	118	97	114	218
Haryana	27	24	0	0	0	0	0	0	46	55	34	61	35	24	29
Kerala	103	129	14	15	15	25	23	29	358	525	410	694	463	534	401

(continued)

Table 2 (continued)

State/UT	wom11–12	wom12–13	wom13–14	wom14–15	wom15–16	avg10–11	avg11–12	avg12–13	avg13–14	avg14–15	avg15–16	100_10.11	100_11.12	100_12.13
Assam	88	44	74	59	163	26	26	20	24	22	32	44681	15701	1201
Bihar	189	152	302	131	274	34	38	33	42	34	45	186588	162940	59812
Goa	2	0	1	1	1	23	28	13	23	24	18	413	143	0
Gujarat	139	78	101	78	104	45	38	31	40	35	40	67651	41759	23158
Haryana	40	28	49	26	22	37	39	33	36	28	29	8858	13762	5976
Kerala	587	436	809	542	677	39	45	31	57	43	49	52599	124865	21233

The queries run on MongoDB are given below:

1. MongoDB query to display the state with minimum number persondays in lakhs (total) for the year 2010–11.

```
> db.mnrega.find().sort({"tot10_11":1}).limit(1)
{ "_id" : ObjectId("5a5b34006873d8f6a61c0591"), "State" : "Lakshwadeep", "Dem10_11" : "4507", "Dem11_12" : "3866", "De
m12_13" : "841", "Dem13_14" : "926", "Dem14_15" : "549", "Dem15_16" : "147", "Prov10_11" : "4507", "Prov11_12" : "3855
", "Prov12_13" : "841", "Prov13_14" : "841", "Prov14_15" : "614", "Prov15_16" : "477", "tot10_11" : "1", "tot11_12" :
"2", "tot12_13" : "0", "tot13_14" : "0", "tot14_15" : "0", "tot15_16" : "0", "SC10_11" : "0", "SC11_12" : "0", "SC12_1
3" : "0", "SC13_14" : "0", "SC14_15" : "0", "SC15_16" : "0", "ST10_11" : "1", "ST11_12" : "2", "ST12_13" : "0", "ST13_
14" : "0", "ST14_15" : "0", "ST15_16" : "0", "Oth10_11" : "0", "Oth11_12" : "0", "Oth12_13" : "0", "Oth13_14" : "0", "
Oth14_15" : "0", "Oth15_16" : "0", "wom10_11" : "1", "wom11_12" : "1", "wom12_13" : "0", "wom13_14" : "0", "wom14_15"
: "0", "wom15_16" : "0", "avg10_11" : "30", "avg11_12" : "43", "avg12_13" : "31", "avg13_14" : "24", "avg14_15" : "26"
: "avg15_16" : "22", "awaili10_11" : "71", "awaili11_12" : "134", "awaili12_13" : "16", "awaili13_14" : "11", "awaili14_15"
: "4", "awaili15_16" : "3" }
```

2. MongoDB query to display the development in Chhattisgarh for the year 2010–11 to 2015–16.

```
> db.mnrega.find({state:"Chhattisgarh"});
{ "_id" : ObjectId("5a5b4fca6873d8f6a61c059a"), "State" : "Chhattisgarh", "Dem10_11" : "2470926", "Dem11_12" : "273817
9", "Dem12_13" : "2203344", "Dem13_14" : "2748721", "Dem14_15" : "2042939", "Dem15_16" : "261192@", "Prov10_11" : "244
8745", "Prov11_12" : "2724228", "Prov12_13" : "2203344", "Prov13_14" : "2512379", "Prov14_15" : "1748266", "Prov15_16"
: "2174153", "tot10_11" : "1097", "tot11_12" : "1287", "tot12_13" : "690", "tot13_14" : "1299", "tot14_15" : "556", "
tot15_16" : "1014", "SC10_11" : "158", "SC11_12" : "116", "SC12_13" : "68", "SC13_14" : "117", "SC14_15" : "60", "SC15
_16" : "85", "ST10_11" : "397", "ST11_12" : "452", "ST12_13" : "229", "ST13_14" : "521", "ST14_15" : "178", "ST15_16"
: "431", "Oth10_11" : "541", "Oth11_12" : "639", "Oth12_13" : "393", "Oth13_14" : "661", "Oth14_15" : "318", "Oth15_16
" : "498", "wom10_11" : "535", "wom11_12" : "545", "wom12_13" : "326", "wom13_14" : "630", "wom14_15" : "277", "wom15_
16" : "497", "avg10_11" : "45", "avg11_12" : "44", "avg12_13" : "31", "avg13_14" : "52", "avg14_15" : "32", "avg15_16"
: "47", "await10_11" : "182113", "await11_12" : "208146", "await12_13" : "44933", "await13_14" : "346287", "await14_1
5" : "48087", "await15_16" : "242583" }
>
```

3. MongoDB query to display the average persondays per household for the year 2015–16 in Kerala.

```
> db.mnrega.find({State:"Kerela"},{avg15_16:1});
{ "_id" : ObjectId("5a5b3c376873d8f6a61c0593"), "avg15_16" : "49" }
```

4 Conclusion and Future Work

Big data techniques have been used in the past and will continue to be used in the future since big data is the heart of engineers and help them in solving big challenges and design more user-friendly equipments. Second, big data can be used in any field be it agriculture or weather forecasting. Big data is a vast topic and important for the development, some means of encouragement must be provided either financially or by some other means to attract expertise in this field. In the future, own data mining techniques and their implementations in Hadoop will be proposed. To the best of our knowledge and understanding, big data are not a market but a technology that is still evolving and its power and potential still needs to be harnessed.

References

1. https://www.impactradius.com/blog/7-vs-big-data/
2. http://blog.softwareinsider.org/2012/02/27/mondays-musings-beyond-the-three-vs-of-big-data-viscosity-and-virality/
3. https://bigdata.cioreview.com/cxoinsight/the-other-five-v-s-of-big-data-an-updated-paradigm-nid-10287-cid-15.html
4. https://tdwi.org/articles/2017/02/08/10-vs-of-big-data.aspx
5. Vera-Baquero A, Colomo-Palacios R, Molloy O (2014) Towards a process to guide big data based decision support systems for business processes. Procedia Technol 16:11–21
6. Gandomi A, Haider M (2015) Beyond the hype: Big data concepts, methods, and analytics. Int J Inf Manage 35(2):137–144
7. Özköse H, Arı ES, Gencer C (2015) Yesterday, today and tomorrow of big data. Procedia Soc Behav Sci 195:1042–1050
8. Ziora, ACL (2015) The role of big data solutions in the management of organizations. Review of selected practical examples. Procedia Comput Sci 65:1006–1012
9. Kościelniak H, Puto A (2015) BIG DATA in decision making processes of enterprises. Procedia Comput Sci 65:1052–1058
10. Kubina M, Varmus M, Kubinova I (2015) Use of big data for competitive advantage of company. Procedia Econ Finan 26:561–565
11. Jin X et al (2015) A domain knowledge based method on active and focused information service for decision support within big data environment. Procedia Comput Sci 60:93–102
12. Portela F, Luciana L, Manuel FS (2016) Why big data? Towards a project assessment framework. Procedia Comput Sci 98:604–609
13. Elgendy N, Elragal A (2016) Big data analytics in support of decision making process. Procedia Comput Sci 65:1071–1084

14. Braganza A et al (2017) Resource management in big data initiatives: processes and dynamic capabilities. J Bus Res 70:328–337
15. Marjani M et al (2017) Big IoT data analytics: architecture, opportunities, and open research challenges. IEEE Access 5:5247–5261
16. Bajaj S, Rahul J (2016) Big data: a boon or bane-the big question. In: 2016 Second international conference on computational intelligence & communication technology (CICT). IEEE
17. https://www.mongodb.com/download-center#atlas
18. https://data.gov.in/resources/state-ut-wise-implementation-report-under-mgnrega-2010-11-2015-16

Fuzzy Risk Assessment Information System for Coronary Heart Disease

Pankaj Srivastava and Neeraja Sharma

Abstract It is a well-known fact that Coronary Heart Disease (CHD) is spreading worldwide due to the lavish lifestyle of people and this one is verified by the report of WHO which declared South Asian Subcontinent a hub of cardiac disease people. Several known and vague factors play the major role which is responsible for CHD. In the present paper, we made an attempt to develop Fuzzy Informative System for risk assessment scheme for CHD by making use of fuzzy relational features and assessment function to assess the different phases of the cardiac patient.

Keywords Softcomputing · Fuzzy tools · Coronary heart disease (CHD)
Risk assessment function · Blood pressure (BP) · Cholesterol · Triglyceride

1 Introduction

According to the Indian Heart Watch (IHW), risk factors for Coronary Heart Disease (CHD) in Indian are at higher levels than in developed countries and regions such as the US and Western Europe [1, 2]. There is no single factor that causes CHD, but a large number of factors such as blood pressure, cholesterol, smoking, physically inactive, obesity, diabetes, and so on affects the CHD which give uncertain, vague information to the medical experts for the prediction and diagnosis of CHD. Since vague and imprecise features are handled by the fuzzy theory [3], which calibrates experts reasoning and intuition in risk assessment for CHD. Klire and Yuan [4] gave detailed theory and application of soft computing.

A rule-based system for cardiac analysis was developed by Pandey et al. [7] in which they diagnosed ECG graphs by using the principles of fuzzy logic and knowl-

P. Srivastava
Motilal Nehru National Institute of Technology, Allahabad, Uttar Pradesh, India
e-mail: drpankaj23@gmail.com

N. Sharma (✉)
Government Girls Inter College, Lalganj, Raebareli, Uttar Pradesh, India
e-mail: 06.neeraja@gmail.com

© Springer Nature Singapore Pte Ltd. 2019
S. Bhattacharyya et al. (eds.), *International Conference on Innovative Computing and Communications*, Lecture Notes in Networks and Systems 56,
https://doi.org/10.1007/978-981-13-2354-6_18

edge of medical experts for cardiac analysis. Srivastava and Srivastava [5] designed Fuzzy Risk Assessment system to determine coronary heart disease (CHD) besides it, they [6] also designed a user-friendly, intelligent, and effective diagnosis system for risk assessment of hypertension. Gayathri and Jaisankar [7] gave a comprehensive study of Heart Disease Diagnosis using Data Mining and Soft Computing Techniques. Rajalakshmi and Latha [8] analysis of the uniqueness of medical data mining with soft computing techniques used for classification, diagnosis, prediction, and prognosis in Coronary Artery Diseases (CAD). Uyara and Ilhana [9] used a genetic algorithm based on trained recurrent fuzzy neural networks to the diagnosis of Heart Diseases. Wiharto et al. [10] used Intelligence systems for diagnosis CHD with K-Star Algorithm.

Davidson et al. [11] proposed a fuzzy risk assessment tool for early-stage risk assessment of microbial hazards in food systems and it motivated an attempt in the present paper to design and development of a Fuzzy Risk Assessment Information System to assess Coronary Heart Disease risk grade in men and women.

2 Material and Methodology

In the design and development of fuzzy risk assessment system for CHD, we have used trapezoidal fuzzy number [4] and their multiplication (1) for the imprecise factors affecting CHD.

Two trapezoidal fuzzy numbers X = (a, b, c, d) and Y = (p, q, r, s) with α-cuts and the membership function of the product (2) are as follows:

$$X^\alpha = [a + (b - a)\alpha, d(d - c)\alpha] \text{ and } Y^\alpha = [p + (q - p)\alpha, s(s - r)\alpha]$$

$$X \otimes Y = (min(ap, as, dp, ds), min(bq, br, cq, cr), max(bq, br, cq, cr), max(ap, as, dp, ds)) \tag{1}$$

$$\mu_{X \otimes Y}(x) = \begin{cases} 0, & x \leq ap \\ \frac{-(b-a)p+(q-p)a)+\sqrt{((b-a)p+(q-p)a)_2-4(b-a)(q-p)(ap-x)}}{2(b-a)(q-p)} & ap < x < bq \\ 1, & bq \leq x \leq cr \\ \frac{((s-r)d+(d-c)s)+\sqrt{((s-r)d+(d-c)s)_2-4(d-c)(s-r)(sd-x)}}{2(d-c)(s-r)} & cr < x < sd \\ 0, & x \geq sd \end{cases} \tag{2}$$

We have developed a risk assessment function that estimates early risk assessment of the proposed risk assessment informative system for coronary heart disease, which is as follows:

$$RA = \frac{1}{1 + exp^{\frac{-\sum_{k=1}^{n} u_k x_k}{2}}} \tag{3}$$

where n is the number of parameters, u_k and x_k are fuzzy knowledge of parameters that govern the proposed information system.

2.1 Key Rule for the Proposed Fuzzy Information System

The diagnosing and analyzing behavior of the proposed algorithm for the Information System is mentioned in the following three steps.

STEP 1: Algorithm for Fuzzification Mechanism

    1. Input—Information System with suitable "N" parameters A_i, i = 1, 2, …, N
    2. Initialize $i \leftarrow 1$
while $i \leq N$ **do**
    1. Categorize in M_i fuzzy sets X_j in terms of linguistic variables, j = 1, 2, …, M_i
    while $j \leq M_i$ **do**
        1. Design Trapezoidal membership function $\mu_{X_j}(a, b, c, d)$ using Eq. (2)
    end while
end while

STEP 2: Algorithm for Fuzzy Relation Mechanism

    1. Input—All important pairs of parameters A_i and A_j for calculating Fuzzy Relation
    2. Initialize $i \leftarrow 1$
while $i \leq N$ **do**
    while $j \leq N$ **do**
        1. Develop fuzzy relation R_{ij} in consultation with medical experts
    end while
end while

STEP 3: Algorithm for Risk Assessment Mechanism

    1. Input—All pairs of parameters A_i and A_j used in K calculating Fuzzy Relation
    2. Initialize $k \leftarrow 1$
while $k \leq K$ **do**
    1. Calculate product $u_k = a_i a_j$ using trapezoidal arithmetic operations using (1).
    2. Determine membership grade of product u_k, using (2).
end while
    3. Assess risk of the patient by the proposed Information system using the product u_k and $x_k = r_{ij}$ corresponding to parameters A_i and A_j in (3).

3 Risk Factors for Risk Assessment Fuzzy Information System of Coronary Heart Disease (CHD)

It is a known fact that CHD is mainly caused by Age (Men and Women), Systolic Blood Pressure (SBP), Diastolic Blood Pressure (DBP), Low-Density Lipoprotein Cholesterol (LDL), High-Density Lipoprotein Cholesterol (HDL), Total cholesterol (TC), and Triglyceride due to which we have taken these features as risk factors to calculate risk estimate for CHD. The age of men and women has been categorized into five fuzzy sets, where in women, the CHD risk appears at higher level after postmenopause than premenopause [12]. Smoking is categorized on the basis of number of cigarettes per day a person consumes. These risk factors are categorized into different fuzzy sets which are given below.

We have also considered diabetes, family history, and obesity factors whose membership grades are only 1 and 0, as they prevail or not prevail, respectively.

4 Fuzzy Relations for Risk Assessment Information System of Coronary Heart Disease (CHD)

In this section, we develop fuzzy relational matrices among the risk factors selected for men and women in consultation with cardiologists Dr. Geeta Shukla (MD) and Dr. Omar Hasan (MD, DM) (Nazareth Hospital, Allahabad, India) and as per selected reports of Framingham Heart Study [13] to assess CHD risk.

We have categorized elements of fuzzy relation matrices using fuzzy set = {(Low(L), 0.25), (Medium, 0.55), (High, 0.75)} and hedges given as: Little Bit (LB) = $(\mu_A(x))^3$, Slightly (S) = $(\mu_A(x))^2$, Above (A) = $(\mu_A(x))^{1/2}$, Very (V) = $(\mu_A(x))^{1/3}$, Extremely (E) = $(\mu_A(x))^{1/6}$.

The fuzzy relations for men and women are as follows.

5 Result and Discussion

In this section, we would like to check the output for various cases by the proposed Fuzzy Information System (Sect. 2.1) as explained.

Case 1:

The patient's Age = 47 year old, Gender = Male, SBP = 143 mm/Hg, DBP = 88 mm/Hg, TC = 180 mg/dl, LDL = 100 mg/dl, HDL = 30 mg/dl, Triglyceride = 250 mg/dl, Diabetic = Prediabetes, Smoker = No, Family History = Yes, Obese = No.

As per first step proposed in Information System (Sect. 2.1), we develop corresponding trapezoidal fuzzy numbers for Age, SBP, DBP, LDL, HDL, and Triglyceride which are as follows:

Table 1 Fuzzy set of age

Age	Ranges (yrs)
Aged young (AY)	28–35
Slightly aged (SA)	35–45
Aged (A)	45–55
Very aged (VA)	55–65
Old (O)	≥65

Table 2 Fuzzy set of SBP

SBP	Ranges (mm/Hg)
Normal (N)	110–120
Above normal (AN)	119–140
Moderate (M)	139–160
High (H)	159–170
Very high (VH)	≥169

Table 3 Fuzzy set of DBP

DBP	Ranges (mm/Hg)
Normal (N)	70–80
Above normal (AN)	79–85
Moderate (M)	84–90
High (H)	89 to100
Very high (VH)	≥99

Table 4 Fuzzy set of LDL

LDL	Ranges (mg/dl)
Optimal (O)	≤100
Above optimal (AO)	100–130
Borderline high (M)	129–160
High (H)	159–190
Very high (VH)	≥189

Age = [40, 44, 45, 49], SBP = [140, 144, 145, 149], DBP = [78, 79, 80, 81], TC = [175, 179, 180, 184], LDL = [95, 99, 100, 104], HDL = [25, 29, 30, 34], Triglyceride = [241, 249, 251, 259].

Making use of Eqs. (1) and (2), we evaluate membership value of product u_k of trapezoidal fuzzy numbers for corresponding fuzzy relations (in Tables 1, 2, 3, 4, 5, 6, 7, 8, 9, and 10 for Men) which are as follows:

$\mu_{Age \times SBP} = 0.8685$, $\mu_{Age \times DBP} = 0.7125$, $\mu_{Age \times TC} = 0.7278$, $\mu_{Age \times LDL} = 0.7834$, $\mu_{Age \times HDL} = 0.9271$, $\mu_{Age \times Triglycerides} = 0.8775$, $\mu_{Age \times Diabetes} = 1$, $\mu_{Age \times Smoking} = 0$, $\mu_{Age \times FamilyHistory} = 1$, $\mu_{Age \times Obesity} = 0$.

Table 5 Fuzzy set of HDL

HDL	Ranges (mg/dl)
Very low (VL)	≤ 35
Low (L)	35–40
Borderline high (M)	39–60
Optimal (O)	≥ 59

Table 6 Fuzzy set of total cholesterol

TC	Ranges (mg/dl)
Very low (VL)	≤ 160
Low (L)	160–200
Desirable (D)	199–240
High (H)	239–260
Very high (VH)	≥ 259

Table 7 Fuzzy set of Triglyceride

Triglyceride	Ranges (mg/dl)
Normal (N)	≤ 150
Mildly high (MH)	150–200
High (H)	199–500
Very high (VH)	≥ 499

Table 8 Fuzzy set of smoking

Smoking	Ranges
Low smoking (LS)	≤ 3 per day
Moderate smoking (MS)	3–7 per day
Above moderate smoking (AMS)	7–12 per day
High smoking (HS)	12–18 per day
Very high smoking (VHS)	≥ 18 per day

Table 9 Age(Men) between SBP

VVL	L	M	H	VH
SL	M	AM	AH	VH
L	M	H	VH	EH
L	AM	AH	EH	EH
L	AM	VH	EH	EH

Now, we compute $\sum_{k=1}^{10} u_k x_k$ of Risk Assessment Function (3) by making use of evaluated values of u_k and fuzzy relational value for x_k and obtained risk grade as per procedure mentioned in third step of (Sect. 2.1) as follows:

Table 10 Age(Men) between DBP

SL	L	M	H	VH
SL	L	M	H	VH
SL	M	AM	AH	VH
L	SM	AM	VH	EH
L	AM	AM	EH	EH

Table 11 Age(Men) between LDL

SL	L	M	H	AH
L	SM	M	AH	VH
L	M	AH	VH	EH
L	AM	VH	EH	EH
L	AM	VH	EH	EH

Table 12 Age(Men) between HDL

H	M	SL	VVL
AH	M	L	VVL
AH	H	SM	SL
VH	H	AM	SL
EH	VH	AM	SL

$\sum_{k=1}^{10} u_k x_k$ = (0.7500)0.4839 + (0.7416)0.5896 + (0.3025)0.6545 + (0.5500)0.7500 + (0.8660)0.7203 + (0.8660)0.8474 + (0.3025)1 + 0 + (0.2500)1 + (0.3025)0 = 3.3049.

Risk Grade = 0.8392.

The result shows that the concerned person has 83.92% risk estimate for coronary heart diseases. This risk estimate may be reduced by adopting a balanced cardiovascular suitable diet and exercises.

Case 2:

The patient's Age = 32 year old, Gender = Female, SBP = 156 mm/Hg, DBP = 90 mm/Hg, TC = 256 mg/dl, LDL = 185 mg/dl, HDL = 30 mg/dl, Triglyceride = 203 mg/dl, Diabetic = No, Smoker = 10 cigarettes per day, Family History = Yes, Obese = No.

We follow the same mechanism as quoted in case 1 to evaluate risk grade.

Trapezoidal fuzzy numbers for Age, SBP, DBP, LDL, HDL, Triglyceride and membership values of products u_k, k = 1, 2, ..., 10 corresponding to fuzzy relations (in Tables 11, 12, 13, 14, 15, 16, 17, 18, 19 and 20 for Women) are given below: Age = [30, 34, 35, 39], SBP = [150, 154, 155, 159], DBP = [88, 89, 90, 91], TC = [250, 254, 255, 259], LDL = [180, 184, 185, 189], HDL = [30, 34, 35, 39], Triglyceride = [201, 209, 211, 219].

Table 13 Age(Men) between TC

SL	L	SM	H	AH
SL	L	M	H	VH
L	SM	AM	AH	VH
L	SM	H	EH	EH
L	AM	AH	EH	EH

Table 14 Age(Men) between triglyceride

SL	SM	H	VH
SL	AM	AH	VH
L	AM	AH	VH
L	H	VH	EH
L	AH	EH	EH

Table 15 Age(Men) between diabetes

Age	Prediabetes	Type-1 diabetes	Type-2 diabetes
Aged young	L	SM	M
Slightly aged	L	SM	M
Aged	SM	M	AM
Very aged	SM	AM	H
Old	SM	AM	H

Table 16 Age(Men) between smoking

SM	M	AM	H	AH
SM	M	AM	AH	VH
M	AM	H	AH	VH
M	AM	AH	VH	EH
AM	H	VH	VH	EH

Table 17 Age(Men) between family history

Age	Family history
Aged young	SL
Slightly aged	L
Aged	L
Very aged	SM
Old	M

Risk Grade = 0.7887.

This shows 78.87% risk estimate of the concerned smoker woman. It can be lowered by removing or reducing modifiable risk factors like smoking, high cholesterol,

Table 18 Age(Men) between obesity

Age	Obesity
Aged young	L
Slightly aged	L
Aged	SM
Very aged	M
Old	AM

Table 19 Age(Women) between SBP

SL	SL	M	H	AH
SL	L	M	AH	VH
L	SM	AH	AH	VH
L	M	H	VH	EH
L	AM	AH	EH	EH

Table 20 Age(Women) between DBP

SL	L	M	H	VH
L	M	AM	AH	VH
L	AM	AH	VH	EH
SM	AM	VH	EH	EH
SL	AH	VH	EH	EH

Table 21 Age(Women) between LDL

SL	L	M	AM	AH
SL	L	M	H	AH
L	SM	M	AH	VH
L	M	H	VH	EH
L	AM	AH	EH	EH

high blood pressure, so on and by adopting balanced cardiovascular suitable diet (Tables 21, 22, 23, 24, 25, 26, 27, and 28).

Case 3:

Age = 32 year old, Gender = Female, SBP = 156 mm/Hg, DBP = 90 mm/Hg, TC = 256 mg/dl, LDL =185 mg/dl, HDL = 30 mg/dl, Triglyceride = 203 mg/dl, Diabetic = No, Smoker = 10 cigarettes per day, Family History = Yes, Obese = Yes.

Following the same mechanism as quoted in case 1 and 2 to evaluate risk grade which is = 0.8088.

This shows 80.88% risk estimate of the concerned chain smoker woman when in Case 2, we include the obesity factor in the assessment of risk of CHD.

In all the three cases, we have taken in this paper gives the current status of Coronary Heart Diseases for the respective patients and these results were verified with satisfaction by the medical experts.

Table 22 Age(Women) between HDL

H	SM	SL	VVL
H	M	L	VVL
AH	AM	L	SL
VH	AM	SM	SL
EH	VH	AM	SL

Table 23 Age(Women) between TC

SL	L	SM	AM	H
SL	L	M	H	AH
L	SM	M	AH	VH
L	M	AM	VH	EH
L	AM	H	EH	EH

Table 24 Age(Women) between triglyceride

SL	L	M	H
SL	SM	AM	AH
SM	M	AH	VH
SM	AM	AH	EH
SM	H	VH	EH

Table 25 Age(Women) between diabetes

Age	Prediabetes	Type-1 diabetes	Type-2 diabetes
Aged young	SL	L	SM
Slightly aged	L	SM	M
Aged	SM	M	AM
Very aged	SM	AM	H
Old	SM	AM	H

Table 26 Age(Women) between smoking

L	SM	M	AM	H
L	SM	AM	H	AH
M	AM	AH	VH	EH
M	AM	AH	VH	EH
AM	H	VH	VH	EH

Table 27 Age(Women) between family history

Age	Family history
Aged young	SL
Slightly aged	L
Aged	M
Very aged	M
Old	AM

Table 28 Age(Women) between obesity

Age	Obesity
Aged young	L
Slightly aged	SL
Aged	M
Very aged	AM
Old	H

6 Conclusion

The risk assessment scheme developed in the present research article plays a key role in designing a user-friendly Fuzzy Information System which is helpful in sharpening medical diagnostic skill of medical experts as well as guide patients about current status of their health.

Acknowledgements We express our sincere thanks to the prominent Cardiologists of Allahabad district namely Dr. Omar Hasan (MD, DM) and Dr. Geeta Shukla (MD) for their valuable comments and suggestions in designing fuzzy relational matrices as well as for critical examination of the evaluated results of various cases taken in the present research article.

References

1. We have weaker heart than Americans. Times News Network, 23 April, 2012
2. 81% Indian inactive: study. Times News Network, 22 April, 2012
3. Zadeh LA (1976) The concept of linguistic variable and its application to approximate decision making. Mir, Moscow
4. Klire, GJ, Yuan B (2009) Fuzzy sets and fuzzy logic: theory and applications. PHI Learning Private Limited, New Delhi (2009)
5. Srivastava P, Srivastava A (2012) A note on soft computing approach for cardiac analysis. J Basic Appl Sci Res 2(1):376–385
6. Srivastava P, Srivastava A (2012) Spectrum of soft computing risk assessment scheme for hypertension. Int J Comput Appl (0975-8887) 44(17): 23–30 (2012)
7. Gayathri P, Jaisankar N (2013) Comprehensive study of heart disease diagnosis using data mining and soft computing techniques. Int J Eng Technol 5(3)
8. Rajalakshmi R, Latha R (2016) Diagnosis of CAD and risk factors using DM and soft computing techniques. A survey. Middle-East J Sci Res 24(2):298–305

9. Uyara K, Ilhana A (2017) Diagnosis of heart disease using genetic algorithm based trained recurrent fuzzy neural networks. Proc Comput Sci 120:588–593

10. Wiharto W (2016) Intelligence system for diagnosis level of coronary heart disease with K-Star algorithm. Healthc Inf Res 22(1):30–38

11. Davidson VJ, Ryks J, Fazil A (2006) Fuzzy risk assessment tool for microbial hazards in food systems. Fuzzy Sets Syst 157:1201–1210

12. Matthews KA (1989) Menopause and risk factors for coronary heart disease. New Engl J Med 321:641–646

13. Sullivan LM et al (2004) Tutorial in biostatistics presentation of multivariate data for clinical use: the Framingham study risk score functions. Stat Med 23:1631–1660

A Soft Computing-Based Approach to Group Relationship Analysis Using Weighted Arithmetic and Geometric Mean

Poonam Rani, M. P. S. Bhatia and D. K. Tayal

Abstract Relationships patterns between social entities in the social network are the main attribute that plays important role in our lives. They are mostly complex in nature and uncertain to find out. To quantify these relationships, patterns is a very potent issue in social networks analysis. This paper proposes a robust function that finds the relationships between groups of finite size based on fuzzy graphs theory. The relationship among elements in-group is found out by using the arithmetic mean or geometric mean. This paper has taken advantages of both weighted arithmetic and geometric mean, which combines the advantage of both arithmetic and geometric mean. The weights taken are the function of the importance of both the social elements participating in a term. These weights can be the parameters like the betweenness centrality or closeness centrality.

Keywords Social networks · Social network analysis · Fuzzy graphs
Arithmetic mean · Geometric mean · Betweenness centrality
Closeness centrality

1 Introduction

Social networks [1, 2] is defined as a circle of the finite number of similar or dissimilar kinds of social elements linked-up with single or multi-relations between them. The social element may range from an individual person to any organization. The link between social elements defines the relationships that help greatly in the diffusion

P. Rani (✉) · M. P. S. Bhatia
NSIT – Delhi University, Dwarka, New Delhi, India
e-mail: poonam.rani.nsit@gmail.com

M. P. S. Bhatia
e-mail: bhatia.mps@gmail.com

D. K. Tayal
IGDTUW University, New Delhi, India
e-mail: dev_tayal2001@yahoo.com

© Springer Nature Singapore Pte Ltd. 2019
S. Bhattacharyya et al. (eds.), *International Conference on Innovative Computing and Communications*, Lecture Notes in Networks and Systems 56,
https://doi.org/10.1007/978-981-13-2354-6_19

of information, knowledge, and technology in the network. Social network analysis (SNA) [1] is "a distinct research perspective, which focuses on *relationships* among the social elements, and on the patterns and implications of these relationships, mainly within social and behavioral sciences". It mainly studies social elements structure patterns rather than the attributes of individual social elements. But Yager [3] introduced the concept of "Intelligent Social Network Modeling", which included the attributes of the nodes. It helps intelligently in handling the analysis, with uncertainty in both the node attributes and relationships, based on two techniques like the graph theory and concepts from the fuzzy set theory. So, today, it has become an astute paradigm, for the representation and analyses of the complexes of social relations between social elements. The relationships patterns are essential parameters of SNA. They need to be explored and mapped carefully at each step. The quantification of these relationships patterns in a social network is one of the potential challenging tasks as discussed in [4]. This can also be used in comparing the social networks. The comparison task is done by using OWA [5] and Fuzzy approach method [6] in the previous work.

To address this potential challenge in the social network, the paper proposes an astute robust function based on the weighted arithmetic and geometric means. It acts as a tool in finding the relationships between groups of finite size using fuzzy graphs theory. It helps in representing the uncertainty both in attributes and connections. The other part of this paper is structured in consecutive sections. Section two discusses the related work with its limitation. Section three gives the overview of the fuzzy graphs. Section four presents the proposed robust function with all concepts employed in an algorithm. In the end, section five briefly concludes the paper.

2 Related Work

In order to quantify the relationship of whole or subgroup of a social network, some researchers [7–9] have used OWA operator. They have calculated m-dimensional relationships, using the primary pairwise relations between social elements. They have increased the extent of analysis from binary to m-dimensions relations. For this, they have explored the fuzzy adjacency matrix. They have used fuzzy membership degree for representing the intensity of the relationship between social elements of a network. It has an advantage of preserving more information as compared to a classical binary relationship. They have explored OWA-based aggregation approach for finding the relationships between m social elements of a social network.

So, this section has explored the research gap in their work. It has made an endeavor to fill that research gap. The authors have explored a single type of relationship. But, in the real-life problems, there exist simultaneous multiple kinds of relationships between the elements in the network. These multi-kinds of relationships between entities are too difficult to find and to address. The authors have exploited the fuzzy set theory to address the uncertainty only in relation. Moreover, there is dynamism in a social network so there will be uncertainty in nodes attributes also. The authors have developed a recursive solution for finding the relationships between elements in social networks. But finding a recursive solution is not always feasible in case of presence of uncertainty. They have used OWA operator for aggregation of the relationship between all social elements in social networks, the weights used here are taken randomly. Although "Fuzzy set network Model with OWA" [8] relate well to the real-world problems in social networks, but it has a number of limitations as discussed above. So, it may not be efficiently applied to real-life social networks in presence of high level of uncertainty in the system. Thus, modeling social networks in terms of fuzzy sets may seem to an effective solution in real-life social networks to some extent. But it is actually not mapping the attributes of social elements in the real world. The fuzzy graph can handle uncertainty in the attribute as well in the relationship. So, this paper explored the fuzzy graph concept instead of the fuzzy adjacency matrix.

3 Fuzzy Graphs

In 1965, Zadeh [10] had given the big idea of fuzzy sets along with fuzzy relations in order to capture uncertainty in the real-world problems. After that, in 1975, Rosenfeld [11] introduced the Fuzzy graph theory by experimenting with fuzzy relations on fuzzy sets. A fuzzy graph has some uncertainty value assigned to every node and edge that accounts for the participation value of every member (node) and also the value of a relationship. The definition of a fuzzy graph is better understood only when we understand the definition of the graph and fuzzy set.

Definition 1 A simple graph [12] G_s is represented by two-tuple < vertices set, edge set >, and is given as $G_s = (V_s, E_s)$. The V_s set is containing a fixed set of n elements. The E_s is containing a fixed set of ordered pairs of different vertices called edges.

Definition 2 A fuzzy set [10], S_f on a set Z_s is represented by two-tuple $\langle Z_s, \mu_f \rangle$, and is distinguished by a mapping μ_f *called* membership function and is given as follows

$$\mu_f : Z_S \rightarrow [0, 1]$$

Definition 3 A fuzzy graph [11], G_ξ is described by three-tuple $\langle V_s, \sigma_V, \mu_E \rangle$. It consists of fixed finite set V_s of n nodes, along with two membership mapping $\sigma_V : V_s \rightarrow [0, 1]$ and $\mu_E : V_s \times V_s \rightarrow [0, 1]$, such that $\forall v_i, v_j \in V_s$, $\mu_E(v_i, v_j) \leq \sigma_V(v_i)^\wedge \sigma_V(v_j)$, and μ_E is a symmetric fuzzy relation on σ_V. Here, $\sigma_V(v_i)$ and $\mu_E(v_i, v_j)$ represents the membership mapping values of the vertex v_i and of the edge (v_i, v_j) in G_ξ, respectively. Where $(1 \leq i, j \leq n)$

The other two definitions which have been explored in the proposed work are given below:

Definition 4 Strength of a Path:

The strength of a path is given as -

$$\min\{\mu(x1, xi) : i = 1, 2, \ldots\ldots, n\}.$$

In other words, the strength of a path is the weight (membership value) of the weakest arc of the path.

Definition 5 Binary Relationship Strength:

The strength of a relationship between any two social elements x and y are defined as the maximum of the strengths of all paths between x and y. It is depicted by $C(x, y)$.

4 Proposed Work

Mathematically, a social network is defined as a two-tuple of nodes and edges as <nodes, edges>, where nodes are set of the social elements belonging to a universal set $U = \{u_1, u_2, \ldots, u_n\}$. The edges depict the relationships between elements. They are the set of the pair of social elements, having a relationship between them and is defined as $R \subseteq UXU$.

In literature [1, 2, 9], the binary scale of 0 and 1 have been explored to represent the presence and absence of a relationship between social entities u_i and u_j. It consists of a binary relation on the pair of social elements, is a relation, $R \subseteq UXU$. Its characteristics function $\mu_R : UXU \rightarrow \{0, 1\}$ is given as follows:

$$\mu_R(u_i, u_j) = \begin{cases} 1, & \text{if } u_i \text{ has a relation with } u_j \\ 0, & \text{if } u_i \text{ has no relation with } u_j \end{cases}$$

But, this binary scale is not efficient to map with the real relationship between entities, as it is lacking in representing the facts regarding the degree of relationship between elements. Bruneli and Fedrizzi [7] proposed the fuzzy relationship concept. They have mapped the value of a degree of relationship between u_i and u_j elements by membership degree (μ_R) in the interval [0,1] and is defined as follows:

$$\mu_R(u_i, u_j) = \begin{cases} 1, & \text{if } u_i \text{ has a relation with } u_j \\ \varphi \in (0, 1), & \text{if } u_i \text{ has a relation of some degree with } u_j \text{ .} \\ 0, & \text{if } u_i \text{ has no relation with } u_j \end{cases}$$

This fuzzy relationship contains more information than a crisp one.

In this work, the fuzzy graph is explored for finding the relationship between n social entities $\{1,2,3, \ldots, n\}$. The total number of pairs is given by k. The k is given by the binomial formula

Fig. 1 Social network 1
graph with four nodes a, b, c,
and d

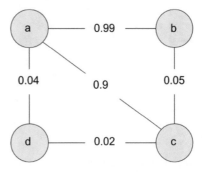

$k = {}^nC_2 = \{(1, 2), \{1, 3\}, (1, 4), .., (2, 1)...., (n, 1), (n, 2), .., (n, (n-1))\}$

The relationship between n elements can be found out by using the arithmetic mean or geometric mean or combining both with some weights.

1. **Using Arithmetic Mean**:

The relationship of n social elements using arithmetic mean function CAM (1,2 … n), is given by

$$CAM(1, 2, 3 \ldots n) = \frac{\sum_{i=1}^{n} \sum_{j=1, j \neq i}^{n} C(i,j)}{k} \tag{1}$$

The "C" in Eq. 1 is the function name means connectedness, which this paper is trying to estimate. The "AM" represents the arithmetic mean that we are using to calculate the connectedness. The above technique is an OR-like operation as if any relationship between any one arbitrary pair is very strong but, all other k-1 pairs have very weak relationships, the output of the function will be reasonably high, equivalent to the case when the relationship between all the k pairs has an average value. Lets take an example of social network 1 having four social elements (a, b, c, and d) with given connectedness values among pairs is shown in Fig. 1.

Here, AM gives a value of nearly 0.4, but the network should have a lower value as only two pairs have good relationships, others have negligible. So, this is its limitation.

2. **Using Geometric Mean**

The relationship of n social elements using geometric mean function $CGM(1, 2, \ldots, n)$ is given as

$$CGM(1, 2, 3 \ldots n) = \left(\prod_{i=1}^{i=n} \prod_{j=1, j \neq i}^{j=n} C(i,j) \right)^{\frac{1}{k}} \tag{2}$$

The "C" again here is the function name. The "GM" represents the geometric mean that is used to calculate the connectedness. The above technique is an AND like operation as if the relationship between any one arbitrary pair is very weak but, all other k-1 pairs have very strong relationships, the output of the function will be

Fig. 2 Social network 2
graph with four nodes a, b, c,
and d

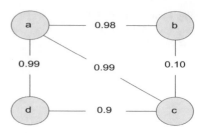

very low, equivalent to the case when the relationship between all the k pairs has a
low value as shown in social network 2 graph in Fig. 2.

In the worst case, suppose when c and d elements are not connected than in that
case, GM gives a value of 0.

3. Combined Approach

As the above two approaches have their own limitations. So, this section proposes to
combine the advantages of both the AM and GM method, to get a robust function,
which finds the relationships between n social elements. The relationship between n
social entities $(1, 2, \ldots n)$ is defined as follows:

$$C(1, 2, 3, \ldots, n) = (\alpha CAM(1, 2, 3, \ldots, n) + (1 - \alpha)CGM(1, 2, 3, \ldots, n)) \quad (3)$$

where α is called the *OR-ness* coefficient $(0 \leq \alpha \leq 1)$. If no specific purpose is
specified α should be taken as 0.5, which means an equal blend of AM and GM. If
$\alpha = 1$, the proposed function becomes identical to the AM function. Similarly, if
$\alpha = 0$, the proposed function becomes identical to the GM function.

Any intermediate value will result in blending of both the functions CAM (1, 2,
3, ..., n) and CGM (1, 2, 3, ..., n), which will make the function more robust to
any kind of network. Instead of using simple arithmetic and geometric means, the
weighted arithmetic means, and weighted geometric means can be used.

Here, one possible choice of the weights can be taken as a function of the importance of both the social elements participating in a term, i.e., the weight of C(x,y) is
a function of $\sigma(x)$ and $\sigma(y)$, either AM or GM of these two factors. These weights
can be the parameters like the betweenness centrality or closeness centrality.

Now, we define

$$W(x, y) = \frac{\sigma(x) + \sigma(y)}{2} \quad (4)$$

or

$$W(x, y) = \sqrt{\sigma(x) * \sigma(y)} \quad (5)$$

We can use these weights given in Eqs. (4) and (5) into Eqs. (1) and (2), respectively, and get values as follows:

$$CWAM\,(1, 2, 3 \ldots n) = \frac{\sum_{i=1}^{n} \sum_{j=1, j \neq i}^{n} W(i, j) * C(i, j)}{k} \qquad (6)$$

$$CWGM\,(1, 2, 3 \ldots n) = \prod_{i=1}^{i=n} \prod_{j=1, j \neq i}^{j=n} W(i, j) * C(i, j)^{1/k} \qquad (7)$$

By using above equations in Eq. 3, the relationship of a group or subgroup can be found out by given function as follows:

$$C(1, 2, 3, ..n) = \propto \frac{\sum_{i=1}^{n} \sum_{j=1, j \neq i}^{n} W(i, j) * C(i, j)}{k}$$
$$+ (1 - \propto)(\prod_{i=1}^{i=n} \prod_{j=1, j \neq i}^{j=n} W(i, j) * C(i, j))^{1/k}$$

This function will be the most robust among all the functions mentioned and will give the best possible relationships between n numbers of social elements. The choice of $\propto$ depends on the kind of application.

5 Conclusion

The paper has discussed various methods of analyzing the relationship of a group or subgroup of social networks along with their limitations. The best and most robust function is a blending version of both weighted arithmetic and weighted geometric means, where weights assigned to the social elements can be taken as proportional to the means of betweenness or closeness centrality of an element. In the future work, this proposed robust function with betweenness centrality as a weight to social elements can be applied to some real-world data. Hence, the fuzzy graph is the suitable model for modeling the social networks with uncertain parameters.

References

1. Faust, K, Wasserman S (1994) Social network analysis: methods and applications, 8th edn. Cambridge University Press
2. Hanneman RA, Riddle M, Robert A (2005) Introduction to social network methods. University of California Riverside
3. Yager RR (2008) Intelligent social network modeling and analysis. In: 3rd international conference on intelligent system and knowledge engineering-2008. ISKE 2008, pp 5–6
4. Rani P, Bhatia MPS, Tayal DK (2018) Different aspects & potential challenges in social networks. In: 3rd international conference on information, communication and computing technology (ICICCT-2018), Springer, 2018, pp 1–12
5. Rani P, Bhatia M, Tayal DK (2017) An astute SNA with OWA operator to compare the social networks. Int J Inf Technol Comput Sci

6. Rani P, Bhatia M, Tayal DK (2018) Qualitative SNA methodology. In: Proceedings of the 12th INDIACom and 5th international conference on computing for sustainable global development
7. Brunelli M, Fedrizzi M (2009) A fuzzy approach to social network analysis. In: ASONAM'09. international conference on advances in social network analysis and mining, 2009, pp 225–240
8. Brunelli M, Fedrizzi M, Fedrizzi M (2011) OWA-based fuzzy m-ary adjacency relations in social network analysis. Springer
9. Brunelli M, Fedrizzi M, Fedrizzi M (2014) Fuzzy m-ary adjacency relations in social network analysis: optimization and consensus evaluation. Inf Fusion, Elsevier, 17, 36–45
10. Zadeh LA (1965) Fuzzy sets. Inf Control 8(3):338–353
11. Rosenfeld A (1975) Fuzzy graphs. In: Zadeh LA, Fu KS, Shimura M (eds) Fuzzy sets and their applications. Academic Press, New York
12. Graph_theory @en.wikipedia.org. https://en.wikipedia.org/wiki/Graph_theory
13. Scott J (2012) Social network analysis Sage Publishers

Hand Gesture Recognition Using Convolutional Neural Network

Savita Ahlawat, Vaibhav Batra, Snehashish Banerjee, Joydeep Saha and Aman K. Garg

Abstract Nowadays, Hand Gesture Recognition can be used for human interaction with electronic devices and computers easily. In this paper, a hand gesture recognition application is proposed to recognize hand gestures using a webcam. The application uses Convolutional Neural Network (ConvNet) for learning and classifying the hand gestures. Contour formation takes place around the palm and Hue Saturation Values (HSV) are extracted from the hand for detection. The application classifies eight hand gestures and provides an accuracy of above 90%.

Keywords Hand gesture recognition · ConvNet · HSV · Max pooling

1 Introduction

Hand gestures are widely used by humans to convey messages. Therefore, it is normal for us to interact with machines in a similar way. For example, safety and comfort in vehicles can be improved using touch-less human–computer interfaces. Computer vision systems can be very helpful for designing such interfaces [1]. Hand gestures provide an easy and innovative way to interact with computers to do a task [2, 3].

There are two types of hand gesture recognition approaches that are used most often, i.e., Data-Glove-based approach and Vision-based approach. In the first approach, sensors are used to collect the data. These sensors are attached on gloves, which the users wear on their hands. Only the essential information is gathered using this method, which helps in minimizing the need for data preprocessing which in turn helps in reducing the amount of junk data [4]. But this approach is not feasible and it can lead to various issues such as sensor sensitivity, connectivity, synchronization, etc.

S. Ahlawat (✉) · V. Batra · S. Banerjee · J. Saha · A. K. Garg
Computer Science and Engineering Department, Maharaja Surajmal
Institute of Technology, New Delhi, India
e-mail: savita.ahlawat@gmail.com

© Springer Nature Singapore Pte Ltd. 2019
S. Bhattacharyya et al. (eds.), *International Conference on Innovative Computing and Communications*, Lecture Notes in Networks and Systems 56,
https://doi.org/10.1007/978-981-13-2354-6_20

Vision-based approaches are simpler and convenient due to the fact that these systems only need a camera or scanner. This kind of approach provides biological human vision by artificially illustrating the visual field [5]. Though this type of approach is more economical than the former approach; it generates an enormous amount of data which needs to be processed carefully for extracting only the significant information. This problem is handled making such a recognition system that is not affected by the lighting conditions, does not vary with the background details, and independent of the subject matter and type of camera. These systems also need to provide real-time interaction and hence, they need to be extremely fast and efficient [6].

According to various researchers, it has been found that ConvNets are very effective when they are used to solve problems related to computer vision. The biological similarity between the ConvNets and the human visual system is the main reason for the compatibility of ConvNet. The processing units of ConvNets consist of multiple layers of neurons that are distributed hierarchically. The parameters are shared among various neurons that are present in different levels in the network. This results in various connection patterns with the connections having varying weights. These weights are then used to complete the process of classification [7].

Real-time automatic gesture recognition is an important area for developing intelligent vision systems [8]. There are many applications of vision systems that include process control, automatic inspection, robot guidance, etc. A ConvNet can be used in such cases to develop applications to produce better outcomes.

A ConvNet is a deep learning artificial neural network technology that is used to analyze images and to recognize different features in the images. ConvNet acts as a robust classification tool, especially with max pooling [9]. A ConvNet consists of one or more fully connected layers which contain various filters that extract the most important features that are used for image analysis and recognition. The layers contain multiple convolution layers and filtering layers that compress the image matrix continuously and give a small matrix as output which contains only the most important features that the neural network uses for training itself.

2   Proposed Method

The development method of the application for hand gesture recognition is divided into five parts.

2.1   Experimental Setup

The user interface of the recognition application is developed using Java and the ConvNet model is developed in Python. A minimum processing speed of 2.2 GHz and a GPU is required for running the recognition application. The application uses an HD Webcam to scan the hand gestures of the user and to capture the images.

The recognition application has been developed and tested in a Windows Operating System.

The ConvNet model is developed using the Keras Library and Theano backend. Virtual environment for Python 3.4 is created using the Anaconda IDE. The user interface has been developed using IntelliJ IDEA to link the libraries and JAR files with the application.

2.2 Feature Extraction

In this step, the interface which will capture the image and return the required features of the image is developed in Java. The interface takes the image from the webcam and looks for HSV features of skin color. It filters out all the other colors and only displays skin color. Then, it extracts the largest collection of skin color to get the hand gesture. In earlier applications, RGB-based skin detection was used which was not robust enough to extract the features properly. There used to be a high amount of noise and distortion associated with detection with RGB [10]. Hence, HSV was chosen over RGB for skin detection and feature extraction.

2.3 Dataset Collection

The algorithm clicks a lot of images to collect the dataset that will be used to train the neural network. The application clicks images via webcam when it is set to data collection mode. 1000 images, 500 positional, and 500 rotational, are used to train the neural network for each hand gesture. Since there are 9 gestures (one being "none"), there are 9000 images in the dataset. Steps to collect dataset:

1. Open the Java file for hand gesture recognition in the editor.
2. Set the dataCollectionMode Boolean variable to true. Then set the gestureIndex integer to the value for which the image is to be collected.
3. Save the Java file. Then, compile the Java file using Java compiler and then run the Java file.
4. The image will be captured using a webcam. After that, click on the applet screen to start collecting the images corresponding to that hand gesture.
5. 1000 images will be collected for each gesture: 500 rotational images and 500 positional images in one go.
6. The images will be stored in the source location where the Java file is present.
7. Run the Java file again after changing the gestureIndex for capturing other gestures.

2.4 Training the Neural Network

After the collection of the dataset is completed, it is fed to the application for training the neural network. Figure 1 shows the basic ConvNet architecture that is used to develop the neural network. The neural network consists of multiple convolution layers and pooling layers which are used to compress the image and extract the important features which are used for training the neural network.

In the sequential convolution neural model, a 50×50 gesture image (50-pixel height and 50-pixel width) is input into the ZeroPadding2D layer which outputs a 54×54 image which is further fed into the Convolution2D layer. Convolution2D layer applies eight filters to the input image and then the output is sent to ZeroPadding2D layer. After this layer, there is another Convolution2D layer which filters the input again with eight filters. Then, the MaxPooling2D layer pools the input matrix and reduces it to the image of size 25×25. The same process continues two more times (with various filters) where the image is compressed and the neural network is trained with those obtained outputs.

In every convolution layer, ReLU activation is applied. The ReLU activation function is as follows:

$$Relu(x) = maximum(0, x) \tag{1}$$

That is, if x < 0, Relu(x) = 0 and if x ≥ 0, Relu(x) = x.

Afterward to prevent the problem of overfitting the model, a dropout layer is added. Flatten layer helps in flattening the input which does not affect the input batch size. Activation layers apply an activation function (tanh) to an output to increase the complexity of the output signal from a linear function because a neural network will not be able to perform complex computations with a linear function only.

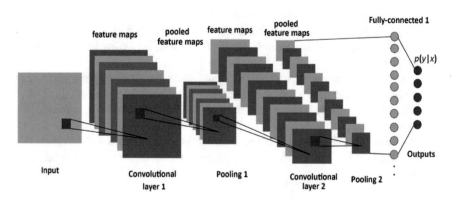

Fig. 1 ConvNet architecture using alternating convolution and pooling layers

Hyperbolic Tangent function—Tanh: Its mathematical formula is

$$func(x) = \frac{1 - e^{(-2x)}}{1 + e^{(-2x)}}$$

(2)

The output of this function is zero centered because its range is in between -1 and 1, i.e., $-1 < \text{output} < 1$.

The entire training of the neural network takes approximately 60 min. The steps to train the model are:

1. Store the images at a particular location in the Python model directory.
2. Add the directory path to the hand detection file. Set the training mode variable to true and prediction mode variable to false.
3. Check whether the path of image folder is given correctly and then run the hand detection file.
4. Each counter will run for almost 4 min reducing loss at every counter.
5. Total training time will be around 1 h. New file "hand_detection_weights_3.h5" will be created in the python model directory.

2.5 Testing the Application for Recognition

The application is tested using user's hand gestures after the training of the neural network has been completed. The application is tested using a variety of hand gestures and the accuracy of the recognition application is calculated.

3 Observations

The recognition application is provided with the user input for various hand gestures and the accuracy of the system is calculated. It was observed that the highest level of accuracy was obtained when the amount of light was within a specific range. The accuracy was neither directly proportional to the illumination nor inversely proportional. The optimum conditions lied within a specific range.

Figure 2 shows the collection of images of the recognition of each hand gesture by the application. The bars in the images in Fig. 2 represent the percentage of images in the dataset of a hand gesture that matches with the user input of that hand gesture. Each gesture has some features that are similar to another gesture. For example, when the "hand" gesture is taken far away from the camera, it appears like the "palm" gesture. Similarly, the "thumb" gesture and "swing" gesture have similar features. So, a few images of one gesture might match with the user input for another gesture. The matching percentage of each hand gesture is shown in Table 1.

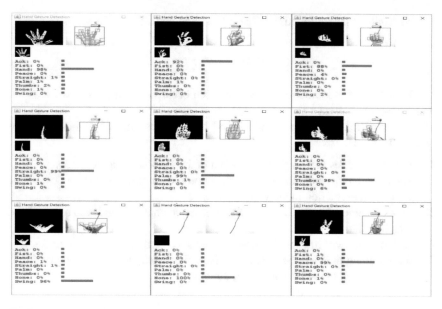

Fig. 2 Recognition of hand gestures by the application

Table 1 Matching percentage

	Ack	Fist	Hand	Peace	Straight	Palm	Thumbs	None	Swing
Ack	88	0	0	0	0	1	0	0	0
Fist	0	85	0	4	0	0	3	0	2
Hand	0	0	92	0	1	1	2	1	0
Peace	0	1	0	94	0	0	0	1	0
Straight	0	0	0	0	94	0	0	1	0
Palm	8	0	0	0	0	88	0	0	0
Thumbs	0	0	0	0	0	0	90	0	6
None	0	0	0	0	0	0	0	100	0
Swing	0	0	0	1	1	0	3	0	90

The matching percentage in the table is taken less than the matching percentage in the images because the images were taken in optimum light conditions and a stable environment.

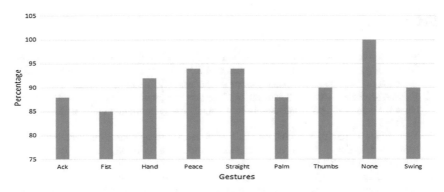

Fig. 3 Matching percentage

4 Results

The accuracy of the recognition application is calculated from the observed values of matching percentage. The accuracy is the mean of the correctly matched percentage of each hand gesture (except "none" gesture). The accuracy comes out to be 90.125% as shown in Fig. 3.

5 Conclusion

This research paper proposes a hand gesture recognition system to recognize various hand gestures using the PC Webcam. The ConvNet is being used by the application for learning and classifying the hand gestures.

In the current implementation, eight different gestures are being recognized but the number of gestures can be extended to as many as we require. A total of 9000 images were fed to the application to train the hand gestures. The application provides a recognition accuracy of 90.125% and works for simple webcams like the laptop webcam.

Acknowledgements The hand gestures employed for the recognition task in this work has been contributed by the authors. The authors are thankful to the anonymous reviewer for their review comments.

References

1. Molchanov P, Kim K, Gupta S, Tyree S, Yang X, Kautz J (2016) Online detection and classification of dynamic hand gestures with recurrent 3D convolutional neural networks. In: IEEE conference on computer vision and pattern recognition (CVPR), pp 4207–4215
2. Vafadar M, Behrad A (2008) Human hand gesture recognition using spatio-temporal volumes for human-computer interaction. In: International symposium on telecommunications, pp 713–718
3. Yang F, Shi H (2016) Research on static hand gesture recognition technology for human computer interaction system. In: International conference on intelligent transportation, big data & smart city (ICITBS), pp 459–463
4. Abhishek KS, Ho D, Qubeley Lee CF (2016) Glove-based hand gesture recognition sign language translator using capacitive touch sensor. In: IEEE international conference on electron devices and solid-state circuits (EDSSC), pp 334–337
5. Zou Y, Liu Z, Hu C, Chen W (2012) Regions of interest extraction based on HSV color space. In: 10th international conference on industrial informatics, pp 481–485
6. Agrawal R, Gupta N (2016) Real time hand gesture recognition for human computer interaction. In: IEEE 6th international conference on advanced computing (IACC), pp 470–475
7. Strezoski G, Dimitrovski I, Stojanovski D, Madjarov G (2016) Hand gesture recognition using deep convolutional neural networks. In: ICT innovations, pp 49–58
8. Chen W-K, Hsu M-H, Lin H-I (2014) Human hand gesture recognition using a convolutional neural network. In: IEEE international conference on automation science and engineering (CASE), pp 1038–1043
9. Ducatelle F, Meier U, Schmidhuber J, Nagi J, Ciresan D, Di Caro GA, Giusti A, Gambardella L, Nagi F (2011) Max-pooling convolutional neural networks for vision-based hand gesture recognition. In: IEEE international conference on signal and image processing applications (ICSIPA) 342–347
10. Muhammad B, Rahman Abu-Baka SA (2015) A hybrid skin color detection using HSV and YCgCr color space for face detection. In: IEEE international conference on signal and image processing applications (ICSIPA), pp 95–98

Arrhenius Artificial Bee Colony Algorithm

Sandeep Kumar, Anand Nayyar and Rajani Kumari

Abstract The foraging behavior of real honey bees inspired D. Karaboga to develop an algorithm, namely Artificial Bee Colony (ABC) Algorithm. The ABC performs well in comparison to other swarm-based algorithms but has few drawbacks also. Similar to other stochastic techniques, the step size during the position update play a very imperative part in the potential of ABC. The ABC is very good in the exploration of search space but not fine in exploitation. So, as to improve balancing between diversification and intensification process of ABC algorithm, a novel variation of ABC proposed termed as Arrhenius ABC (aABC) algorithm. The suggested algorithm tested over eight unconstrained global optimization functions and two constrained problems. The results prove that aABC algorithm performs better for considered low dimensional problems in comparison to basic ABC and its current variants.

Keywords Swarm intelligence · Nature-inspired algorithm · Computational intelligence · Exploration and exploitation

1 Introduction

It is human nature to find the optimum solution for a problem in order to minimize efforts and requirement of resources. Many disciplines and domains have various optimization problems with different specifications and different degree of complexity. In these problems, it is required to discover optimal or adjacent to optimal results

S. Kumar (✉)
Amity University Rajasthan, Jaipur, India
e-mail: sandpoonia@gmail.com

A. Nayyar
Graduate School, Duy Tan University, Da Nang, Vietnam
e-mail: anand_nayyar@yahoo.co.in

R. Kumari
JECRC University, Jaipur, India
e-mail: rajanikpoonia@gmail.com

© Springer Nature Singapore Pte Ltd. 2019
S. Bhattacharyya et al. (eds.), *International Conference on Innovative Computing and Communications*, Lecture Notes in Networks and Systems 56,
https://doi.org/10.1007/978-981-13-2354-6_21

with respect to different constraints. Some optimization problems are not solvable in a fixed number of steps but it is required to follow some process that leads to optimal results. The nature has great capability to deal with complex problems in a very easy manner. A number of algorithms are available that are inspired by natural phenomenon and categorized as Nature-Inspired Algorithms (NIAs). The NIAs are very popular in the field of optimization and giving best results. The nature-inspired algorithms are stimulated by extraordinary conduct of some natural phenomenon or intelligent behavior of natural insects in a group. Like ABC algorithm [1], Ant Colony Optimization (ACO) [2, 3], Firefly Algorithm (FA) [4], etc.

The NIAs are further classified into different classes based on their source of inspirations. Like Bio-inspired algorithm, chemistry-based algorithm and physics-based algorithms. Classification of NIAs is not limited to these categories only, one may classify them into different classes based on some other criteria. The natural entities living or nonliving that collectively shows intelligent behavior is cooperative with each other and follow swarming behavior. They have no any central controlling unit and most of the times work in distributed manner but they have a division of labor and self organization. Researchers and scientists are continuously trying to understand their behavior and simulated in terms of some algorithm. The ABC simulates rummaging demeanor of real honey bees. It is first identified by D. Karaboga in the year 2005 [1]. It is stochastic in nature and also depicts swarming behavior. Honey bees shows extraordinary behavior while probing for proper food sources and then exploiting nectar from best food sources. The ABC algorithm is very efficient in comparison to other competitive algorithm but it has few drawbacks also similar to other algorithms, like premature convergence and stagnation. That is the reason that since its inception it has been modified by a number of researchers. The ABC algorithm is very simple in implementation as it has only three control parameters and follows three simple steps while moving towards optimum.

The remaining article is prepared as follows: Sect. 2 explain basic ABC algorithm with some variants. Section 3 deliberates the Arrhenius ABC algorithm. Investigational outcomes are evaluated in Sect. 4. Section 5 presents the conclusion.

2 ABC Algorithm

The practical implementation of ABC is easy, and has only three parameters. The complete ABC algorithm spliced into three phases. The Algorithm 1 revealed the core steps of ABC algorithm.

Algorithm 1: Artificial Bee Colony Algorithm

Initialize all parameters
Repeat step 1 to step 3 till the termination criteria meet
    Step 1: Apply employed bee phase to generate new food sources using (2)
    Step 2: Onlooker bee phase to update food sources according to their nectar quality
    Step 3: Scout bee phase to generate new solution in place of rejected solutions
Remember the finest solutions established until now

2.1   Steps of ABC Algorithm

Initialization: The first phase in ABC is initialization of parameters (Colony Size, Limit for scout bees, and maximum number of cycles) and set up an initial population randomly using Eq. 1.

$$p_{ij} = LB_j + rand \times (UB_j - LB_j) \tag{1}$$

where $i = 1, 2, \ldots, (Colonysize/2)$ and $j = 1, 2 \ldots, D$. Here, D represent dimension of problem. p_{ij} denotes location of ith solution in jth dimension. LB_j and UB_j denotes lower and upper boundary values of search region correspondingly. $rand$ is a randomly selected value in the range (0, 1).

Employed Bee Phase: This phase try to detect superior quality solutions in proximity of current solutions. If the quality of fresh solution is enhanced than present solution, the position is updated. The position of employed bee updated using Eq. 2.

$$V_{ij} = p_{ij} + \overbrace{\phi_{ij} \times (p_{ij} - p_{kj})}^{s} \tag{2}$$

where $\phi_{ij} \in [-1, 1]$ is an arbitrary number, $k \in 1, 2, \ldots (Colonysize/2)$ is a haphazardly identified index such that $k \neq i$. In this equation, s denotes step size of position update equation. A larger step size leads to skipping of actual solution and convergence rate may degrade if step size is very small.

Onlooker Bee Phase: The selection of a food source depends on their probability of selection. The probability is computed using fitness of solution with the help of Eq. 3.

$$Prob_i = \frac{fitness_i}{\sum_{i=1}^{colonysize/2} Fitness_i} \tag{3}$$

Scout Bee Phase: An employed bee become a scout bee when the solution value not updated till the predefined threshold limit. This scout bee engenders new solution instead of rejected solution using Eq. 1.

2.2   Recent Modifications in ABC Algorithm

The ABC is very popular in the midst of academicians and researchers operational in the arena of optimization due to its outstanding performance and simplicity. The performance of ABC algorithm is critically analyzed and concluded that it is better than other NIAs most of the times [5, 6]. The ABC algorithm hybridized with other nature-inspired algorithms and provided good results for benchmark problems. Recently, TK Sharma et al. [7] hybridized ABC with shuffled frog-leaping. In this hybrid algorithm, the population is divided into a couple of parts (superior and inferior) according to the fitness of individuals and solved five real-world problems from field of chemical engineering. Akay and Karaboga [8] proposed eight variants of ABC algorithm by modifying all phases and introduced some new control parameters and analyzed for constrained optimization problems. Bansal et al. [9] anticipated an innovative variant of ABC with global and local neighborhoods for optimal power flow and also performed stability and convergence analysis of ABC [10]. Sharma et al. anticipated several alternatives of ABC like Lbest-Gbest ABC [11], fully informed [12], and Levy flight [13] ABC to balance intensification and diversification competencies of ABC algorithm. Tiwari et al. [14] introduced a weight-driven approach to update the location of solution in ABC algorithm. Bhambu et al. [15] also introduced a new position update process in ABC and named it as modified gbest as they used information about global best solutions. The detailed study of ABC algorithm is available in [16, 17].

3   Arrhenius ABC Algorithm

JC Bansal et al. [17] found that ABC algorithm does not have proper balancing between exploration and exploitation of optimal solution in search region. This balancing may be achieved with proper step size while updating the position of individuals. As shown in Eq. 2, step size is defined as combination of a random number ϕ_{ij} and difference of ith and kth solution. A large step size leads to skipping of true solutions and tiny step size reduce the rate of convergence. To avoid these problems it is highly desirable to develop some new strategy that is able to maintain proper step size. Here, this paper has proposed two modifications in position update equation. First, the value of coefficient ϕ_{ij} is decided by a new process that is inspired by Arrhenius equation [18]. Second, it identifies the global best solution in the current population and uses it in place of randomly selected kth solution. In 1989, Arrhenius [18] established a relation between temperature and reaction rate, which is equally valid for forward and reverse reactions. It makes use of activation energy which is the least possible energy essential to start the reaction. The formula given to computer rate constant (k) is as follows:

$$k = Ae^{\frac{-E_a}{RT}} \tag{4}$$

where A denotes a constant that is defined as the frequency of collisions and preset for every chemical reaction, E_a denotes the activation energy for the reaction, R indicates the universal gas constant, The absolute temperature is denoted by T [18].

Algorithm 2: Solution update strategy in Arrhenius ABC   Algorithm

for $j \in 1, 2, 3, ..., D$ do

    if $rand(0,1) > prob_j$ then

$$V_{ij} = p_{ij} + (e^{-Fitness_i \times D/k}) \times (p_{ij} - p_{Bestj})$$

    else

$$V_{ij} = p_{ij}$$

    endif

end for

The proposed algorithm use this constant in modified form, the ϕ_{ij} and V_{ij} is engendered using Eq. 5.

$$V_{ij} = p_{ij} + \phi_{ij} \times (p_{ij} - p_{bestj}) \tag{5}$$

where

$$\phi_{ij} = e^{\frac{-Fitness_i \times D}{t}} \tag{6}$$

Here, $Fitness_i$ denotes fitness of ith solution. D denotes the dimension of selected problem and t denotes generation number of solution. The modified position update equation is defined in Eq. 6. Where ϕ_{ij} is a coefficient generated using Arrhenius equation and $Pbest$ denotes best feasible solution in jth direction. The newly introduced algorithm selects the solution with higher fitness for the next iteration and tries to balance exploration and exploitation with a step size. It assumes that the extremely fitted solution has good solutions in their closeness. Algorithm 3 list the major steps of aABC.

Algorithm 3: Arrhenius ABC Algorithm

Initialize the population using (1)
Reiterate step 1 to 3 till the termination criteria meet
1. Apply employed bee phase to produce new food sources using (8)
2. Onlooker bee phase to modernize food sources as per their nectar quality
3. Scout bee phase to generate new solution in place of rejected using (1)
Remember the best solutions initiated up to now

---

4   Experimental Settings and Result Analysis

The proposed Arrhenius ABC Algorithm is tested on eight unconstrained optimization problems. The CEC benchmark problems [19] that are considered for experiments are Beale (f_1), Colville (f_2), Branins (f_3), Six-hump Camel Back (f_4), Dekkers and Arts (f_5), Hosaki (f_6), McCormick (f_7), and Mayer and Roth problem (f_8). The newly anticipated algorithm is applied to solve Compression Spring Problem and Parameter Estimation for FM Sound Wave, two well known real-world constrained optimization problems.

Compression Spring Problem: It diminishes the weight of a compression spring with four restrictions (minimum surge wave frequency, shear stress, minimum deflection, and bounds on diameter (outside) of the spring and on design variables). Three design variables are deliberated while designing compression spring: d_1, d_2, and d_3 are mean diameter of coil wires diameter and count of active coils, respectively. This problem is described Eq. 7 [20].

$$f_9(X) = \pi^2 \times \frac{(d_1 + 2)d_2 d_3^2}{4} \tag{7}$$

The optimum solution is $f(7, 1.386599591, 0.292) = 2.6254$ and $1.0E - 04$ is passable error for this function. Details about the constraint available is available in [20].

Parameter Estimation for FM Sound Wave: It is a six- dimensional function. The six parameters that are to be optimized are denoted by a vector $X = p_i, w_i$, where $i = 1, 2, 3$. The Eqs. 8 and 9 denotes the predictable sound and the objective sound waves, respectively [15].

$$y(t) = p_1 \times sin(t\omega_1\theta + p_2 \times sin(t\omega_2\theta + p_3 \times sin(t\omega_3\theta))) \qquad (8)$$

$$y_0(t) = sin(5\tau\theta - (\frac{3}{2})sin((4.8)\tau\theta + 2 \times sin((4.9)\tau\theta))) \qquad (9)$$

Table 1 Outcomes for test problems

TP	Algorithm	ME	SD	AFE	SR
f_1	aABC	4.93E−06	3.12E−06	51365.21	97
	ABC	8.10E−06	1.83E−06	16004.86	100
	MFABC	8.39E−06	6.47E−06	100436.1	73
f_2	aABC	6.67E−03	1.91E−03	92250.05	98
	ABC	1.39E−01	7.81E−02	200019.83	0
	MFABC	7.31E−03	1.98E−03	151363.13	43
f_2	aABC	5.74E−06	6.51E−06	63618.9	89
	ABC	5.66E−06	6.58E−06	21877.03	90
	MFABC	8.52E−06	1.21E−05	94049.52	79
f_4	aABC	1.64E−05	1.28E−05	120583.3	46
	ABC	2.21E−03	2.26E−03	200021.2	0
	MFABC	1.30E−05	1.29E−05	154243	23
f_5	aABC	4.49E−01	1.19E−02	107220.6	78
	ABC	2.64E+01	2.48E+01	200026.2	0
	MFABC	5.11E−01	3.18E−02	148462.3	52
f_6	aABC	5.93E−06	6.21E−06	41894.47	93
	ABC	2.73E−02	1.14E−02	200026.1	1
	MFABC	5.78E−06	6.31E−06	52480.92	93
f_7	aABC	8.95E−05	7.48E−06	70871.26	96
	ABC	1.98E−02	2.08E−02	200023.1	0
	MFABC	9.58E−05	1.26E−05	120556	70
f_8	aABC	1.36E−03	2.87E−06	39275.3	100
	ABC	1.17E−03	2.86E−06	22724.49	99
	MFABC	1.95E−03	3.20E−06	65008.61	98
f_9	aABC	8.29E−03	1.81E−02	181660.6	29
	ABC	1.65E−02	1.76E−02	190788.4	9
	MFABC	1.71E−02	1.30E−02	188964.2	14
f_{10}	aABC	5.66E+00	5.26E+00	141286.4	46
	ABC	6.83E+00	5.15E+00	200012.1	0
	MFABC	3.91E+00	5.62E+00	177092.8	24

$$f_{10}(X) = \sum_{i=1}^{100} (y(t) - y_0(t))^2 \tag{10}$$

where $\theta = 2\pi/100$ and the parameters typically assessed in the search range $[-6.4, 6.35]$. The square of the difference of predictable sound and the objective sound are denoted as fitness function and is shown in Eq. 10. The global minimum value is $f(x) = 0$. Tolerable fault is $1.0E - 05$.

4.1 Experimental Settings

The proposed Arrhenius Artificial Bee Colony Algorithm is implemented in C programming language and compared with basic ABC and Levy Flight Memetic Search in ABC (MFABC) [20]. In order to perform these experimentations, the colony size taken is 50, the limit for foraging is 5000 cycles, the experiment repeated 100 times, and the results are average of 100 run. The termination criteria, i.e., limit in this case is, $Limit = D \times 25$. A food source discarded if it is not updated after "limit" trials in employed bee phase and become scout.

4.2 Result Analysis

The efficiency of newly advised variant of ABC is equaled with ABC and MFABC [20] based on success rate (SR), mean error (ME), average function evaluations

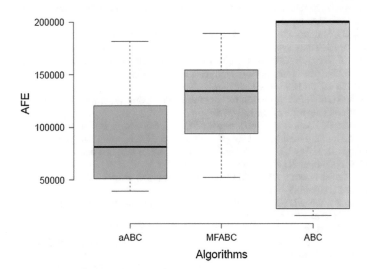

Fig. 1 Boxplots for average function evaluations

(AFE), and standard deviation (SD). Experimental outcomes are displayed in Table 1. A nonparametric analysis, namely boxplot analysis of AFE is also done. Boxplots for aABC and other measured techniques are depicted in Fig. 1. The exploration exposes that the interquartile range and the median of aABC are proportionately low in contrast to basic ABC and MFABC, thus it takes less time to find optimum solutions.

5 Conclusions

This paper propositions a novel alternative of ABC, namely Arrhenius Artificial Bee Colony which introduce a new parameter and modify the position update process. To appraise the potential of the proposed strategy, it is evaluated for eight unconstrained global optimization functions and two constrained problems. The evaluation of the results leads to the conclusion that aABC may be a better alternative for timely solution of the complex optimization problems. The aABC more efficiently explore the feasible search space as it selects the solution with higher fitness for the next iteration and tries to balance exploration and exploitation with a balanced step size. It assumes that the extremely fitted solution has good solutions in their closeness. It is more reliable in comparison to other measured algorithms as median for aABC is very low.

References

1. Karaboga D (2005) An idea based on honey bee swarm for numerical optimization. Technical report, Technical report-tr06
2. Dorigo M, Maniezzo V, Colorni A (1996) Ant system: optimization by a colony of cooperating agents. IEEE Trans Syst Man Cybern Part B (Cybernetics) 26(1): 29–41
3. Nayyar A, Singh R (2016) Ant colony optimization computational swarm intelligence technique. In: 2016 3rd International conference on computing for sustainable global development (INDIACom), pp 1493–1499
4. Yang X-S (2009) Firefly algorithms for multimodal optimization. In: International symposium on stochastic algorithms, pp 169–178
5. Karaboga D, Basturk B (2007) A powerful and efficient algorithm for numerical function optimization: artificial bee colony (abc) algorithm. J Glob Optim 39(3):459–471
6. Karaboga D, Basturk B (2008) On the performance of artificial bee colony (abc) algorithm. Appl Soft Comput 8(1):687–697
7. Tarun Kumar Sharma and Millie Pant (2017) Shuffled artificial bee colony algorithm. Soft Comput 21(20):6085–6104
8. Akay B, Karaboga D (2017) Artificial bee colony algorithm variants on constrained optimization. Int J Optim Control 7(1):98
9. Bansal JC, Jadon SS, Tiwari R, Kiran D, Panigrahi BK (2017) Optimal power flow using artificial bee colony algorithm with global and local neighborhoods. Int J Syst Assur Eng Manag 8(4): 2158–2169
10. Bansal JC, Gopal A, Nagar AK (2018) Stability analysis of artificial bee colony optimization algorithm. Swarm Evolut Comput

11. Sharma H, Sharma S, Kumar S (2016) Lbest gbest artificial bee colony algorithm. In: 2016 International conference on advances in computing, communications and informatics (ICACCI), pp 893–898
12. Sharma K, Gupta PC, Sharma H (2016) Fully informed artificial bee colony algorithm. J Exp Theor Artif Intel 28(1–2):403–416
13. Sharma H, Bansal JC, Arya KV, Yang X-S (2016) Lévy flight artificial bee colony algorithm. Int J Syst Sci 47(11):2652–2670
14. Tiwari P, Kumar S (2016) Weight driven position update artificial bee colony algorithm. In: International conference on advances in computing, communication, & automation (ICACCA)(Fall), pp 1–6
15. Bhambu P, Sharma S, Kumar S (2018) Modified gbest artificial bee colony algorithm. In: Soft computing: theories and applications, pp 665–677. Springer
16. Karaboga D, Gorkemli B, Ozturk C, Karaboga N (2014) A comprehensive survey: artificial bee colony (abc) algorithm and applications. Artif Intell Rev 42(1):21–57
17. Bansal JC, Sharma H, Jadon SS (2013) Artificial bee colony algorithm: a survey. Int J Adv Intel Paradig 5(1–2): 123–159
18. Arrhenius S (1889) Über die dissociationswärme und den einfluss der temperatur auf den dissociationsgrad der elektrolyte. Zeitschrift für physikalische Chemie 4(1):96–116
19. Montaz Ali M, Khompatraporn C, Zabinsky ZB (2005) A numerical evaluation of several stochastic algorithms on selected continuous global optimization test problems. J Glob Optim 31(4):635–672
20. Kumar S, Kumar A, Sharma VK, Sharma H (2014) A novel hybrid memetic search in artificial bee colony algorithm. In: Seventh international conference on contemporary computing (IC3), pp 68–73

Analysis of Refactoring Effect on Software Quality of Object-Oriented Systems

Ruchika Malhotra and Juhi Jain

Abstract Software industry primarily recognizes the significance of high quality, robust, reliable, and maintainable software. The industry always demands for efficient solutions that can improve the quality of software. Refactoring is one such potential solution; however, literature shows varied results of the application of refactoring techniques on software quality attributes. There are number of refactoring techniques that still needs to be empirically validated. This paper focuses on analyzing the effect of four unexplored refactoring techniques on different software quality attributes like coupling, cohesion, complexity, inheritance, reusability, and testability on object-oriented softwares. Impact analysis is performed by calculating Chidamber-Kemerer (CK) metrics of the projects, both before and after applying the refactoring techniques, and the results are statistically validated. Empirical analysis of results revealed that different refactoring techniques have different effects on internal and external quality attributes of object-oriented systems.

Keywords Refactoring · Object-oriented metrics · Software quality

1 Introduction

With the evolution of software industry resulting in growth of size and complexity of software day by day, software practitioners are acknowledging the significance of good quality software. Refactoring provides an effective way to improve source code efficiently. Refactoring involves changes in software internal code while keeping the external behavior intact [1, 2]. As per IEEE standards, software quality is defined as the extent to which any software or process adheres to requirements specified [3].

R. Malhotra · J. Jain (✉)
Computer Science and Engineering Department,
Delhi Technological University, New Delhi, Delhi, India
e-mail: erjuhijain@gmail.com

R. Malhotra
e-mail: ruchikamalhotra2004@yahoo.com

© Springer Nature Singapore Pte Ltd. 2019
S. Bhattacharyya et al. (eds.), *International Conference on Innovative Computing and Communications*, Lecture Notes in Networks and Systems 56,
https://doi.org/10.1007/978-981-13-2354-6_22

Software quality attributes can be further categorized into internal or external quality attributes [4].

Though refactoring is supposed to improve various software quality attributes like correctness, reusability, compatibility, and efficiency [5], however, it is necessary to explore refactoring techniques in terms of their effect on different internal and external quality attributes. Many researchers have empirically proved positive effect of refactoring techniques on different software quality attributes but some researchers empirically raise doubt on this fact. Majority of the related work is carried out for the maintainability [6–9] but other software quality attributes need more empirical studies for their validation.

The aim of the study is to apply less explored refactoring techniques on software projects and major concerns addressed in this study are: (a) To analyze the individual effect of Replace Constructor with Builder, Replace Constructor with Factory, Wrap Return Value, and Encapsulate Field refactoring techniques on internal quality attributes like complexity, coupling, cohesion, and inheritance of object-oriented (OO) projects using CK metrics. (b) To explore the individual effect of aforesaid refactoring techniques on maintainability, adaptability, completeness, understandability, reusability, and testability of OO projects.

In this research, four OO projects are used to analyze the effect of selected refactoring techniques on software quality attributes by analyzing CK metrics. Experimental results recommend applying Replace Constructor with Builder, Replace Constructor with Factory method, and encapsulate field refactoring techniques as they improve adaptability, maintainability, completeness, and understandability but these techniques increase testing effort. Wrap return value refactoring technique shows an adverse effect on all addressed quality attributes. Investigated refactoring techniques can improve software quality, if applied properly based on their behavioral properties.

Organization of remainder paper is classified in the following sections: Sect. 2 includes literature review, Sect. 3 outlines the refactoring process acquired in this paper and describes addressed internal and external software quality attributes, Sect. 4 explains the methodology followed in this research, Sect. 5 comprises results with detailed investigation, Sect. 6 covers validity threats of this research, and Sect. 7 highlights inferences drawn from this study with the future directions.

2 Related Work

In 1990, Opdyke [1] introduced the term refactoring which aims at the restructuring of OO code such that it does not affect its external behavior.

Detailed investigations of existing studies have revealed that there is a varied effect of different refactoring techniques on the quality attributes. Summary of related studies is provided in Table 1. Closest to this study is Elish and Alshayeb [6, 10] but they did not perform statistical validation which is crucial for validation of results. Also, a set of refactoring techniques applied is totally different. Only common refactoring technique in [6] and in this study is encapsulate field refactoring technique. Work

Table 1 Literature studies related to work

Study	Projects	No. of RT	Software quality attributes	
			Internal	External
Bavota et al. [12]	Five open source	01	Cohesion	–
Elish and Alshayeb [6]	Three open source an one self-made	09	CK metrics, FOUT, NOM, LOC	Adaptability, completeness, maintainability, reusability, testability, understandability
Alshayeb [7]	Three open source	09	CK metrics, FOUT, NOM, LOC	Adaptability, maintainability, reusability, testability, understandability
Stroggylos and Spinellis [13]	Four open Source	Not mentioned	CK metrics, Ca, NPM	–
Moser et al. [14]	One commercial	Not mentioned	CK metrics, MCC, LOC	Reusability
Du Bois et al. [8]	One open source	05	Coupling, cohesion	Maintainability
Kataoka et al. [9]	A small C++ program	02	Coupling	Maintainability
Elish and Alshayeb [10]	Four open source	07	CK metrics, FOUT, NOM, LOC	Adaptability, maintainability, reusability, testability, understandability

done on analyzing the impact of refactoring techniques on design patterns by Elish and Alshayeb [10] do not include the design patterns addressed in this study. Yu and Ramaswamy [11] have studied Replace Constructor with Factory in context with complexity and class structural quality only. Other internal and external quality attributes still needs to be explored. In the best of the authors' knowledge, none of the researchers have studied the impact of Replace Constructor with Builder and Wrap Return Value refactoring techniques on any of software quality attributes.

3 Refactoring and Software Quality

3.1 Refactoring Techniques

Four refactoring techniques are selected Replace Constructor with Builder, Replace Constructor with Factory, Wrap Return Value, and Encapsulate Field as they are not

Table 2 Refactoring techniques (RTs)

S. no.	Refactoring technique	Usage
RT1	Replace_constructor with factory	It hides the constructor and can return already created object. Used if the constructor is complex
RT2	Replace constructor with builder	Same usage as RT1 but used when the object cannot be created in single step
RT3	Wrap return value	Used to allow the method to return more data than planned. It is executed by either creating a new or using an existing wrapper class for return values of selected method
RT4	Encapsulate field	This is performed by changing the access modifier from public to private and creating the get–set accessors

yet explored properly in the context of software quality. A recent survey conducted by Dallal and Abdin [15] also supports this fact. Though there are 23 and 17 refactoring patterns identified by Gamma [16] and Kereivsky [17], respectively; we covered, for now, two of the most popularly used creational design patterns. Addressed refactoring techniques are summarized in Table 2.

3.2 Software Quality Attributes

Software quality aims at determining how well the software is designed and how well it conforms to that design [18]. Software quality can be measured by different attributes. According to Pressman [4], software quality attributes can be represented at two levels: (i) Micro level—Internal quality attributes and (ii) Macro level—External quality attributes.

Micro level—Internal Quality Attributes
CK metric suite is selected for this study as Basili et al. [19] found on investigation that CK metrics are good and useful quality indicators. CK metrics consists of six metrics described in Table 3. CKJM tool [20] is used for the calculation of CK metrics.

Macro level—External Quality Attributes
Quality attributes, viz. testability, maintainability, adaptability, completeness, understandability, and reusability are addressed in this study. For estimating reusability, one can rely upon adaptability, maintainability, completeness, and understandability [21].

Table 3 CK metrics

CK metric	Description
CBO	Coupling between object. This is count of other classes to which given class is coupled
RFC	Response for a class. This is number of methods in that class and methods called by that class
LCOM	Lack of cohesion in methods. This is lack of cohesion in metrics
DIT	Depth of inheritance tree. This is number of ancestor classes
NOC	Number of children. This is number of direct subclasses of a class
WMC	Weighted methods per class. This is number of methods that belong to the class

Table 4 Project details

Project name	Github URL	#Classes excluding test classes
RxJava	https://github.com/ReactiveX/RxJava	1449
Fastjson	https://github.com/alibaba/fastjson	157
Hollow	https://github.com/Netflix/hollow	284
Scott	https://github.com/dodie/scott	27

4 Research Methodology

An association between refactoring techniques and software quality are established by providing systematic solutions for the following research questions:

RQ1: What is the effect of investigated refactoring techniques on internal quality attributes of software projects?

Statistical validations need to be done for concluding this investigation.

RQ2: What is the effect of investigated refactoring techniques on selected external quality attributes of software projects?

4.1 Dataset Collection

Four projects were selected from GitHub repository (https://github.com/) on the context of number of classes. Number of classes guides in defining projects as small-, mid-sized, and large projects. Table 4 summarizes projects' names, source, number of classes excluding the test sources and size of project.

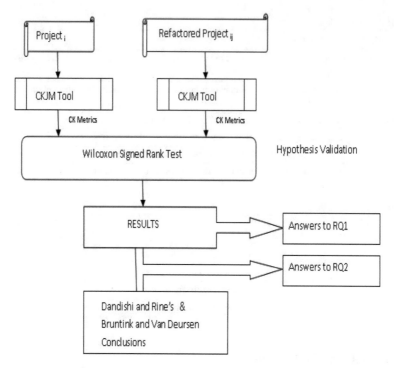

Fig. 1 Experimental design

4.2 Experimental Design

Experimental design is based on performing metric-based refactoring on chosen projects and statistically validating the differences between metrics of original and refactored projects. Diagrammatically, the complete methodology is depicted in Fig. 1.

Project$_i$ are projects undertaken where $i = 1$–4. Refactored project$_{ij}$ are projects after refactoring where $j = 1$–4 symbolizes applied refactoring technique. For each project, four refactoring techniques are applied individually. Metrics are then calculated for original as well as each refactored version of project using CKJM tool [20] for common classes. Refactoring opportunities are found manually for particular refactoring technique and then, refactoring technique is applied on projects using IntelliJ IDEA [22].

Next, the Wilcoxon signed-rank test is performed to find statistically significant difference between CK metrics of original and each refactored version. Statistical results help in understanding how individual refactoring technique affects the internal quality of object-oriented projects by relating CK metrics with coupling, cohesion, inheritance, and complexity.

This study aims at finding the effect of refactoring techniques on internal quality attributes by setting null hypothesis as

H_{0i}: Investigated refactoring techniques do not affect internal quality attribute$_i$.

where $i = 1$–4, internal quality attribute$_1$ = coupling, internal quality attribute$_2$ = inheritance, internal quality attribute$_3$ = cohesion, and internal quality attribute$_4$ = complexity.

To answer RQ2, we need to establish a correlation between external and internal software quality metrics. Dandashi and Rine [21] by qualitative survey found that calculated CK metrics can be well correlated with external quality attributes like adaptability, completeness, maintainability, reusability, and understandability of OO software. Bruntink and van Deursen [23] investigated the methodology to estimate the testing effort and found a correlation between CK metrics and testing effort. Findings of these researchers provide the guidelines to establish a correlation between individual refactoring technique and external quality attributes. Behavioral analysis of refactoring techniques [2, 16, 17] guided experimental findings to derive conclusions.

5 Results and Discussions

5.1 Analysis of Impact of Refactoring Techniques on Internal Quality Attributes

RQ1: What is the effect of investigated refactoring techniques on internal quality attributes of software projects?

To answer this research question, statistical validation is performed using Wilcoxon signed-rank test. Two levels of significance are defined: Highly significant—99% significant and Significant—95% significant. L, M, and S notations with the names of projects in Table 5 denotes large-, mid-sized, and small project. ++ indicates highly significant increase in metric value, + indicates significant increase in metric value, x indicates no change in metric value, and I indicates insignificant change in metric value. Thus, Table 5 provides the precise information regarding change in metric values of different sized projects when considered refactoring techniques are used for these projects.

From Table 5, the correlation between CK metrics and applied refactoring techniques can be deduced which are summarized in Table 6. ↑ signifies statistically significant increase in CK metric value, ↓ signifies statistically significant decrease in CK metric value, I signifies insignificant change, and x signifies no impact. L, M, and S denotes large-, mid-sized, and small project. The constructed hypotheses can be validated or refuted with help of Tables 5 and 6.

Table 5 Statistical validation of Wilcoxon signed-rank test results

Refactoring Technique	Project with size	CBO^R-CBO	RFC^R-RFC	$LCOM^R$-LCOM	DIT^R-DIT	NOC^R-NOC	WMC^R-WMC
Replace Constructor with Builder	RxJAVA (L)	++	++	x	x	x	x
	HOLLOW (M)	++	++	I	x	x	I
	FASTJSON (M)	++	++	x	x	x	x
	SCOTT (S)	I	I	x	x	x	I
Replace Constructor with Factory Method	RxJAVA (L)	++	++	x	x	x	++
	HOLLOW (M)	+	++	++	x	x	++
	FASTJSON (M)	x	++	++	x	x	++
	SCOTT (S)	I	I	x	x	x	I
Encapsulate Field	RxJAVA (L)	x	++	++	x	x	++
	HOLLOW (M)	x	++	++	x	x	++
	FASTJSON (M)	x	++	++	x	x	++
	SCOTT (S)	x	I	I	x	x	I
Wrap Return Value	RxJAVA (L)	++	++	I	x	x	++
	HOLLOW (M)	+	++	I	x	x	++
	FASTJSON (M)	++	++	x	x	x	I
	SCOTT (S)	I	I	I	x	x	I

$Metric^R$ = metric value of refactored code

metric = metric value of original code

++ = statistically highly significant increase

+ = statistically significant increase

x = no change

I = statistically insignificant change

H_{01}: Investigated refactoring techniques do not affect coupling.

As can be concluded by Table 6, the increase in CBO value and RFC value are either highly significant or significant after applying Replace Constructor with Builder.

Replace Constructor with Factory Method and Wrap Return Value on large- and mid-sized projects. As the number of methods in classes as well as interdependency among classes has increased significantly, null hypothesis will be rejected for these three refactoring techniques. Hence, these three refactoring techniques have negative impact on coupling.

In encapsulate field refactoring, public data members are transformed into private data members by encapsulating it in public method, so this refactoring technique does not have any impact on CBO but increase in public method results in significant increase in RFC which finally conclude to high coupling. Again, this results in the rejection of null hypothesis. Encapsulate field refactoring has significant negative impact on coupling.

Table 6 Refactoring impact on CK metrics

Refactoring technique	CBO			RFC			LCOM			DIT			NOC			WMC		
	L	M	S	L	M	S	L	M	S	L	M	S	L	M	S	L	M	S
Replace constructor with builder	←	←	I	←	←	I	×	×	×	×	×	×	×	×	×	×	I	I
Replace constructor with factory method	←	←	I	←	←	I	←	←	×	×	×	×	×	×	×	←	←	I
Wrap return value	←	←	I	←	←	I	I	I	I	×	×	×	×	×	×	←	←	I
Encapsulate field	×	×	×	←	←	I	←	←	I	×	×	×	×	×	×	←	←	I

H_{02}: Investigated refactoring techniques do not affect cohesion.

LCOM value, measure of cohesion, shows high significant increase for mid-sized projects and large projects on the application of Encapsulate Field refactoring technique; resulting in negative impact on cohesion. Replace Constructor with Factory Method refactoring technique also depicts high significant increase for mid-sized projects. Cohesion acts as a trade-off for assessing software quality. Replace constructor with factory method is used when we have complex constructor to make design flexible [2]. In this case, though new static method is formed resulting in less cohesiveness but the overall software quality is improved. So, the null hypothesis is rejected for abovementioned techniques.

Replace Constructor with Builder and Wrap Return Value, whereas, have either no effect or insignificant increase in LCOM value for all projects, respectively, irrespective of size of projects. Therefore, the null hypothesis is accepted for these RTs.

H_{03}: Investigated refactoring techniques do not affect inheritance.

DIT and NOC metrics are directly related to inheritance. By definitions given in Table 2, none of the investigated refactoring techniques relate to change in number of subclasses or super classes. The experimental results also depict no change in DIT or NOC values for all considered projects. So, we accept the null hypothesis for all four addressed refactoring techniques.

H_{04}: Investigated refactoring techniques do not affect complexity.

Replace Constructor with Builder has insignificant effect or no effect on WMC values of all projects, whereas there is highly significant increase in WMC value for the emaining three refactoring techniques. WMC is a trade-off metric. High WMC values make project more complex but in case of complicated and repeating design problems, high WMC may indicate better object-oriented software that is easy to implement and reuse [2]. So, Replace Constructor with Factory Method refactoring technique, though having high WMC, results in reducing the complexity of object-oriented software. Hypothesis is rejected for all refactoring techniques except for Replace Constructor with Builder.

Based on accepted and rejected null hypotheses, classification of refactoring techniques for internal quality attributes (complexity, coupling, cohesion, and inheritance) are illustrated in Table 7 where L, M, and S are notations for large, mid-, and small-sized projects, respectively. ↑ signifies an increase in addressed internal quality attribute, ↓ signifies decrease in addressed internal quality attribute, I signifies insignificant impact, and x signifies no impact of investigated refactoring technique on addressed internal quality attribute.

Table 7 Classification of refactoring techniques for internal software quality attributes

Refactoring technique	Coupling			Cohesion			Inheritance			Complexity		
	L	M	S	L	M	S	L	M	S	L	M	S
Replace constructor with builder	←	←	I	×	×	×	×	×	×	×	I	I
Replace constructor with factory method	←	←	I	×	→	×	×	×	×	→	→	I
Wrap return value	←	←	I	I	I	I	×	×	×	←	←	I
Encapsulate field	←	←	×	→	→	I	×	×	×	←	←	I

Table 8 Correlation between CK metrics and external software quality attributes [21, 22]

External quality attributes	CBO	RFC	LCOM	DIT	NOC	WMC
Adaptability	–	–	NA	–	–	+
Completeness	–	–	NA	–	–	+
Maintainability	–	–	NA	–	–	+
Understandability	–	–	NA	–	–	+
Reusability	–	–	NA	–	–	+
Testability (Testing Effort)	0	+	0	0	0	+

5.2 Analysis of Impact of Refactoring Techniques on External Quality Attributes

Dandashi and Rine [21] & Bruntink and van Deursen [23] results are provided in Table 8 where + signifies positive correlation, − signifies negative correlation, 0 signifies no correlation, and NA signifies not applicable. For example, high WMC (+) indicates improvement in adaptability, completeness, maintainability, reusability, and understandability and reusability but demands for more testing effort, i.e., high testability.

Based on Tables 7 and 8, refactoring techniques' impact on external software quality attributes is interpreted and is summarized in Table 9.

For all investigated refactoring techniques, small-sized projects show insignificant impact on all external software quality attributes. The reason for this may be that small projects are manageable by nature and may seldom require refactoring.

The impact of individual refactoring technique is summarized below:

Replace constructor with factory method

This refactoring technique results in increase in RFC, CBO, and WMC values. Increase in number of classes and methods are not always related to increased complexity. It may also be the results of simplifying complex design structure of object-oriented software [2]. Keeping in consideration that replaces the constructor with factory method is used to restructure code in case there are complex constructors, and therefore, statistically increase in RFC, CBO, and WMC indicates reduced complexity of software. This technique aims at building software that can embrace changes easily [2].

Therefore, it results in increase in adaptability, completeness, maintainability, reusability, testability, and understandability. So, it has positive impact on reusability.

Table 9 Classification of refactoring techniques for external software quality attributes

Refactoring technique	Adaptability			Completeness			Maintainability			Understandability			Reusability			Testability (Testing Effort)		
	L	M	S	L	M	S	L	M	S	L	M	S	L	M	S	L	M	S
Replace constructor with builder	←	←	I	←	←	I	←	←	I	←	←	I	←	←	I	←	←	I
Replace constructor with factory method	←	←	I	←	←	I	←	←	I	←	←	I	←	←	I	←	←	I
Wrap return value	→	→	I	→	→	I	→	→	I	→	→	I	→	→	I	←	←	I
Encapsulate field	←	←	I	←	←	I	←	←	I	←	←	I	←	←	I	←	←	I

But high testability accounts for more testing effort, therefore, has negative impact on testability.

Replace constructor with builder

Builder is more complicated than factory method and has less usage than factory method. This refactoring technique shows an increase in CBO and RFC values of large- and medium-sized projects. WMC is insignificantly increased. Therefore, for similar reasons as of Replace constructor with factory method refactoring technique, this refactoring technique also leads to improvement in adaptability, completeness, maintainability, reusability, and understandability but it degrades testability as more testing effort is required.

Wrap return value

This technique results in either creation of new class (wrapper class) or it uses already existing wrapper class which is most compatible with method. Hence, it results in increase in CBO, RFC, and WMC values which is statistically validated also. From the following prevailing studies' conclusions [21, 23], this refactoring technique has negative impact on addressed software external quality attributes.

Encapsulate Field

Similarly, conclusions can be drawn for encapsulate field refactoring technique. It has no effect on CBO value but increases RFC and WMC values. So, trade-off results in positive impact of encapsulate field refactoring technique on adaptability, completeness, maintainability, reusability, and understandability but negative impact on testability.

6   Threats to Validity

One of the obvious validity threats to conduct study is that we totally relied on prevailing studies [21, 22] for analyzing effect of refactoring techniques on external software quality attributes and did not validate them. Also, there are many other factors than project metrics that affects internal or external quality attributes like developer's expertise and gut feeling which are not considered in this study. Further, the results cannot be generalized as only four Java projects are considered for investigated refactoring techniques. As none of the researchers has considered these refactoring techniques so far, there is a need of replication of this study with more projects to generalize results for the selected refactoring techniques. Though we considered projects of different sizes to accumulate indefinite behavior of refactoring techniques, still the replication of study with more projects of different types is needed to generalize results.

7 Conclusions and Future Work

Results of the study support the investigated refactoring techniques for improvement of selected software quality attributes. For small projects, impact of all refactoring techniques is insignificant as small projects, by nature, are simple and easy to handle. So, generally, developers structure them well. But for medium and large projects, it is important to wisely choose refactoring techniques before applying them. This is experimentally proved that inheritance is not affected by any of the investigated refactoring techniques and other internal quality attributes have varied impact of refactoring techniques on them. Wrap return value refactoring technique shows adverse effect on addressed external quality attributes. The remaining three refactoring techniques improve addressed external quality attributes except for testability of mid-sized and large object-oriented software. Developers and software practitioners can predict positive or negative change in their software's quality before application of aforesaid refactoring techniques.

Future directions include replicating the study with measured external quality attributes and validating the results of [21, 22]. Other future directions will certainly be the reinvestigation of these refactoring techniques with more projects of different nature and sizes that can ascertain or disprove our findings. In addition to this, if we incorporate qualitative aspect of refactoring with our quantitative approach, we can get better and accurate refactoring solutions.

References

1. Opdyke WF (1992) Refactoring object-oriented frameworks
2. Fowler M, Beck K (1999) Refactoring: improving the design of existing code. Addison-Wesley Professional
3. Bansiya J, Davis CG (2002) A hierarchical model for object-oriented design quality assessment. IEEE Trans Softw Eng 28(1):4–17
4. Pressman RS (2005) Software engineering: a practitioner's approach. Palgrave Macmillan
5. Mens T, Tourwé T (2004) A survey of software refactoring. IEEE Trans Softw Eng 30(2):126–139
6. Elish KO, Alshayeb M (2011) A classification of refactoring methods based on software quality attributes. Arab J Sci Eng 36(7):1253–1267
7. Alshayeb M (2009) Empirical investigation of refactoring effect on software quality. Inf Softw Technol 51(9):1319–1326
8. Du Bois B, Demeyer S, Verelst J (2004) Refactoring-improving coupling and cohesion of existing code. In: Proceedings of 11th working conference on reverse engineering. IEEE, pp 144–151
9. Kataoka Y, Imai T, Andou H, Fukaya T (2002) A quantitative evaluation of maintainability enhancement by refactoring. In: Proceedings of international conference on software maintenance. IEEE, pp 576–585
10. Elish KO, Alshayeb M (2012) Using software quality attributes to classify refactoring to patterns. J Softw 7(2):408–419
11. Yu L, Ramaswamy S (2018) An empirical study of the effect of design patterns on class structural quality. In: Application development and design: concepts, methodologies, tools, and applications. IGI Global, pp 315–334

12. Bavota G, De Lucia A, Marcus A, Oliveto R (2014) Automating extract class refactoring: an improved method and its evaluation. Empir Softw Eng 19(6):1617–1664
13. Stroggylos K, Spinellis D (2007) Refactoring–does it improve software quality?. In: Proceedings of the 5th international workshop on software quality. IEEE Computer Society, vol 10
14. Moser R, Sillitti A, Abrahamsson P, Succi G (2006) Does refactoring improve reusability?. In: Proceedings of international conference on software reuse, vol 4039, pp 287–297
15. Al Dallal J, Abdin A (2017) Empirical evaluation of the impact of object-oriented code refactoring on quality attributes: a systematic literature review. IEEE Trans Softw Eng
16. Gamma E (1995) Design patterns: elements of reusable object-oriented software. Pearson Education India
17. Kerievsky J (2005) Refactoring to patterns. Pearson Deutschland GmbH
18. Malhotra R (2016) Empirical research in software engineering: concepts, analysis, and applications. CRC Press
19. Basili VR, Briand LC, Melo WL (1996) A validation of object-oriented design metrics as quality indicators. IEEE Trans Softw Eng 22(10):751–761
20. CKJM Tool. https://www.spinellis.gr/sw/ckjm/
21. Dandashi F, Rine DC (2002) A method for assessing the reusability of object-oriented code using a validated set of automated measurements. In: Proceedings of ACM symposium on applied computing, pp 997–1003
22. Jetbrains IntelliJ IDEA Version 5.0. http://www.jetbrains.com/idea/
23. Bruntink M, van Deursen A (2006) An empirical study into class testability. J Syst Softw 79(9):1219–1232

Rumor Detection Using Machine Learning Techniques on Social Media

Akshi Kumar and Saurabh Raj Sangwan

Abstract Information overload on the Web has been a well-identified challenge which has amplified with the advent of social web. Good, bad, true, false, useful, and useless are all kinds of information that disseminates through the social web platforms. It becomes exceedingly imperative to resolve rumors and inhibit them from spreading among the Internet users as it can jeopardize the well-being of the citizens. Rumor is defined as an unverified statement initiating from a single or multiple sources and eventually proliferates across meta-networks. The task for rumor detection intends to identify and classify a rumor either as true (factual), false (nonfactual), or unresolved. This can immensely benefit the society by preventing the spreading of such incorrect and inaccurate information proactively. This paper is a primer on rumor detection on social media which presents the basic terminology and types of rumors and the generic process of rumor detection. A state-of-the-art depicting the use of supervised machine learning (ML) algorithms for rumor detection on social media is presented. The key intent is to offer a stance to the amount and type of work conducted in the area of ML-based rumor detection on social media, to identify the research gaps within the domain.

Keywords Rumor · Machine learning · Social media

1 Introduction

With the inception of Web 2.0 and the increasing ease of access methods and devices, more and more people are getting online, making Web indispensable for everyone. The focal point of Web 2.0 is social media. Active participation is a key element

A. Kumar (✉) · S. R. Sangwan
Department of Computer Science & Engineering, Delhi Technological University,
New Delhi, Delhi, India
e-mail: akshikumar@dce.ac.in

S. R. Sangwan
e-mail: saurabhsangwan2610@gmail.com

© Springer Nature Singapore Pte Ltd. 2019
S. Bhattacharyya et al. (eds.), *International Conference on Innovative Computing and Communications*, Lecture Notes in Networks and Systems 56,
https://doi.org/10.1007/978-981-13-2354-6_23

213

that builds the social web media. Numerous social networking sites like Twitter, YouTube, and Facebook have become popular among the masses. It allows people to build connection networks with other people and share various kinds of information in a simple and timely manner. Today, anyone, anywhere with the internet connection can post information on the Web. But like every coin has its two sides, this technological innovation of social media also has some good as well as bad aspects. We are really benefited by social media but we cannot oversee its negative effects in society. Most people admire it as a revolutionary invention and some seem to take it as a negative impact on the society. As a positive case, these online communities facilitate communication with people around the globe regardless of your physical location. The perks include building connection in society, eliminating communication barriers, and helping as effective tools for promotion, whereas on the flip side, privacy is no more private when sharing on social media.

Due to the ubiquitous and overdependence of users on social media for information, the recent trend is to look and gather information from online social media rather than traditional sources. But there are no means to verify the authenticity of information available and spreading on these social media platforms thus making them rumor breeding sources. A rumor is defined as any piece of information put out in public without sufficient knowledge and/or evidence to support it thus putting a question on its authenticity. It may be true, false, or unspecified and is generated intentionally (attention seeking, self-ambitions, finger pointing someone, prank, to spread fear, and hatred) or unintentionally (error). Further, these can be personal as well as professional. Knapp [1] classified rumors into three categories, namely pipe dream, bogy, and wedge driving for describing intentional rumors.

Rumors are circulated and believed overtly. And due to the increasing reliance of people on social media, it is inevitable to detect and stop rumors from spreading to reduce their impact. It takes only little time for a single tweet or post to go viral and affect millions. Thus, rumor detection and mitigation have evolved as a recent research practice where the rumor has to be recognized and its source has to be identified to limit its diffusion. It is essential not just to detect and deter, but to track down the rumor to its source of origin. Various primary studies with promising results and secondary studies [2, 3] have been reported in this direction. The work presented in this paper is a primer on rumor detection on social media to explicate the what, why, and how about the rumor detection on online social media. The intent is to aid novice researchers with a preliminary introduction to the area and at the same time, offer background work to the experts. The types of rumors and the typical process of rumor detection are discussed followed by a state-of-the-art review of supervised ML-based rumor detection on online social media. The research gaps have been identified as issues and challenges within the domain which make it an active and dynamic area of research.

2 Rumor Detection on Social Media

Social media has the power to make any information, be it true or false, go viral, and reach and affect millions. Due to the speed of information spread, even rumors are spread. Hence, it is necessary to detect and restraint these rumors before they have a serious impact on people's lives.

2.1 Types of Rumors

A rumor is defined as information whose veracity is doubtful. Some rumors may turn out to be true, some false, and others may remain unverified. Not all false information can be classified as a rumor. Some are honest mistakes by people and are referred to as misinformation. On the other hand, there may be intentional rumors put to mislead people into believing them. These are labeled as disinformation and are further classified based on the intent of the originator. The following Fig. 1 depicts the classification of rumors.

We define a rumor as any information put out in public without sufficient knowledge and/or evidence to support it. It is misleading, either intentionally or unintentionally. If some information has been put out in public erroneously without authentic or complete information with no ulterior motive of hurting or causing any disturbance to anyone whatsoever, it is called misinformation. It is an honest mistake. Disinformation, on the other hand, is information that is intentionally put out in public view to mislead people and start a false rumor. Disinformation depending on the motive of the writer and nature of the post can be classified as humorous, hoax, finger pointing, tabloids, and yellow press. The most harmless type of rumor is the humorous ones.

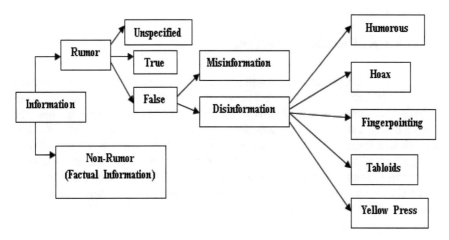

Fig. 1 Classification of rumors

Sources spreading this type of information fabricate news and stories to give it an amusing side. The motive is usually to entertain people. The information is predeclared to be false and intended only for comical purposes. The best examples of such sources include news satires and news game shows.

The next form of disinformation is a hoax. A hoax is intentional fake news spread to cause panic among people and cause trouble to people at whom it is aimed. A hoax can also be an imposter. Examples include fabricated stories, false threats, etc. In 2013, a hoax stating Hollywood actor "Tom Cruise to be dead" started doing the rounds. Social messaging apps like WhatsApp worsen the situations when it comes to hoaxes. Currency ban of Indian rupees 500 and 1000 was done in November 2016. Soon after a hoax message went viral on WhatsApp stating that the government will release a new 2000 rupee denomination that would contain a GPS trackable nano chip that would enable to locate the notes even 390 feet buried underground. The government and bank spokespersons had to finally issue an official statement stating it was false. Still, many people found the official statement hard to believe as they were so brainwashed by the hoax message.

Another form of disinformation is finger pointing. Finger pointing always has an associated malicious intent and personal vested interest. It blames a person or an organization for some bad event that is happening or happened in the past. It aims at political or financial gain by tarnishing the image of the target person/organization/party/group, etc. Tabloids have a bad name for spreading rumors from since when they started. It is the type of journalism that accentuates sensational stories and gossips about celebrities that would amount to spicy page 3 stories. Yellow press journalism is a degraded form of journalism which reports news with little or no research at all. Journalists' only aim is to catch attention using catchy headlines with no regards whatsoever to the authenticity of news. They do not bother to delve deep into a story but just publish it to sell as many stories as possible and make money. It is the most unprofessional and unethical form of journalism.

3   Process of Rumor Detection

Rumor detection is primarily a four-step process, where the task begins by collecting data from various social media sites under consideration. This collected data needs to be in a uniform structured format so that relevant features can be extracted. The preprocessing task includes consolidation, cleaning, transformation, and reduction. The relevant features (including content-based, pragmatic, and network-specific features) are then extracted and each dataset is then classified as being a rumor or not a rumor using various machine learning techniques like Naïve Bayesian, Support Vector Machines, etc. The typical rumor detection process which essentially consists of four steps as described in Fig. 2.

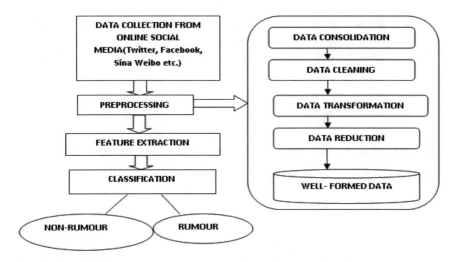

Fig. 2 Process of rumor detection

3.1 Data Collection

The first and foremost step is to collect data from various social media platforms like Facebook, Twitter, and SinaWeibo. This is the data that we have to analyze for rumor detection. Each social media platform offers application programming interfaces (APIs) which allows extracting data from them. Each API is accompanied with detailed documentation elaborating the methods, features, limitations, etc., of the API.

3.2 Preprocessing

The data gathered must be transformed into a form suitable for tools to extract features and perform classification. Preprocessing involves the following steps:

- **Data Consolidation**: In this step, the data collected is integrated together and combined into one data source. It also includes converting all collected data into a single format.
- **Data Cleaning**: The data collected contains a lot of noise (data irrelevant to us) which needs to be removed. Data cleaning includes removing noise, inconsistencies and any missing data.
- **Data Transformation**: In this, the data is transformed by adding certain attributes, aggregating and normalizing the data.

- **Data Reduction**: Here, the number of variables and cases are reduced for ease. The skewed data is also balanced in this step.
- After preprocessing, the data is well formed and suitable for feature extraction.

3.3 Feature Extraction

The following figure depicts the various features that can be extracted (Fig. 3).

3.4 Classification

In the final step, the data is analyzed and classified into two pre-defined categories, namely rumor and non-rumor by using various classification techniques. Rumor detection has been an interesting area of research since the start of Web 2.0 era. Various techniques have been tried and tested on social media for rumor detection. Among various techniques, machine learning methods have emerged as a key player when applied to rumor detection on online social media. We describe these machine learning techniques and review their use in rumor detection in the next section.

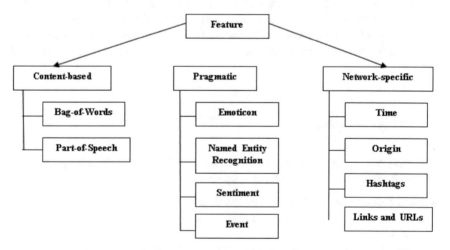

Fig. 3 Typical features extracted in rumor detection on social media

Table 1 Research on rumor detection on social media using ML techniques

Author	Year	Technique	Dataset	Evaluation parameters
Zhao et al. [4]	2015	SVM, Decision Trees	Gardenhose	Precision @ N
Takahashi and Igata [5]	2012	Natural Language Processing	Twitter	Accuracy
Wang [6]	2014	Social Spam Analytics and Detection Framework (SPADE)	Email, Twitter	Tp, Fp, F-measure, Accuracy
Liu et al. [7]	2015	SVM	Twitter	Accuracy
Cai et al. [8]	2014	SVM	SinaWeibo	P, Recall, F-measure
Lukasik et al. [9]	2015	SVM	Twitter	Accuracy
Kwon et al. [10]	2013	Decision tree, SVM, Random forest	Twitter	Accuracy, Precision, Recall, F1
Yang et al. [11]	2012	SVM	SinaWeibo	Precision, Recall, F-score
Jin et al. [12]	2013	SEIZ	Twitter	SEIZ compartment and RSI
Yang et al. [13]	2015	Logistic, Naïve Bayes, random forest	News, Twitter	Precision, Recall, F-measure
Liu et al. [14]	2016	Information propagation model based on heterogeneous user	SinaWeibo	Precision, recall, F-rate, Accuracy
Wang and Terano [15]	2015	Graph-based pattern matching algorithm	Twitter	Tf–idf
Wu et al. [16]	2015	SVM	SinaWeibo	Accuracy

4 Machine Learning-Based Techniques for Rumor Detection on Social Media

Machine learning-based techniques have emerged as promising viable approach for detecting rumors on social media. This section summarizes the state-of-the-art review to account for the work done within the domain of rumor detection on social media using various supervised ML techniques. The following Table 1 summarizes the primary research carried out.

As seen from the table, most of the research for rumor detection have been carried out on Twitter and SinaWeibo (Chinese micro-blogging site) while some have also incorporated data sets from news websites and Gardenhose. The most popular machine learning techniques used includes SVM, Naïve Bayes, and Decision Trees. Techniques like k-Nearest Neighbor, Clustering, Bernoulli Naïve Bayes, Random Forests, Logistic Naïve Bayes, Natural Language Processing, social spam analytics and detection framework (SPADE), SEIZ, etc. have also been used. Parameters such as precision, recall, accuracy, number of true and false positives, and F-measure [17, 18] have been used as evaluation metrics.

4.1 Issues and Challenges

Rumor detection comes with its share of issues and challenges. The main challenge for carrying out the rumor detection task is the collection of data. Even the most popular social media sites, namely Twitter and Facebook do not provide full freedom to users for extracting data. Most of the data posted on Facebook is private in nature, hence inaccessible. Only data posted on Facebook pages can be collected. Twitter, on the other hand, these days does not allow data older than seven days to be fetched. Another issue faced by researchers is the detection of new rumors from real-time data. It is easier to detect old posts regarding a rumor that we know of because we know the keywords. But with emerging rumors we are in a fix as we do not know what to look out for. Also, some rumors remain unspecified and there is no conformation or debunking for them. Hence, detecting rumor veracity is very challenging. Another aspect that needs to be taken care of is the detection of origin of a rumor as it is difficult to identify the user who started a particular rumor. These issues need to be addressed to improve the quality and speed of rumor detection.

5 Conclusion

This paper presented the primary concepts of rumor detection. As much as social media has become an invaluable source for sharing real-time and crucial information, it is also a breeding platform for rumors. Timely rumor detection is essential to prevent panic and maintain peace in society. This paper explains the rumor detection process and reviews the research carried out for rumor detection using various ML techniques. The scope of this review is limited to a single level classification task where we predict whether given online information is a rumor or not. This task can be extended to a multi-level, fine-grain classification where rumors can be detected for being a misinformation or a disinformation, hoaxes, etc. Various novel and hybrid machine learning techniques such as fuzzy, neuro fuzzy can also be used for detecting rumors.

References

1. Knapp R (1944) A psychology of rumor. Public opinion. Quarterly 1:22–37
2. Serrano E, Iglesias CA, Garijo M (2015) A survey of Twitter rumor spreading simulations. In: Núñez M, Nguyen N, Camacho D, Trawiński B (eds) Computational collective intelligence. Lecture notes in computer science, vol 9329. Springer, Cham, pp 113–122
3. Zubiaga A, Aker A, Bontcheva K, Liakata M, Procter R (2017) Detection and resolution of rumours in social media: a survey. arXiv:1704.00656
4. Zhao Z, Resnick P, Mei Q (2015) Enquiring minds: Early detection of rumors in social media from enquiry posts. In: Proceedings of the 24th international conference on world wide web international world wide web conferences steering committee. Florence, pp 1395–1405
5. Takahashi T, Igata N (2012) Rumor detection on twitter. In: Joint 6th international conference on soft computing and intelligent systems (SCIS) and 13th international symposium on advanced intelligent systems (ISIS). IEEE, Kobe, pp 452–457
6. Wang D (2014) Analysis and detection of low quality information in social networks. In: 2014 IEEE 30th international conference on data engineering workshops (ICDEW). IEEE, Chicago, pp 350–354
7. Liu X, Nourbakhsh A, Li Q, Fang R, Shah S (2015) Real-time rumor debunking on twitter. In: Proceedings of the 24th ACM international on conference on information and knowledge management. ACM, Mebourne, pp 1867–1870
8. Cai G, Wu H, Lv R (2014) Rumors detection in Chinese via crowd responses. In: 2014 IEEE/ACM international conference on advances in social networks analysis and mining (ASONAM). IEEE, Beijing, pp 912–917
9. Lukasik M, Cohn T, Bontcheva K (2015) Classifying tweet level judgements of rumors in social media. arXiv:1506.00468
10. Kwon S, Cha M, Jung K, Chen W, Wang Y (2013) Prominent features of rumor propagation in online social media. In: 2013 IEEE 13th international conference on data mining (ICDM). IEEE, Dallas, pp 1103–1108
11. Yang F, Liu Y, Yu X, Yang M (2012) Automatic detection of rumor on Sina Weibo. In: Proceedings of the ACM SIGKDD workshop on mining data semantics. ACM, Beijing, p 13
12. Jin F, Dougherty E, Saraf P, Cao Y, Ramakrishnan N (2013) Epidemiological modeling of news and rumors on twitter. In: Proceedings of the 7th workshop on social network mining and analysis. ACM, Chicago, p 8, Aug 2013
13. Yang Z, Wang C, Zhang F, Zhang Y, Zhang H (2015) Emerging rumor identification for social media with hot topic detection. In: 2015 12th web information system and application conference (WISA). IEEE, Jinan, pp 53–58
14. Liu Y, Xu S, Tourassi G (2015) Detecting rumors through modeling information propagation networks in a social media environment. In: Agarwal N, Xu K, Osgood N (eds) Social computing, behavioral-cultural modeling, and prediction. SBP 2015. Lecture notes in computer science, vol 9021. Springer, Cham, pp 121–130
15. Wang S, Terano T (2015) Detecting rumor patterns in streaming social media. In: 2015 IEEE International Conference on Big Data (Big Data). IEEE, Santa Clara, pp 2709–2715
16. Wu K, Yang S, Zhu KQ (2015) False rumors detection on Sina Weibo by propagation structures. In: 2015 IEEE 31st international conference on data engineering (ICDE). IEEE, Seoul, pp 651–662
17. Bhatia MPS, Kumar A (2008) A primer on the web information retrieval paradigm. J Theor Appl Inf Technol 4(7):657–662
18. Kumar A, Jaiswal A (2017) Empirical study of Twitter and Tumblr for sentiment analysis using soft computing techniques. In: Proceedings of the World Congress on Engineering and Computer Science, vol 1, pp 1–5

Empirical Analysis of Supervised Machine Learning Techniques for Cyberbullying Detection

Akshi Kumar, Shashwat Nayak and Navya Chandra

Abstract Cyberbullying is utilization of digital technology for targeting a person or a group in order to bully them socially and psychologically. Real-time social media platforms such as Instagram, Twitter, and YouTube have a large viewership, which serves as a fertile medium for such bullying activities. Instances of such harassment or intimidation are maximally found in the comments of a dynamic and an expressive medium like YouTube. This necessitates an adequate requisite to take relevant steps to find solutions for the detection and prevention of cyberbullying. The work presented in this paper focuses on the implementation of four supervised machine learning methodologies, namely Random Forest, k-Nearest Neighbor, Sequential Machine Optimization, and Naive Bayes in order to identify and detect the presence or absence of cyberbullying in YouTube video comments. The experimentation was carried out expending the Weka toolkit and utilizing the data gathered from comments obtained from YouTube videos involving core sensitive topics like race, culture, gender, sexuality, and physical attributes. The results are analyzed based on the measures like precision, accuracy, recall, and F-score and amongst the four techniques implemented, k-Nearest Neighbor is able to recognize the true positives with highest accuracy of around 83%. We also discuss various future research prospects for detection of cyberbullying.

Keywords Cyberbullying · Social media · Supervised machine learning YouTube

A. Kumar (✉) · S. Nayak · N. Chandra
Delhi Technological University, New Delhi, Delhi, India
e-mail: akshikumar@dce.ac.in

S. Nayak
e-mail: shashwatnayak@outlook.com

N. Chandra
e-mail: navya.chandra2010@gmail.com

© Springer Nature Singapore Pte Ltd. 2019
S. Bhattacharyya et al. (eds.), *International Conference on Innovative Computing and Communications*, Lecture Notes in Networks and Systems 56,
https://doi.org/10.1007/978-981-13-2354-6_24

1 Introduction

With the advent of Web 2.0 technologies, the use of the internet has increased tremendously [1] and its pervasive reach has some unintended consequences. One of the negative outcomes is cyberbullying, which is a major concern for society. Bullying taking place over devices like cell phones, tablets, and computers are known as cyberbullying. It can happen through messages and apps, or online through social networks, forums, or gaming portals where anyone can look at, participate in, or share information and content [2]. It comprises of but is not limited to posting, sending, or sharing undesirable, damaging, dishonest, or degrading content about someone else. It might include sharing personal details of a person causing embarrassment or humiliation. Some cases can lead to unlawful or criminal behavior. Harassment, denigration, flaming, impersonation, masquerading, pseudonyms, trolling, and cyberstalking are some of the common ways in which cyberbullying takes place [3].

Cyberbullying has an adverse effect both on the minds of the victim and the bully. A bully is someone who uses strength, force, or threat to intimidate or abuse another. This happens or is capable of happening over an extended period of time. A cyberbully, quite unlike a traditional bully can make use of the varied resources available on the internet to render himself anonymous. This boosts his confidence and many a times, he is also not able to gage the reaction of his victims who might similarly be unknown to him. This is also known as "online disinhibition effect" [4]. The range of his victims can increase considerably creating a larger group of affected individuals. Cyberbullying can produce psychological and emotional repercussion for the victim which includes lowering of self-esteem and self-confidence, depression, loneliness, anger, sadness, stress, degradation of health, and academic achievement and in the worst cases, self-injury and suicidal tendencies [5]. The experience of being a cyberbully has been frequently linked to psychological impacts that go all the way to the bully's childhood years and external difficulties. Cyberbullies lack personal awareness, have low self-esteem, feel the need to have power over someone, portray depressive symptoms, have low self-efficacy, low empathy level, paranoia, phobic anxiety, and poor psychological well-being sometimes leading to suicide ideation.

Many people try to change themselves physically and mentally to be accepted into this fake world created by social media. That is, with the evolution of technology and uninhibited access to any information worldwide, the rate of cyberbullying has increased manifold. Cyberbullying Research Center in 2016 had published a study [6] which highlights that around 33.8% of the students between the ages of 12–17 years were prime victims of cyberbullying. Conversely, 11.5% of students between the age group of 12–17 accepted that they had been involved in cyberbullying. A study by McAfee, found that 87% of teens have witnessed cyberbullying.

In light of cyberbullying concerns, there is an urgent need for the detection and prevention of cyberbullying. The recent trend of shifting from reading to watching videos has led to a large diversity in the ages of people viewing YouTube, demanding that greater attention should be paid to prevent cyberbullying on this social media giant in comparison to others. Currently, social media platforms rely only on users alerting network software to remove the comments of bullies. The performance of alerting software can be improved by detecting instances and then removing the comments automatically. In this paper, four supervised machine learning [7] algorithms, namely random forest, sequential minimal optimization, k-nearest neighbor, and Naïve Bayes are implemented in "Waikato Environment for Knowledge Analysis" tool (Weka) to detect instances of cyberbullying across YouTube related to aspects like race, culture, gender, and physical attributes. The corpus for analysis comprises of comments gathered from arbitrary YouTube channels. The data is preprocessed and the results are evaluated based on measure like TPR, TNR, precision, recall, F-measure, and accuracy [8] for each selected algorithm. The following sections briefly represent the related work done in this area followed by a discussion of the proposed approach.

2   Background Work

Previously, a tremendous amount of work has been done for the detection of cyberbullying across various social media sites with Twitter being the most popular choice among researchers. A survey of existing research based on cyberbullying detection on well-liked social media giants like Twitter, YouTube, and Instagram have been done and tabulated below in Table 1.

From this table, we can infer that the most well-known social media sites are Twitter, YouTube, and Instagram when examined for detection of cyberbullying. As YouTube is the world's most notable user-generated content website with a viewership comprising almost the entire global Internet population, it is quite vulnerable to bullying through videos, the comments following, and most importantly, the replies to those comments. With about a 100 hours of videos uploaded every minute, the social networking site's acclaim, anonymity and almost negligible publication barriers allow users to upload content profane in nature. Thus, we can say that YouTube is highly susceptible to cyberbullying.

3   Proposed Approach

In this paper, we investigate the comment-based features to detect instances of cyberbullying in YouTube. Most of the researches in this area involve comments as a single attribute. However, the source of such profanity might be a part of some conversation or a response to a comment. Therefore, we consider the entire discussion or the reply comments as instances and manually label them as positive or negative instances of cyber bullying.

We extract the dataset for statistical machine learning using the social networking site's API for comments and their following conversations manually. Initially, we scraped comments from random YouTube channels. A manual inspection of the

Table 1 Survey of existing research in detecting cyberbullying

Social media	Author	Year	Journal/ Publisher	Technique
Twitter	Singh et al. [9]	2016	IEEE/ACM	Naïve Bayes
	Zhang et al. [10]	2016	IEEE	Pronunciation-based convolutional neural network (PCNN)
	Huang et al. [11]	2014	ACM	Bagging, J48, sequential minimal optimization, naive Bayes
	Santos et al. [12]	2014	Springer	Random Forest, J48 (decision tree), k-nearest neighbor, sequential minimal Optimization, naive Bayes
	Mangaonkar et al. [13]	2015	IEEE	naive Bayes, support vector machine, logistic regression
	Dadvar et al. [14]	2012	ACM	Support vector machine
YouTube	Dadvar et al. [15]	2012	ACM	Support vector machine
	Dinakar et al [16]	2011	AAAI	Naive Bayes, Rule based J-Rip, tree based J48, Support vector machine
	Dadvar et al. [17]	2014	Springer	Naive Bayes+ multi-criteria evaluation system, decision tree+ multi-criteria evaluation system, support vector machine+ multi-criteria evaluation system
	Dadvar et al. [18]	2013	BNAIC	Rule-based approach
Instagram	Miller et al. [19]	2016	IJCAI	Convolutional neural network (CNN) for clustering of Images
	Mattson et al. [20]	2015	University of Colorado	Naive Bayes, linear support vector machine

channel's comments showed that cyberbullying instances snowballed in case of controversial YouTube videos involving sensitive subjects like race (attacks on racial minorities like African American), culture (comments on the stereotypes attached to cultural traditions like Islam, Jewish), gender, sexuality (comments on LGBT Community), and physical attributes (topics related to the inherent characteristics of an individual).

Thus, for the detection of cyberbullying, YouTube videos on such sensitive topics are taken into account. The video IDs of the offensive videos acquired are annotated according to the type of offense. The video IDs generated and the respective offense category under which they fall has been listed in Table 2.

Table 2 Sample snippet of offensive videos

Video ID	Category	Comments	Rating	Date
2O7IW1CP0jA	LGBT	21,474	4.1	10/08/10
FrbC0ZGZ_pA	Culture	865	4.2	17/05/14
gQp0UugrpTk	Race	2942	4.1	20/11/15
KMU8TwC652 M&	Race	31,549	3.7	24/10/16
qXHPh3ecZEI	Physical attribute	1772	4	14/10/16

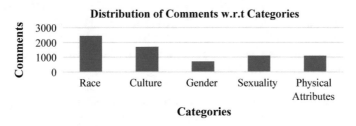

Fig. 1 Distribution of number of comments in each category

In totality, 7962 comments are scraped and labeled from 60 videos, which are roughly around 116 comments per video. A distribution representing the number of comments procured per category has been portrayed in Fig. 1.

The data obtained is preprocessed. The first step of preprocessing involves the correction of spelling mistakes in the comments. People tend to make short abbreviations for words such as "he's fy9" was converted to "He's fine". All the uppercase letters of comments are converted to lowercase characters. All the URL content, unwanted spaces and symbols such as "@$%#" are removed. Stopwords are also removed. These are words that are basically considered unwanted such "he, she, want, for, etc." The words are also stemmed to their basic form such as "hunting", "hunted", etc., reduce to the stem "hunt". They are then tokenized by breaking words into small segments called tokens. The next step is to select the features of the experimentation. The features are TF-IDF [21] (Term Frequency–Inverse Document Frequency); the Ortony lexicon of words [21] denoting negative abstract number density of foul words; frequently occurring part of speech unigram; and bi-gram tags observed in the training set across the dataset.

Using these features, different weights are assigned to the words in comments. 80% of the compiled dataset obtained is utilized for training data and the rest 20% as testing data. After being subjected to preprocessing to clean and organize the data, small datasets were evaluated using machine learning techniques, namely Random forest, Sequential Minimal Optimization (SMO), k-nearest neighbor, and Naïve Bayes.

4   Evaluation and Results

The results were analyzed using measures like TPR, TNR, precision, recall, F-measure, and accuracy. The terms negative and positive indicates the classifier's prediction while the terms false and true indicates whether that prediction corresponds to the user's judgment. 20-fold cross validation is used All these measures are calculated and portrayed in Table 3.

The results show that the best accuracy of 83% is given by KNN. Random Forest, Naïve Bayes, and Sequential Minimal Optimization closely follows KNN in decreasing order of accuracy with Random Forest giving an accuracy of 82.1%, Naïve Bayes of 81.4%, and SMO of 80.8%. The higher Precision and Recall Value indicates that a few false positives were detected. The graph in Fig. 2 compares the accuracy, precision, F-measure, and recall for all the training classifiers with the obtained results plotted on the y-axis and the selected classifiers on the x-axis.

According to the results, we deduce that among the above-applied algorithms, KNN has produced the highest accuracy of 83% when applied to cyberbullying detection of YouTube videos.

Table 3 Results obtained from testing the test dataset

Classifier	TPR	TNR	Precision	Recall	F-measure	Accuracy
KNN (k = 1)	0.830	0.308	0.824	0.830	0.826	0.830
Random Forest	0.822	0.324	0.815	0.822	0.817	0.821
SMO	0.809	0.389	0.799	0.809	0.797	0.808
Naïve Bayes	0.814	0.329	0.808	0.814	0.810	0.814

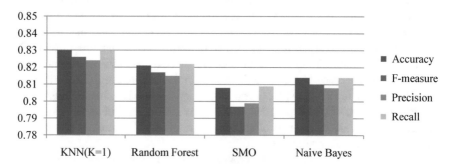

Fig. 2 Comparison of different measures for each of the training classifiers

5 Conclusion and Future Scope

This study empirically contrasted the analysis of cyberbullying detection on YouTube videos using four supervised machine learning algorithms, namely k-Nearest Neighbor, Random Forests, Sequential Minimal Optimization, and Naïve Bayes. 7962 comments were scraped and labeled from around 60 videos on YouTube, which are approximately 116 comments per video. These comments were based on sensitive topics like race, culture, gender, sexuality, and physical attributes which were then analyzed and the results were evaluated based on the various performances. The best accuracy is achieved using k-Nearest Neighbor, followed by Random Forest, Naïve Bayes, and Sequential Minimal Optimization. The experiments can be modified to include the use of comments and social networking graphs for better modeling of the problem. Various soft computing and deep learning techniques can also be used for cyberbullying detection.

References

1. Kumar A, Sebastian TM (2012) Sentiment analysis: a perspective on its past, present and future. IJISA 4:1–14
2. Agatston P-W, Kowalski R, Limber S (2007) Student perspective of cyberbullying. J Adolesc Health. Official publication of the Soc Adolesc Med 41:59–S60
3. Colette L (2013) Cyberbullying, associated harm and the criminal law. PhD Thesis, University of South Australia, pp 1–352
4. Suler J (2004) The online disinhibition effect. Cyberpsychology Behav 7:321–326
5. Foody M, Samara M, Carlbring P (2015) A review of cyberbullying and suggestions for online psychological therapy. Internet Inventions Elsevier 2:235–242
6. Robers S, Kemp J, Truman J, Synder T-D (2013) Indicators of school crime and safety: 2012. Bur Justice Stat 1–211
7. Bhatia MPS, Khalid AK (2008) Information retrieval and machine learning: supporting technologies for web mining research and practice. Webology 5
8. Bhatia MPS, Kumar A (2008) A primer on the web information retrieval paradigm. J Theor Appl Inf Technol 4(7):657–662
9. Huang Q, Singh V-K, Atrey PK (2014) Cyberbullying detection using probabilistic socio-textual information fusion. In: ACM international conference on advances in social networks analysis and mining (ASONAM), pp 3–6
10. Zhang X, Tong J, Vishwamitra N, Whittaker E, Mazer J-P, Kowalski R, Hu H, Luo F, Macbeth J, Dillon E (2016) Cyberbullying detection with a pronunciation based convolutional neural network. In: 15th IEEE international conference on machine learning and applications (ICMLA), pp 740–745
11. Huang Q, Singh V-K, Atrey P-K (2014) Cyber bullying detection using social and textual analysis. In: Proceedings of the 3rd international workshop on socially-aware multimedia ACM
12. Santos I, Miñambres-Marcos I, Laorden C, Galán-García P, Santamaría-Ibirika A, Bringas PG (2013) Twitter content-based spam filtering. In: Herrero Á et al (eds) International joint conference SOCO'13-CISIS'13-ICEUTE'13. Advances in intelligent systems and computing, vol 239, pp 449–458. Springer
13. Mangaonkar A, Hayrapetian A, Raje R (2015) Collaborative detection of cyberbullying behavior in Twitter. In: 2015 IEEE international conference on electro/information technology (EIT), pp 611–616

14. Dadvar M, Jong F (2012) Cyberbullying detection: a step toward a safer internet yard. In: Proceedings of the 21st international conference on World Wide Web ACM, pp 121–126
15. Dinakar K, Jones B, Havasi C, Lieberman H, Picard R (2012) Common sense reasoning for detection, prevention, and mitigation of cyberbullying. ACM Trans Interact Intell Syst 2
16. Karthik D, Roi R, Henry L (2011) Modeling the detection of textual cyber bullying. In: Cyber-bullying social mobile web workshop at 5th international AAAI conference on weblog and social media, pp 11–17
17. Dadvar M, Trieschnigg D, de Jong F (2014) Experts and machines against bullies: a hybrid approach to detect cyberbullies. In: Canadian conference on artificial intelligence. Springer
18. Dadvar M, de Jong F, Trieschnigg D (2014) Expert knowledge for automatic detection of bullies in social networks. In: Canadian conference on artificial intelligence, pp 275–281
19. Zhong H, Li H, Squicciarini A-C, Rajtmajer S-M, Griffin C, Miller D-J, Caragea C (2016) Content-driven detection of cyberbullying on the instagram social network. In: Proceedings of the twenty-fifth international joint conference on artificial intelligence (IJCAI-16), pp 3952–3958
20. Hosseinmardi H, Mattson S-A, Rafiq R-I, Han R, Lv Q, Mishra S (2015) Detection of cyber-bullying incidents on the instagram social network. Technical report by Association for the advancement of artificial intelligence, pp 1–9
21. Ortony A, Clore GL, Foss MA (1987) The referential structure of the affective lexicon. Cogn Sci 11:341–364

Implementation of MapReduce Using Pig for Election Analysis

Anup Pachghare, Aniket Jadhav, Sharad Panigrahi and Smita Deshmukh

Abstract The main goal of this project focuses on providing another way to conduct elections by highlighting the fundamental flaws present in the current system. The present election system considers the number of votes gained by a candidate as the only parameter to select the winner. Our system will be selecting the deserved candidate by considering, parameters such as the educational qualifications, criminal records, previous term record, past social work, personality, popularity, etc. Our proposal also aims at highlighting the distinction between the most deserving candidate to win the election and the candidate who is anticipated to win based on his or her popularity. The final output will clearly reflect the loopholes present in the current system. We want the current election process to not be a contest of popularity only. We aim to bring about an evolutionary change in the current election process to make it better and unprejudiced.

Keywords Hadoop · Election system · Sqoop · Flume · Hive · Cloudera · Pig
Qlik sense · MySQL

A. Pachghare · A. Jadhav (✉) · S. Panigrahi · S. Deshmukh
IT Department, Mumbai University, Mumbai, India
e-mail: aniketjadhav71196@gmail.com

A. Pachghare
e-mail: anupvpachghare@gmail.com

S. Panigrahi
e-mail: sharadpani786@gmail.com

S. Deshmukh
e-mail: deshmukhsmita17@yahoo.com

A. Pachghare · A. Jadhav · S. Panigrahi · S. Deshmukh
Terna Engineering College, Plot No 12, Sector-22, Phase 2,
Nerul (W), Navi Mumbai 400706, India

© Springer Nature Singapore Pte Ltd. 2019
S. Bhattacharyya et al. (eds.), *International Conference on Innovative Computing
and Communications*, Lecture Notes in Networks and Systems 56,
https://doi.org/10.1007/978-981-13-2354-6_25

231

1   Introduction

The existing election system follows the simplest approach of considering the number of votes gained by candidates to decide the winner ignoring all other different
important elements. An enormous group of the population chooses not to vote as
they believe that the entire system is defective. Wrong candidates get selected for
excessive and crucial positions in Government offices and other organizations. We
agree with that votes given to a candidate just on the idea of the spurious claims made
at some point of their Election campaigns without knowing his or her qualifications,
previous work and eligibility are not justifiable. Subsequently, we came up with an
altogether innovative concept that exploits the strength of Hadoop for an in-depth
evaluation of elections for overcoming the issues present in the existing system [1,
2]. Our proposed system for accomplishing elections is designed to increase people's
participation in elections and to make it greater stringent and impeccable.

The following bar charts bolster the claims made in the abstract. This is depicted
on the basis of the survey we undertook as a part of our project (Figs. 1 and 2).

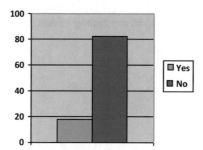

Fig. 1 Bar chart of people's opinion about the effectiveness of the election process

2   Proposed Methodology

2.1   Problem Statement

The election protocols used by the Government of our country should aim to enhance
People's participation in elections efficiently so that the cornerstone of democracy
is maintained. The process of voting should be optimized and simplified as much

Do you agree with the idea 'there should be a change in the Election Process'?

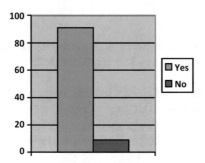

Fig. 2 Bar chart of people's opinion regarding the change in the election process

as possible but should also be straightforward enough to help people choose their representative. People should have access to maximum information about the qualifications, eligibility and records of the candidates contesting elections.

However, the present election system is inefficient and flawed. People vote the candidates without having any foundation for judging the candidates. Based on the survey we conducted it was clear that a huge percentage of the population is unhappy and unsatisfied with the current election process. Hence, we aim to bring a revolution in the election system by introducing a new system and at the same time highlighting the loopholes in the present system.

2.2 Objectives

1. To provide a better and fastidious way of conducting elections.
2. To analyze existing data to understand the importance of various parameters needed to elect the appropriate candidate.
3. To give proper weightage to parameters based on their importance.
4. To develop an algorithm to choose the best candidate from the provided pool of candidates.
5. To help people choose a right person to represent them by performing a meticulous analysis of the achievements, records and other aspects.

2.3 *Working*

The detailed description of our proposed system is divided into following two steps:

Step 1: Analysis

The Analysis step focuses on determining the deserving candidate to win the election from the given list of candidates. The candidates will be judged in this step with following parameters:

Educational Qualifications, Criminal records, Previous social work, Previous term record, Speaking, Motivational skills and personality, Social status, and popularity.

To use the above parameters to decide the candidates it is far important to provide every parameter suitable weightage. To do this we have got performed analysis on two massive datasets:

1. Common people's opinion about these parameters. This information has been accumulated via accomplishing physical surveys as well as the usage of Google forms. Now, after studying this dataset, we have given a preliminary weightage to each of the parameters.
2. Data about the candidates who have or who are serving their term. This dataset contains various fields like tenure, a variety of beneficial tasks completed for the advantage of people, track record, consistency, and data about the above-noted parameters for each candidate, etc. We performed physical surveys and collected most records possible; but, as the political information is exclusive, we have been compelled to make our own dataset. hence, we shaped our personal facts set via coding which consisted of data about applicants serving their term, therefore, after studying this dataset, we were given a quick idea approximately the significance every of the parameter and a secondary weightage changed into given to every one of them regardless of the weightage given within the first step. The final weightage given to parameters is the average of each of them. This has substantially improved the accuracy of evaluation as we are using public opinion in addition to actual information related to the parameters surrounding politicians. When we determined the very final weightage about the parameters, we accepted records about the candidates currently contesting elections through a user interface and found out the deserving person among them by making use of those parameters. A score was generated for every candidate relying on his qualifications on each of the parameters and the result about the deserving candidate was saved inside the database.

Step 2: Prediction

The fundamental goal of prediction step is to predict the candidate to win the election from the given list of candidates. We are going to give this step a real-time emulation, hence the prediction graphs and pie-charts will change at discrete intervals depending on the tweets and posts about candidate updated on social media.

1. Data about the candidates will be extracted out from social media [3]. E.g.: Twitter, Facebook, etc.
2. This data will be stored in a local MySQL database.
3. A candidate who has greater chances of winning the election irrespective of whether he merits to win will be predicted by taking into an account the analysis of this data using Pig Programming and Batch processing. Corresponding graphs and charts will be displayed.
4. This phase matches to the exit polls that are conducted during any major elections.
5. We will be using flume to directly import data into Hadoop from MySQL database, thus giving our project a real-time approach.
6. If our proposed system is actually implemented in future, then the Flume tool will be directly connected to the Twitter or Facebook API [3, 4] (since they are paid APIs we cannot use them in our project).

3   Algorithm

Step 1: Conduct surveys to assemble data about individuals' feelings about the parameters. Preprocess the received data and format the data appropriately for transferring it to Hadoop ecosystem.

Step 2: Utilizing Sqoop tool, import the dataset containing individuals' perspectives into Hive tables.

Step 3: Prepare appropriate set consisting positive, negative, and neutral keywords which will be utilized for performing Text mining on the dataset.

Step 4: Making use of pig script, perform Text mining on the dataset and assign a preliminary weightage to each parameter. Store this output into Hadoop Distributed File System (HDFS) to use it later in future.

Step 5: Gather information about the politicians at present serving or has served their term.

Step 6: Perform data preprocessing activities and include required fields helpful for analysis. Use Sqoop tool to load this dataset into Hive tables.

Step 7: Using attributes such as overall projects done by the candidate for people's benefit, clear track record and overall satisfaction score get the top 10%, successful politicians.

This was done by the following Mathematical Formula:

Score = (((hlp + (mlp/2) + (llp/4))/3) * cost_efficiency * overall_performance)/1000

As clearly seen, the fields like the number of projects, cost efficiency, overall performance from the table data of politicians were used. The hlp, mlp, llp were divided by a number suitable to their weightage.

For e.g., Since the High-Level Projects have more average, they were not divided by any number thus contributing more to the score of the candidate. Also, the final product was divided by 1000 just to decrease the range of score. Once, the score for each candidate was found, only the top 10% records were extracted by using the score generated.

Step 8: Analyze the parameters of the top candidates using Pig programming to again assign a secondary weightage to the parameters and Store this result back into HDFS. For each of the parameter, a different scoring pattern was used. The scoring patterns for each parameter are explained:

Possible values of scores 1, 3, 5, 7, 10.

1 **Education**:
   Masters: 10, Graduate: 7, HSC: 5, SSC: 3, NA: -1
2 **Criminal Record**:
   Yes + Major: $*$ by 0.5 (Made Half), Yes + Minor: * by 0.7 (Made ¾), No: No change
3 **Social work**:
   We took into account five different types of social work:
   NGO work, Charity, Medical Camps, Mental Welfare, Women and rural empowerment The candidate having more varied social work will be given more points:
   1/5: 1, 2/5: 3, 3/5: 5, 4/5: 7, 5/5: 10
4 **Previous term record**:
   If the person had already served a term, his record was fetched from the table and a score was given to him:

   Score = (((hlp + (mlp/2) + (llp/4))/tenure)

   Hence, the person who did more useful work in less tenure was given more score. Score ranges:
   1–20: 1, 21–50: 3, No previous term: 5, 50–70: 7, 70–100: 10
5 **Popularity**:
   Numerical score on the scale of 1–10
   1–2: 1, 2–4: 3, 4–6: 5, 6–8: 7, 8–10: 10
6 **Personality**:
   Numerical score on the scale of 1–100
   1–20: 1, 20–40: 3, 40–60: 5, 60–80: 7, 80–100: 10

Step 9: The two weightages obtained in above steps are integrated and a final weightage is assigned to each of the parameters.

Step 10: Receive data about the current candidates now contesting elections. With the help of final weightage generated in step 9, judge the candidates on the basis of the parameters and assign a score to each candidate.

Step 11: The deserving candidate is one who receives the highest score. Display this output onto Qlik Sense (user interface) and add necessary explanations.

Step 12: Extract data from external sources such as social media about the candidates contesting elections to have a brief idea about their popularity. Load this data into Hive tables.

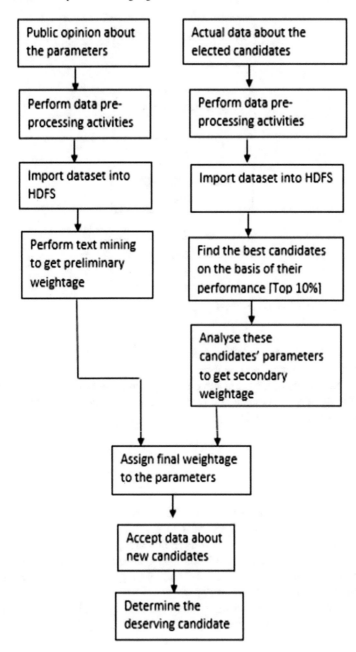

Fig. 3 Flow diagram of analysis phase

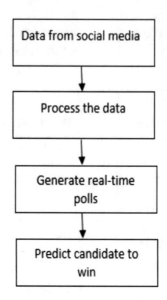

Fig. 4 Flow diagram of prediction phase

Step 13: Utilizing pig programming, Batch processing, and the gathered data generate polls to predict the winner among those with the creation of graphs and charts.
Step 14: The outputs of the corresponding stages highlight the distinction between the actually deserved candidate and the candidate predicted to win thus removing the flaws of the current system.

4   Results

As shown in Fig. 3, when we received public opinion data about the parameters, we performed processing activities on it to give preliminary weightage to these parameters and also decided their rank as shown in Fig. 5 (Fig. 4).

As the actual data about elected candidates were also received in HDFS, we performed processing activities to give secondary weightage to these parameters and also decided their rank as shown in Fig. 6.

The outputs of Figs. 5 and 6 were taken into account to give final weightage to these parameters and also decided their rank as shown in Fig. 7 (Fig. 8).

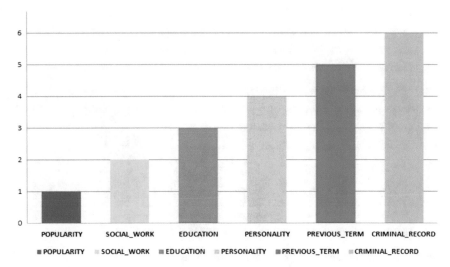

Fig. 5 Ranking of parameters according to preliminary weightage

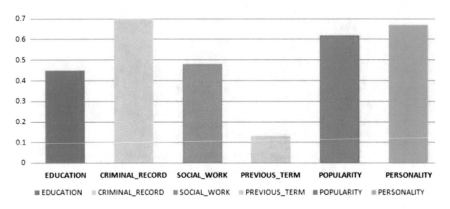

Fig. 6 Ranking of parameters according to secondary weightage

We received the final weightage from Fig. 7 and used the same to judge the new candidates that were eligible for the election.

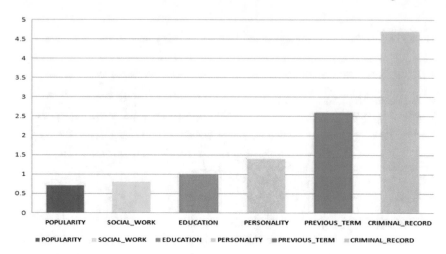

Fig. 7 Ranking of parameters according to final weightage

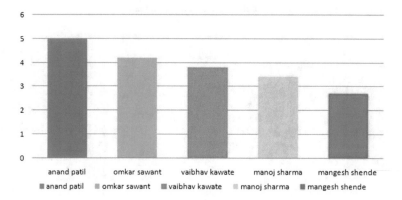

Fig. 8 Output of analysis stage

5 Conclusion

The discrepancy and the loopholes within the existing system are really highlighted with the help of this modern project. After analyzing the result, it became clear that the modern-day election technique is not able to select the maximum suitable candidate of the available people. Subsequently, concerning the modern-day quality of politicians, our proposed system must be implemented by means of the authorities as early as feasible by means of making required changes. This sort of approach primarily based on analytics can be used for any type of election. This approach additionally highlights the strength and scope of information analytics as a whole in the field of Information Technology. The system can bring a revolution in our country if executed properly.

References

1. Zicopoulos P, Eaton C (2011) Understanding big data: analytics for enterprise class Hadoop and streaming data. McGraw-Hill, New York, NY, USA
2. Big data analytics. 2nd edn. Wiley. www.wileyindia.com
3. Sheela LJ (2016) A review of sentiment analysis in twitter data using Hadoop. Int J Database Theory Appl
4. Sharma P, Dwivedi P (2013) Big-data analytics to predict election results. Int J Adv Res Comput Sci Technol
5. Aravinth SS, Haseenah Begam A, Shanmugapriyaa S, Sowmya S (2017) An efficient HADOOP frameworks SQOOP and ambari for big data processing. Int J Innov Res Sci Technol
6. Dhawan S, Rathee S (2016) Big data analytics using Hadoop components like Pig and Hive. Am Int J Res Sci Technol Eng Math

Primary Healthcare Using Artificial Intelligence

**Vaishali D. Khairnar, Archana Saroj, Pooja Yadav, Shraddha Shete
and Neelam Bhatt**

Abstract The right to proper, basic and timely medical attention lies with every individual. Whereas the population of rural India (70%) has been devoid of it since a long time. Some factors causing these problems are poor financial conditions to afford the facilities, lack of means of transportation to reach these services, etc. Our proposed system emphasizes on providing instant help to remote patients. The person who needs medical attention sends a voice message to the system explaining the health condition. The voice message is converted to text message. The system evaluates prescription using Artificial Neural Network. The prescribed medicines are sent to patient in the form of a text message. Our system is efficiently trained for most of the prominent diseases, which prescribes precise medicines within milliseconds excluding communication delay. It is a system, which analyzes the symptoms and can enhance the current medical situation by approximately 20% per year.

Keywords Primary healthcare · Natural language tool kit
Artificial neural network · Back propagation algorithm

V. D. Khairnar (✉) · A. Saroj · P. Yadav · S. Shete · N. Bhatt
Terna Engineering College, Navi Mumbai, India
e-mail: khairnar.vaishali3@gmail.com

A. Saroj
e-mail: archansaroj97@gmail.com

P. Yadav
e-mail: poojayadav0811@gmail.com

S. Shete
e-mail: shraddhashete320@gmail.com

N. Bhatt
e-mail: neelamgbhatt@gmail.com

© Springer Nature Singapore Pte Ltd. 2019   243
S. Bhattacharyya et al. (eds.), *International Conference on Innovative Computing
and Communications*, Lecture Notes in Networks and Systems 56,
https://doi.org/10.1007/978-981-13-2354-6_26

1   Introduction

About 70% of the total population of India sum up to the people living in rural areas. According to recent survey, the most common problem faced in rural areas is the delay in the medical help due to unavailability of the doctors or less number of hospitals. The Health Ministry of India has rural health care as its biggest concern. As 70% population live in the rural areas where quality of health facilities are poor, mortality rates are on a rise. This increasing rate can be controlled if people are provided with the primary aide at the right time [1].

The technology stack for our system includes Neural network, Natural Language Toolkit, Google API, Natural Language Processing, Python as the programming language, Interactive Voice Response, PHP, HTML, R studio. Artificial Neural Networks can be stated as biologically inspired network of artificial neurons designed to perform desired tasks. They are the bio-inspired simulations on the computer to perform certain tasks like clustering, pattern recognition, classification, etc. Using backpropagation algorithm as the learning algorithm ensures minimal error rate which is crucial requirement of our system [2]. The technique used is delta rule or gradient descent. We use neural networks with some inputs, hidden neurons, and output neurons. Hidden and output neurons may contain bias, initial weights are set. Backpropagation aims on optimizing the weights in such a way that mapping of arbitrary inputs towards output is learnt by the neural network [3].

Our system works in four modules—the client, server, doctor machine, and medical machine. The client sends a voice message to the system. This voice message is converted to text message at the server side and this is fed to the neural network which trains itself and provides output. This output is the prescription to be given to the patient. The prescription is sent to the doctor machine for verification. The doctor can modify it, if required. This prescription is then sent to the client as well as the medical machine.

2   Literature Survey

WHO report on India's health workforce states that according to census of 2001, the count of doctors with genuine qualifications in medicine field was 2.7 lakh, 82.6% of which belonged to urban areas. Roughly estimating, in India where about 1.3 billion people reside, the count of doctors is over 9 billion. The WHO recommends having one doctor for every 1,000 people but calculations based on the above statistics show that the actual scenario is a doctor for about 1,450 people. In the year 2011, 31% of Indians resided in urban areas. Assuming it has now reached to around 430 million or one-third. Further assuming that currently 70% of the doctors to be in urban areas. It would still amount to the availability of 6.3 lakh doctors in urban areas whereas the need is just for 4.3 lakh which is about 50% or an excess of 2 lakh. 27% of the overall deaths in India happen due to no medical attention at the time of death [4].

As per National Rural Health Mission Report,

(1) 700 million people live in 6,36,000 Indian Villages.
(2) Cause of death of majority of people are to preventable and curable diseases like measles, typhoid and diarrhea. Critical medicines cannot be accessed by 66% of rural Indians.
(3) 31% of the population travels more than 30 km s to seek health care in rural India.
(4) 8% Primary Healthcare Centre's do not have doctors, 39% do not have lab Technicians and 18% do not have a pharmacist.
(5) There is a critical shortage of medically trained person in Rural health centers.

3 Proposed System

3.1 Motivation

Several organizations are working along with the government and NGOs to help reduce the burden on the public health system using telecommunication technology. India has over 900 million mobile phone users and this fact can be put to advantage to employ better practices in even the remote areas. Leading global organizations of healthcare industry are using this technology to enhance the quality of care and bridge the gaps in healthcare services.

3.2 Problem Statement

The right to proper, basic, and timely medical attention lies with every individual. Whereas the population of rural India have been devoid of it since a long time. Some factors associated with these problems are poor financial conditions to afford the facilities, lack of means of transportation to reach to these services and many more. Why cannot there be a solution to this problem using the current efficient emerging technologies? Our proposed system Primary Healthcare using Artificial Intelligence uses Artificial Neural Network (ANN) to provide a solution for the issues arising in basic primary healthcare.

3.3 Objectives

(1) Disease, Data collection, preprocessing and data analytics to create an environment that promotes high-quality patient care.

(2) To enhance awareness, participation and importance of prevention for diseases easily preventable and treatable (to start with).
(3) To provide an assistance to the doctor, not replace them.
(4) To provide Mobile/PCO compatible system for the villages making doctor available as and when required.
(5) To make Patient-centered healthcare system secure.

3.4 Working

The input data from the user can be received in three ways. A template will be defined to collect user information and symptoms. User can choose any of the below ways as per his/her convenience.

(1) Interactive Voice Response (IVR):

   (a) User can call on a helpline number and live expert agents will be available to attend the calls. Agents will collect the patient information and symptoms as per pre-defined template.

(2) Web Application:
   User can access through a secured web application and share the relevant information in the form provided. Form template will be same for both web-based application and mobile application.
(3) Mobile Application:
   User can also access through a secured mobile application (Android OS) and share the relevant information in the form provided (Fig. 1).

The data collected from above ways will be in the same template and stored in a centralized database on cloud. The location of the user would also be tracked. A User ID and Case number will be generated for each user for better tracking. The medical history through Aadhaar number or User ID would also be kept. A historical dataset containing past patient records will be available. This dataset will have demographic information of patients (Name, Aadhaar number, age, gender, income group, location, height, weight, etc.), observed symptoms and diseases and corresponding medications.

The symptoms data that is collected from user will be preprocessed and matched against symptoms data in historical dataset. Data mining and Predictive modeling techniques will be used and disease from which user is suffering will be predicted, and medication would be prescribed. This prescription along with User ID, Case number and/or Aadhaar number will be send as an SMS directly to the user and also to the pharmacist nearest to user's location.

The entries for new users will be appended to the historical dataset. This aims to bring better decision support, using data from large numbers of successfully treated people, to every new patient. The results predicted for users from above process will

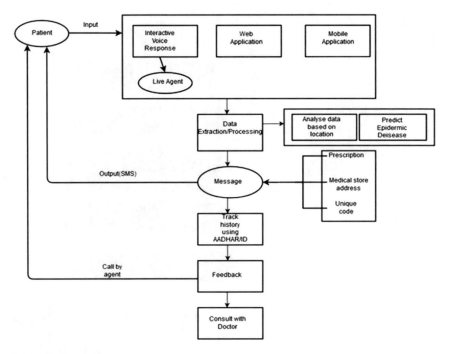

Fig. 1 System architecture

be further analyzed to gather insights. The results will be used to support continuous data analysis and visualization:

(1) find if people from particular areas are affected by any epidemic disease
(2) identify patterns of diseases on seasonal basis
(3) identify types of diseases across people of various age groups, income groups, gender, etc. (Fig. 2)
(4) identify patterns of diseases on seasonal basis
(5) identify types of diseases across people of various age groups, income groups, gender, etc.

3.5 Algorithm

The very first step is to calculate the total net input to each hidden layer neurons using following equation:

$$\text{net} - \text{input} = \text{weight}_1 * \text{input}_1 + \text{weight}_2 * \text{input}_2 \ldots \text{weight}_n * \text{input}_n + \text{bias}_1 * 1$$

(1)

Fig. 2 Flowchart of the system

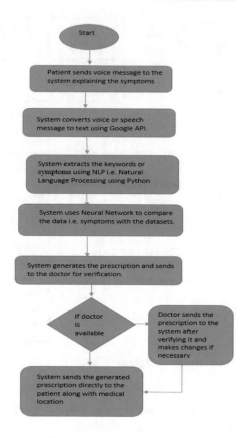

By using an activation function squash the total net input with the help of logistic function as follows:

$$\text{output of hidden layer} = 1/\left(1 + e^{-\text{net input}}\right) \tag{2}$$

Use the output of the hidden layer neurons as an input for the outer layer neurons. Next step would be to calculate the total error using an equation:

$$E_{\text{total}} = \sum 1/2(\text{target} - \text{output})^2 \tag{3}$$

Now based on the total error value we will update each of the weights in the network so that they cause the current output to be closer to the expected output, which minimizes the error for every output neuron; subsequently reducing the error of the network as a whole [3]. By application of chain rule, Calculate the rate of change of error w.r.t change in weight.

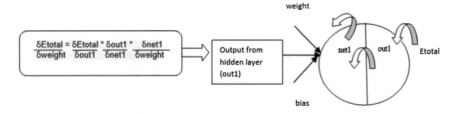

$$\frac{\delta Etotal}{\delta weight} = \frac{\delta Etotal}{\delta out1} * \frac{\delta out1}{\delta net1} * \frac{\delta net1}{\delta weight}$$

Output from hidden layer (out1)

weight

net1   out1   Etotal

bias

Backpropagation Algorithm (Weights, Bias, Input layer, Output layer, Hidden layer).

Initialize weights of each neurons in the network.

Do

(1) Forward propagation

- Find the net total input of input layer.
- Use it as an input for hidden layer.
- Calculate output of hidden layer.
- Take the output of the neurons of hidden layer as inputs and repeat this process for neurons of output layer.

(2) Calculate the total error $= \Sigma\ 1/2\ (expected—output)^2$
(3) Backward propagation

- Calculate the rate of change of error with respect to change in weight.
- Update network weights.

Until all dataset is classified correctly [3].

3.6   Comparative Study

Babylon is a healthcare service that gives medical advice and appointments with doctors. Since it does not have prediction modeling techniques it is not able to find epidemic disease in a particular area. It is completely based on the Internet which is the biggest drawback because Internet facility is not available in rural areas. Our System uses Interactive Voice Response and Live Agent to overcome this drawback (Fig. 3).

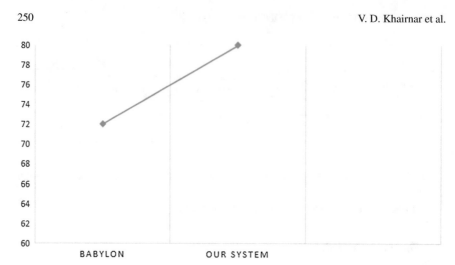

Fig. 3 Relevancy and simplicity of the system when compared with existing system (Babylon)

4 Significance

- Instant medical help anytime, anywhere.
- Increased efficiency of service care with an adequate referral and remote consultation system.
- Valuable lives might be saved even in the absence of a doctor.

5 Future Scope

- Government schemes such as ESIS (Employee service insurance scheme) can be associated with our project.
- Virtual reality can be incorporated in this project by making doctor available virtually for the patients.

6 Conclusion

Our system would prove as an assistant to doctors proving to be a boon in the medical field. It would solve the primary healthcare issues in remote rural areas which are otherwise not easily accessible by the doctor immediately. The mortality rate would decrease significantly. Since we are using ANN the error rate is minimal. It would bridge the gap between doctors and the patients residing in remote/rural areas.

References

1. https://www.quora.com/What-are-the-problems-faced-in-rural-areas-where-proper-chemist-shops-are-not-present
2. https://www.quora.com/What-is-the-major-difference-between-a-neural-network-and-an-artificial-neural-network
3. https://mattmazur.com/2015/03/17/a-step-by-step-backpropagation-example/
4. http://www.gramvaani.org/?p=1629

A Brief Survey on Random Forest Ensembles in Classification Model

Anjaneyulu Babu Shaik and Sujatha Srinivasan

Abstract Machine Learning has got the popularity in recent times. Apart from machine learning the decision tree is one of the most sought out algorithms to classify or predict future instances with already trained data set. Random Forest is an extended version of decision tree which can predict the future instances with multiple classifiers rather than single classifier to reach accuracy and correctness of the prediction. The performances of the Random Forest model is reconnoitered and vary with other models of classification which yield institutionalization, regularization, connection, high penchant change and highlight choice on the learning models. We incorporate principled projection strategies which are aiding to predict the future values. Ensemble techniques are machine learning techniques where more than one learners are constructed for given task. The ultimate aim of ensemble methods is to find high accuracy with greater performance. Ensembles are taking a different approach than single classifier to highlight the data. In this, more than one ensemble is constructed and all individual learners are combined based on some voting strategy. In the current study, we have outlined the concept of Random forest ensembles in classification.

Keywords Machine learning · Classification · Decision tree
Ensembles of decision tree

A. B. Shaik
School of Computing Sciences, VISTAS, Chennai, India
e-mail: anji.sk@gmail.com

S. Srinivasan (✉)
Department of Information Technology, School of Computing Sciences,
VISTAS, Chennai, India
e-mail: ashoksuja08@gmail.com

© Springer Nature Singapore Pte Ltd. 2019
S. Bhattacharyya et al. (eds.), *International Conference on Innovative Computing and Communications*, Lecture Notes in Networks and Systems 56,
https://doi.org/10.1007/978-981-13-2354-6_27

1 Introduction

Classification is one of the machine learning supervised model to be used for predicting the future instances with given training data set [1] Classification algorithms span a wide area of learning methods such as decision trees, Bayesian classification, Rule-based methods, Memory-based Learning and support vector machines to predict the future instances with past data set and all these algorithms are often used in various applications and have proven their capability in terms of accuracy and correctness [2]. Decision tree is a type of classification algorithm which constructs regression model in the form of a tree. By splitting the data into smaller subsets, an associated decision tree is incrementally constructed [3]. Decision tree does not require any domain knowledge and easy to comprehend. To classify a large amount of data set into respective class labels using decision tree may yield inaccuracy of prediction and the results are unknown. Instead of using single classifier we were used multiple classifiers by decision forest known as ensembles [4]. Ensembles are a simple and effective technique for prediction when a large amount of data set has given [5]. In ensemble technique, multiple classifiers are constructed for a given task. Then the different classifiers are combined to form a new classifier. Ensembles are often found to be more accurate than other component learners. In this paper, we present clear understanding about random forest model along with its advantages.

2 An Overview of Random Forest Model

As the name implies the Random Forests Algorithm is a Supervised Classification Algorithm which classifies the data by constructing a number of Classifiers with an aim to achieve a higher accuracy of prediction [6]. The Random Forest techniques are applied to test data set where the trees are constructed while the resultant individuals are combined to predict the class label [6]. To classify a large amount of data with a single classifier is not worthy and may lead to less accuracy in the result. Hence this is often used in many applications where a large amount of data to be classified with Decision Tree Classification Algorithm [7]. Random Forests are easy to comprehend both for computer professionals and end-users without a statistical background. A random forest does not require any cross verification and it is not over-fitting [8]. The Random forest uses Adaboost and Bootstrapping techniques to construct multiple classifiers.

The following Random Forest Algorithm [9] gives the steps in constructing the decision trees.

- Take N as the number of training data instances in the samples. Let M be the number of attributes in given input dataset.
- Let m be the Number of parameter in the input that determines the next attribute to be chosen at each tree node; (where m is lesser than M).

- The training samples are taken and a tree is constructed for each sample with replacement.
- For tree node, arbitrarily select m attributes in that particular node.
- The best split is computed based on the m input attributes of the sample dataset.
- Each tree is grown without pruning.

Visual representation of random forest is explained in the following Fig. 1.

2.1 Visual Representation of Random Forest

See Fig. 1.

2.2 Advantages of Random Forest Algorithm

- Accuracy is more.
- Efficient in handling big databases.
- Easily and efficiently handles thousands of input variables.
- Provides information about variables which are important and which are not in the classification.
- Provides methods for estimating missing data.
- Handles missing data without compromising on accuracy.
- Prototypes are used to give the information or Meta data about the relation between different variables.
- Enables studying variable interactions.

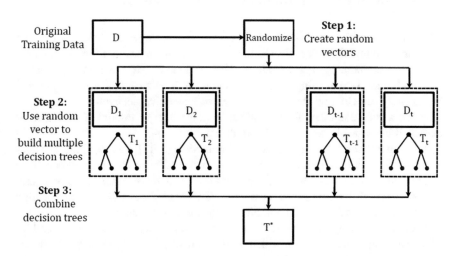

Fig. 1 Representation of random forest

2.3   Random Forest Applications

Banking—In this sector random forest algorithm is used to find the loyal customers and Fraud Customers. Also enables bankers to find out the customers who may not be profitable for the bank. Organizations can also identify defaulters and arbitrators.

Medicine—In the medical field, random forest algorithm is often used to select a correct and optimal combination of the symptoms and components of a certain disease. This is done by analyzing the patient's medical records.

Stock Market—Random Forest algorithm is used to identify the behavior of stock fluctuations and the expected loss or profit for purchasing a particular stock.

E-commerce—Finding the Purchase patterns and also prediction of similar type of customers also called customer segmentation.

2.4   Limitations of Random Forest Model

- Over-fitting to a particular data sample is one of the foremost problems observed in the literature especially in regression tasks [10].
- Random forests have trouble dealing with attributes that are multi-valued and multi-dimensional. They favor categorical variables with many levels [11].

3   A Brief Survey on Random Forest Ensembles

Random forest is classified as a supervised machine learning ensemble algorithm that has come into limelight recently. Decision tree forms the base classifier in a Random forest. As the name suggests randomization is done in two ways in constructing random forests. One is using random sampling for drawing samples and the second is randomly selecting attributes or features for generating decision trees. Decision trees are a good candidate for classification where a large amount of data is to be classified in terms of accuracy and correctness.

In the present context as suggested by the authors [11, 12], Random Forest ensembles from a better approach to classify voluminous and huge volume of data with an aim to achieve highest accuracy both in classification and prediction. They both have proposed various random forest techniques like Bagging and Boosting to construct multiple classifiers and how to combine all the individual tree results.

Breiman [13] suggested that Random Forest is well suited for creating ensembles using randomization approach which works well with bagging on large databases and also provide information about the input variables in order to obtain accuracy of prediction and performance of Classification.

Lebedev et al. [7] and Cohen et al. [14] have carried studies on the Random forests. They have proposed random forest is highly recommended for classifying the data with low error rate and works well even if the data is missing in dataset and maintain accuracy while data size is rapidly growing.

Jampani et al. [11] Proposed Random forest Algorithms in the field of medical diagnosis of breast cancer with proper prediction. He made a comparison of Random forests with all other classification algorithms and concluded that Random Forest Ensembles are good enough to reach the accuracy of prediction.

Banfield et al. [15] broadly discussed the use of the Random forest for Classification. According to him, Random forest refers to split the data into a number of classifiers and each of the Classifiers make some sense with respect to correctness and accuracy. He Proposed Random Forest Algorithms with ensembles and stated that they are well suited in Banking domain to predict the customers who are eligible for getting the loan and who are not eligible for the loan.

Thus the above survey is basically focused on use of the Random Forest in Classification of Big data that are generated by various fields such as in medical diagnosis, Banking Domain and in particular, the detecting frauds in credit card and its associated areas [16].

3.1 Discussion

Table 1 summarizes the survey of Random Forest Ensembles which is used in classification problems using decision tree. It emphasises that using random forest technique in classification can yield better accuracy and correctness than other classification algorithms [17]. As discussed above, [18] studied about various Random Forest ensembles and proposed Bagging and Adaboost techniques for classification to control the over-fitting and noise. Authors like [19] Freund and Schapire have gone through the problems of classification for Image Data set with single classifier and recommended random forest ensembles with MMDT and GASEN-b techniques which can bring accuracy, efficiency and not require domain specific knowledge. A complete study on Random forest by [14, 20] propose random ensembles are better for classification of credit card data set to find out Fraud in the use of credit card. In order to reach accuracy of prediction [11] have made a discussion over the Bootstrap Aggregation in random forest and advised that random forest is good one where prediction and robustness are required.

Table 1 Summary of the survey on random forest for classification

Sl. no	References	Algorithm/ Techniques	Dataset	Results/ Observation	Proposed future enhancements/ Research gaps
1.	Wang et al. [21]	Bagging	Credit card data set	Achieves good effect	A better technique is required to deal with noise in given data set
2.	Du et al. [18]	Adaboost	Medical data set	Low computational complexity and high speed	Optimized algorithm should be developed
3.	Breiman [13]	Mean margins decision tree learning (MMDT)	Image data set	Efficient and simple to implement and parameter less	Accuracy needs to be improved
4.	Zhou and Tang [5]	Genetic algorithm based selective ensembles (GASEN-b)	Fifteen datasets (benchmark datasets from UCI repository)	Domain specific knowledge needs to be incorporated	More powerful algorithms are required that combines component class probabilities and uses ensemble techniques
5.	Freund and Shapire [19]	Bagging with 10-fold cross-validation evaluation method	Bench mark data sets	Flexible, efficiency and parallel construction of trees	Less accurate
6.	Quinlan [22]	Fuzzy decision tree	Iris flower data set	Accuracy, flexibility	This algorithm would be useful in a situation when only a few precedent cases are given
7.	Lou et al. [23]	Generalized additive models	12 datasets from UCI machine learning repository	Reliable	Improved algorithm is desired
8.	Martinović et al. [24]	Structured random forest	Ecole centrale paris facades database (ECP)	Noise free images and high quality	An algorithm that better detects required features is to be developed
9.	Cohen et al. [14], Jampani et al. [11]	Bootstrap aggregation	Toy data set	Robustness and low error rate	Not applicable when data set has too much noise

3.2 List of Research Gaps Identified by the Survey on Random Forests

- A better technique is required to deal with noise in given data set.
- Optimized algorithm should be developed.
- Accuracy needs to be improved.
- More powerful ensemble algorithms that combine class probabilities needs to be developed.
- Domain Knowledge needs to be incorporated.

4 Conclusion

This paper has presented an overview of Random forest and its performance in the Classification model. Random forest is ensemble classifier that includes multiple classifiers to predict class label values with past data set. Random Forests are fast to build and even faster to predict and they do not require any cross-validation and fully parallelizable. Random forest Algorithms are often more accurate than a single classifier. It has an ability to handle the data without preprocessing which means data need not be rescaled and transformed.

References

1. Buczak AL, Guven E (2016) A survey of data mining and machine learning methods for cyber security intrusion detection. IEEE Commun Surv Tutor 18(2):1153–1176
2. Wu X, Zhu X, Wu GQ, Ding W (2014) Data mining with big data. IEEE Trans Knowl Eng 26(1):97–107
3. Kumar GK, Viswanath P, Rao AA (2016) Ensemble of randomized soft decision trees for robust classification. Acad Proc Eng Sci Sadhana 41(3):273–282
4. Mitchell TM (1999) Machine learning and data mining. ACM Commun 42(11):30–36
5. Zhou Z, Tang W (2003) Selective ensemble of decision trees. vol 2639. LNCS, Springer, pp 476–483
6. Gashler M, Carrier CG, Martinez T (2008) Decision tree ensemble: small heterogeneous is better than large homogeneous. In: Proceedings of 7th international conference on machine learning and applications, pp 900–905
7. Lebedev AV et al (2014) Random forest ensembles for detection and prediction of Alzheimer's disease with a good between cohort robustness. NeuroImage Clin 6:115–125
8. Kotsiantis S (2011) Combining bagging, boosting, rotation forest and random subspace methods. Artif Intell Rev 35(3):223–240
9. Dietterich TG (2000) An experimental comparison of three methods for constructing ensembles of decision trees. Mach Learn 40:139–157
10. Louppe G (2014) Understanding random forests: from theory to practice, Cornell University Library, pp 1–225
11. Jampani V, Gadde R, Gehler PV (2015) Efficient facade segmentation using auto-context. In: Proceedings: 2015 IEEE-winter conference on applications of computer vision, WACV'15, Feb 2015, pp 1038–1045

12. Pietruczuk L, Rutkowski L, Jaworski M, Duda P (2017) How to adjust an ensemble size in stream data mining? Inf Sci (Ny) 381:46–54
13. Breiman L (2001) Random forests. Mach Learn 45(1):5–32
14. Cohen L, Schwing AG, Pollefeys M (2014) Efficient structured parsing of facades using dynamic programming. In: Proceedings: IEEE computer society conference on computer vision and pattern recognition, 2014, pp 3206–3213
15. Robert Banfield E, Hall LO, Bowyer KW, Kegelmeyer WP (2007) A comparison of decision tree ensemble creation techniques. IEEE Trans Pattern Anal Mach Intell 29(1):173–180
16. Yin L, Jeon Y (2006) Random forests and adaptive nearest neighbors. J Am Stat Assoc 101(474):578–590
17. Ahn H, Moon H, Fazzari MJ, Lim N, Chen JJ, Kodell RL (2007) Classification by ensembles from random partitions of high-dimensional data. Comput Stat Data Anal 51(12):6166–6179
18. Du YDY, Song ASA, Zhu LZL, Zhang WZW (2009) A mixed-type registration approach in medical image processing. In: Proceedings of 2nd international conference on biomedical engineering and informatics, BMEI'09, 2009, pp 1–4. IEEE
19. Freund. Y, Schapire RRE (1996) Experiments with a new boosting algorithm. In: Proceedings international conference on machine learning, 1996, pp 148–156
20. Lou Y, Caruana R, Gehrke J (2012) Intelligible models for classification and regression. In: Proceedings of KDD'12; 18th ACM SIGKDD international conference on knowledge discovery and data mining, pp 1–9
21. Wang. Y, Zheng. J, Zhou. H, Shen L (2008) Medical image processing by denoising and contour extraction. In: International conference on proceedings: information and automation, ICIA 2008
22. Quinlan JR (1996) Improved use of continuous attributes in C4.5. J Artif Intell Res 4:77–90
23. Lou Y, Caruana R, Gehrke. J (2012) Intelligible models for classification and regression. In: Proceedings of KDD '12; 18th ACM SIGKDD international conference on knowledge discovery and data mining, 2012, pp 1–9
24. Martinovic A, Mathias M, Weissenberg J, Van L (2012) A three-layered approach to facade parsing-supplementary material, pp 1–8

Big Data in Healthcare

Sachet Rajbhandari, Archana Singh and Mamta Mittal

Abstract The evolution of technology and digital devices has qualified a whole new dimension of Big data analytics that can lead to massive developments across the world in every sector including Healthcare. This paper illustrates the challenges and applications of Big Data in healthcare. It is a vast area of study that includes numerous data records, simultaneously; increasing storage of the technology has aided us to store this data. The study is based on different sources of secondary data, thus, is descriptive and qualitative in nature. The details of Big Data analytics in Healthcare has been acknowledged by a methodical literature review, after which selected examples for future benefits has also been evaluated. In the end, based on the assessment, possible ways to minimize the challenges and issues, future prospects and possible suggestions have been provided. The paper covers the challenges, issues, applications, and future scopes of Big Data in Healthcare.

Keywords Big data · Healthcare · Challenges · Applications and analytics

1 Introduction

Digitized information today is ubiquitous and can be found in every sector. The quick development in the utilization of cell phones, note pads, tablets, individual sensors is producing a storm of information. In 2013, 4.4 zettabytes of data were created globally. It is estimated that it will hit 44 zettabytes by 2020. To place this into point of view, this volume of information compares to 500 billion, 5-h long HD motion pic-

S. Rajbhandari · A. Singh
Amity University Uttar Pradesh, Noida, Uttar Pradesh, India
e-mail: sachetr95@hotmail.com

A. Singh
e-mail: archana.elina@gmail.com

M. Mittal (✉)
G. B. Pant Government Engineering College, Okhla, New Delhi, India
e-mail: mittalmamta79@gmail.com

© Springer Nature Singapore Pte Ltd. 2019
S. Bhattacharyya et al. (eds.), *International Conference on Innovative Computing and Communications*, Lecture Notes in Networks and Systems 56,
https://doi.org/10.1007/978-981-13-2354-6_28

tures, which a man would require 115 million years to watch completely. Moreover, this information is foreseen to twofold this and consistently forward. Long back, The Human Genome Project was the foremost dispatch of Big Data in medicinal services, anticipated that would outline nucleotides of a human haploid reference genome. Capable computing, upgrades in IT information administration, analytical tools, utilizing Big Data in research, amusement and marketing, are building an establishment for the expanding utilization of Big Data in healthcare. However, the concept of big data is not new, neither explicit. The way it is defined is constantly changing. Miscellaneous efforts at defining big data fundamentally characterize it as "a term that describes large volumes of high velocity, complex, and variable data that require advanced techniques and technologies to enable the capture, storage, distribution, management, and analysis of the information", by a report in US in 2012. Big Data is more than just size and "Volume", as it additionally envelops attributes, for example, "Variety", "Velocity", and, with deference particularly to medicinal services, "Veracity", regularly alluded to as the "dimensions" or "The 4Vs of Big Data" [1]. The volume of data is growing exponentially all over the world, from 130 exabytes in 2005 to 7,910 exabytes in 2015. In healthcare, development originates from both the current information and from creating new types of information. The current overwhelming volume of information includes individual therapeutic records, genomic arrangements, human hereditary qualities, and so on. New forms of big byte data, fueling this exponential growth are 3D imaging, biometric sensor readings, etc. The assortment of data: structured, unstructured, and semi-structured is a viewpoint that makes healthcare data both captivating, and also difficult. Just 20% healthcare data is structured (proper for PC preparing), with unstructured information (transcribed notes, sound and video documents) growing at around 15 times the rate of organized information. Similarly, as the volume and assortment of data put away has developed, so too has the speed at which it is made. The improvement from cheques to cards is a recognizable case of the update from ease back to quick data processing. Despite all the intrinsic intricacies, there are immense potentials and benefits in developing and implementing big data in healthcare [2, 3].

2 Related Work Done

In paper [1], authors found that the contributions of each company in big data in healthcare by doing extensive survey in various companies. They have given examples and discussed the goals of every company along with the issues and challenges of big data. The authors [4] discussed about various impacts of big data in healthcare and new innovations in healthcare. It also provides information about the new pathways in big data in healthcare. Canada Health Info way (2013), in this paper, the authors have discussed about the economic values and opportunities of big data analytics in healthcare. It further tells about considerations of big data in digital health. The paper [5], authors discussed about the big data in EU healthcare system. It tells about various models, barriers, and applications of big data in European

Union. This paper [6] illustrated about various securities, standards of big data. It discussed mainly about the future scopes of big data in healthcare and how big data can be used to improve the various fragmentations of healthcare. In paper [4], authors recommended the creation of a European wide connected Electronic Health Records Organization (EHRO) empowering patients by enabling the effective collection and usage of patients' health data across member states, revolutionizing healthcare and ultimately leading to better health outcomes for patients and payers irrespective of national borders. In paper [7] gives a lot of examples of big data in healthcare. It discusses about Structured EHR Data, unstructured Clinical Notes, Medical Imaging Data, Genetic Data, Other Data (Epidemiology), etc. To conclude, it gives results depending on these findings. In this paper [8] the authors discuss business intelligence and analytics in big data. Its various characteristics and various capabilities are also discussed. Furthermore, its future scopes and applications of e-commerce have also been highlighted. IBM (2014), this paper informs about big data clinical analytics in action, advanced analytics in action and various steps to becoming a data-driven healthcare organization.

3 Application of Big Data

Genomics has been the bleeding edge of the Big Data agitation in human administrations, one that holds basic affirmation for empowering tweaked solutions. There are various organizations which are genomics-centered, and each of them are adopting different strategies to the information, hoping to progress translational research and inevitably, improve treatment advancement and medicinal practices.

The development of "high-throughput sequencing techniques" has permitted analysts to concentrate hereditary markers over an extensive variety of populace, enhance viability more than five times size since sequencing of the human genome was finished. Examination of the sequencing procedures in genomics is an intrinsically enormous information issue as the human genome comprises of 30–35,000 genes. Big data covers a wide range of application areas. Looking at pathway analysis, where practical effects of genes differentially expressed in an experiment or gene set of specific interests are analyzed [9, 10].

3.1 Pathway Analysis

Resources for reasoning utilitarian impacts for genomics in enormous information for the most part rely on upon factual information of watched quality expression changes and expected useful impacts. Translation of utilitarian impacts needs to join persistent increments in accessible genomic information and comparing documentations of qualities. Different instruments are fused for useful pathway examination of genome scale information. Methods used are divided into three categories or generations;

Fig. 1 Summary of popular methods and toolkits with their applications. *Source* internet

Toolkit name		Selected applications
Onto-Express		Breast cancer
GoMiner	First Generation	Pancreatic cancer
ClueGo		Colorectal tumors
GSEA	⟸ Second Generation	Diabetes
Pathway-Express	⟸ Third Generation	Leukemia

First generation includes "representation and analysis approaches" which determine the part of genes in a certain pathway found within the genes that are expressed unusually. Second generation encompasses "functional class scoring approaches" which consists of expression level changes in separate genes, along with some similar genes. The third generation embraces "pathway topology-based tools", that are easily accessible pathway knowledge databases. It has complete information of gene product interactions, the way specific gene products interact, the location where they interact, etc. (Fig. 1).

A famous Genomic focused company, Genome Health Solutions, applies its proficiency and network of technology providers to incorporate individual genomics and care delivery to initiate a new standard of care for ameliorating patient reports in cancer and other diseases.

3.2 Transforming Data to Information, and Vice Versa

Seeing the developing information storm of human services, it is vital to better sort out this information. The essential viewpoint is to change information to operational data. Changing unstructured information to organized information, for machine administration is a noteworthy venturing stone to empower "information driven human services". Truth be told, in a few cases, turning unstructured information (medicinal notes, diagrams) into data is a required first step. Illustrative examination applies an arrangement of devices in view of information mining, insights, counterfeit consciousness, NLP and so on to compose information for examples and significance. There is numerous product that utilizes cloud as a predictive analytic software to explain designs in hospital datasets to gather re-admissions and maintain a strategic distance from healing facility gained conditions [11].

3.3 Supporting Providers, Improving Patient Care

Care providers undergo intense difficulty: less time and money to do more with an increasing amount of information being concerned and not making errors. Provider support is one of the essential areas where big data is used. Many companies are taking divergent approaches to make provider support systems that are easy to operate, that help save money and improve outcomes, as well as provide providers more time to be concerned. Organizations are combining social and clinical information streams with APIs to produce real-time behavioral wellbeing reports.

3.4 Awareness

Big data is an evident toolbox to build learning to take care of different information-based issues like distinguishing fake drugs, knowing biological worries that trigger asthma, gauge ailment events, helping nations make enhanced arrangement choices, Organizations like Sproxil uses Big Data to recognize fake prescriptions, help pharmaceuticals to track tranquilize appropriation and turn away burglary. PIN codes are printed to each individual solution to confirm whether the prescriptions are genuine or fake, that may hurt patient's well-being [12, 13].

4 Big Data Initiatives Healthcare

Healthcare generates an enormous sum of data every day that requires a technology like Big data. Big Data analytics can execute a real-time data analysis on the extensive data set. Many countries today have taken initiatives and are still into various findings to improve the healthcare sector. Some major steps taken by the countries are: The initiative is a piece of the President's proposed financial 2014 budget plan. The US President divulged striking $100 million research intended to transform our understanding of the human brain. Brain Research through Advancing Innovative Neuro Technologies Initiative.

"*Every dollar we invested to map the human genome returned $140 to our economy... Today, our scientists are mapping the human brain to unlock the answers to Alzheimer's.*"—President Barack Obama, 2013, State of the Union. National Science Foundation, National Institutes of Health, Defense Advanced Research Projects Agency, and other private organizations have come up as a joint venture for this initiative [9, 11]. GE (General Electric, an American multinational conglomerate corporation) and the NFL (National Football League of America) have collaborated to quicken blackout research, determination, and treatment by propelling the Head Health Initiative [6, 12]. This four-year cooperation expects to enhance the wellbeing of athletes, members of the military and society overall. In 2012, Blue Cross and Blue Shield of North Carolina launched a mobile website. The mobile website provides with simple access to their medicinal services data in a hurry, to more than 3.7 million clients. The organization collaborated with Kony Solutions, a main versatile stage supplier, to create and convey the site. The new mobile app enables clients to:

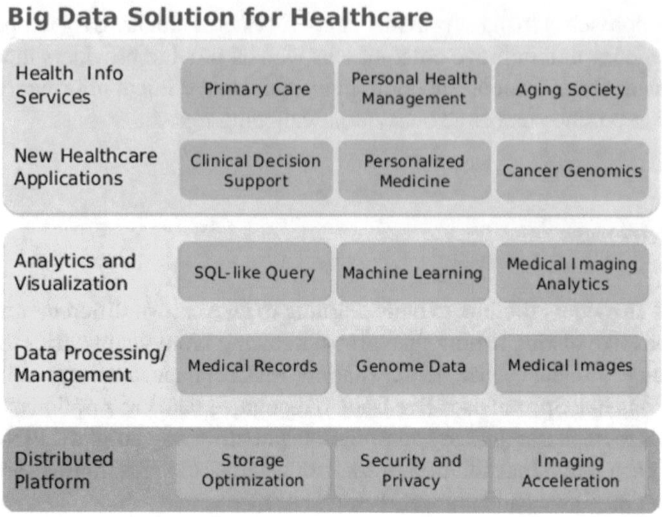

Big Data Solution for Healthcare

Health Info Services	Primary Care	Personal Health Management	Aging Society
New Healthcare Applications	Clinical Decision Support	Personalized Medicine	Cancer Genomics
Analytics and Visualization	SQL-like Query	Machine Learning	Medical Imaging Analytics
Data Processing/ Management	Medical Records	Genome Data	Medical Images
Distributed Platform	Storage Optimization	Security and Privacy	Imaging Acceleration

Fig. 2 Big data solution for healthcare. *Source* IDF 2013 Intel Developer Forum

- View claims (validated clients).
- See current arrangement advantage data (verified clients).
- Find a doctor.
- Get assessed treatment costs (confirmed clients).
- Find the nearest critical care focuses (in light of GPS area).
- Find more affordable physician recommended drugs.
- Compare and search for individual and family under 65 plans.
- Contact customer care [6] (Fig. 2).

5 Challenges Faced by Big Data in Healthcare

Privacy concerns have recently turned out to be aggravating, as Internet exchanges, cloud, online networking, and so forth uncover many individual information to plausible misuse. While different users have been somewhat conflicting about their security issues; expanding mindfulness among individuals makes it clear that a large portion of them are getting to be noticeably worried about ensuring their own information, particularly with regards to well-being and therapeutic information. Exposure of individual wellbeing data to outsiders, pariahs, for example, the media, offenders, and so on is the first motivation behind why healthcare has been dependably a special case [14]. Besides privacy concerns, data security has also been a challenge of Big Data in Healthcare. Inadvertent presentation or information misfortune to unapproved gatherings is the primary risk to big data. Web use, cloud computing and pooling of information all raise the information security stakes. Jason Gilder of IBM

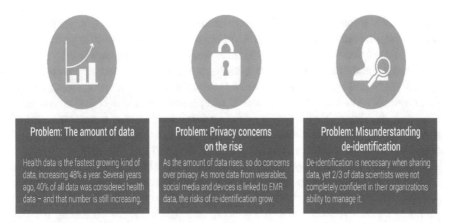

Problem: The amount of data	Problem: Privacy concerns on the rise	Problem: Misunderstanding de-identification
Health data is the fastest growing kind of data, increasing 48% a year. Several years ago, 40% of all data was considered health data – and that number is still increasing.	As the amount of data rises, so do concerns over privacy. As more data from wearables, social media and devices is linked to EMR data, the risks of re-identification grow.	De-identification is necessary when sharing data, yet 2/3 of data scientists were not completely confident in their organizations ability to manage it.

Fig. 3 Challenges in big data analytics in healthcare. *Source* https://privacy-analytics.com/big-data-analytics

Explorys said that healthcare data contains very crucial data of a man's life so we should always be aware and ensure that the data is not leaked out and stays secure in the storage. He also adds that the biggest advantage of the company IBM Explorys is that it has managed to gain the trust from people and the organization also has security audits and also guaranteed consistency and enactment in securing the data [4]. A question that acts as an obstruction for big data is "Who owns the data and who should own the data?" By and large, individuals trust that they ought to claim their own social insurance information; nonetheless, this may not generally be valid. There are stories of patients or their nearby ones attempting to recover fundamental information from report suppliers like Regina Holiday's push to locate her withering spouse's restorative record. These issues have come to fruition "patient advocacy groups" (Fig. 3).

6   Future Prospects of Big Data

Although big data is a nascent field, however, its impact on healthcare is expanding exponentially. Furthermore, big data technologies are anticipated to drive decision-making from a single patient to global population level [14]. One of the chief goals of big data in healthcare is to acquire the facility for every individual with personal care and personal medicines for a precise and an immediate treatment. Big data is possessed to contribute towards this aim in the upcoming years. Some possibilities are: each diabetic, cancer patient has a sequence or a combination of characteristics of genes that can differentiate the patient from others. Analyzing these combinations and drawing inferences from many patients that have similar problem, along with other genetic information, big data may make it possible for service providers to suggest facilities or drugs that are more prone to be helpful to specific patients.

The clinical decision support (CDS) frameworks contain "computerized physician order-entry capabilities" that review patient entries, evaluate them and assess them against medicinal rules to mindful for conceivable mistakes and negative medication responses. With this sort of innovation, suppliers can help decrease unfavorable responses and lessen treatment blunders, especially those, that happen because of clinical missteps. Several companies believe that big data can help healthcare organizations to analyze data from various information sources like EHRs to improve clinical results in global population level. UK's National Health Service (NHS) has come up with techniques to sequence the entire genomes of around hundred thousand patients over 3–5 years to create a national research database. This database will give the researchers a chance to break down and appreciate the hereditary reasons for different ailments, for example, cancer and other uncommon infections. With this information, new solutions, medications, and treatments might be produced. Subsequently, patients can get specific medications that might be more productive. Studies have demonstrated that there exist sharp varieties in healthcare services practices, results, and expenses crosswise over better places, suppliers, and patients. "Dartmouth Atlas project" has said that a few doctors recommend twice the same number of X-rays and CT filters as different specialists in a similar practice. It is likewise recorded that the rates of coronary stents are higher in Elyria and Ohio, than they are in adjacent Cleveland. Basically, investigating various datasets containing understanding qualities and the cost and consequences of medications can find all the more clinically and cost-effective treatments [5].

7 Conclusion

Big data holds millions of varieties of disparate data. It plays a crucial role in the future of Healthcare sector. People can observe a variety of techniques being utilized, to make various decisions and for the performance of healthcare sector. Having all the data of the hospital is impossible to write it down in notebooks or papers, it is necessary to store them digitally so that it can be retrieved and analyzed anytime you want. Storing the numerous data in databases can only be done by technology. There are many hospitals in India where there is extensive use of big data like Apollo Hospitals and Max Hospitals. Thus, many healthcare sectors like hospitals for patient's medical data, pharmacies for medicines data, clinics etc. should use big data for better and safer storage and use of data. Leveraging big data technology will help them expand more and lead themselves towards success.

References

1. Feldman B, Martin EM, Skotnes T (2012) Big data in healthcare hype and hope. Business development for digital health, vol 360
2. Groves P, Kayyali B, Knott D, Kuiken SV (2016) The big data revolution in healthcare: accelerating value and innovation
3. Mittal M, Singh H, Paliwal KK, Goyal LM (2017) Efficient random data accessing in MapReduce. In: 2017 international conference on infocom technologies and unmanned systems (Trends and future directions) (ICTUS), Dubai, 2017, pp 552–556
4. Stril A, Dheur S, Hansroul N, Lode A, Meranto S, Moulonguet M, Ruff I (2015) Report on healthcare
5. Gulamhussen AM, Hirt R, Ruckebier M, Orban de Xivry J, Marcerou G, Melis J (2013) Big data in healthcare
6. Bill H (2012) Big data is the future of healthcare. Cognizant 20-20 Insights (2012)
7. Sun J, Reddy CK (2013) Big data analytics for healthcare. In: Proceedings of the 19th ACM SIGKDD international conference on knowledge discovery and data mining 2013, 11 Aug, pp 1525–1525. ACM
8. Chen H, Chiang RHL, Storey VC (2012) Business intelligence and analytics: from big data to big impact. 36(4):1165–1188
9. Raghupathi W, Raghupathi V (2014) Big data analytics in healthcare: promise and potential. Health Inf Sci Syst 2(1):3
10. Roesems-Kerremans G (2016) Big data in healthcare. J Healthc Commun 1(4):33
11. Arp D (2002) Blue cross and blue shield of North Carolina encourages kids to be active. Healthplan 44(5):37–39
12. https://ninesights.ninesigma.com/web/head-health
13. https://hbr.org/2016/12/how-geisinger-health-system-uses-big-data-to-save-lives
14. Bates DW, Saria S, Ohno-Machado L, Shah A, Escobar G (2014) Big data in healthcare: using analytics to identify and manage high-risk and high-cost patients. Health Aff 33(7):1123–1131

Speedroid: A Novel Automation Testing Tool for Mobile Apps

Sheetika Kapoor, Kalpna Sagar and B. V. R. Reddy

Abstract Automation tools are currently underworked in Quality Assurance and Testing. Nowadays, organizations demand an intellectual tool which can support Agile, DevOps, IoT and mobile software engineering practices. These digital transformations demand smart automation tools to test various applications. Mobile application automation tools present in the market are incapable to meet increased time-to-market pressure due to emerging ways of software development. Digital transformation and third-party relationships have escalated the complexity of testing domain. An innovative tool has been proposed in this paper that provides a smart solution for testing mobile applications. An attempt has been made to develop a mobile application automation tool named as Speedroid. Speedroid aims to provide an intelligent, integrated and automated approach to test continuously changing mobile applications. This automation tool will provide the solution for various challenges in the testing domain like complexity, efficiency, compatibility and portability of tool that are addressed by various organizations. Our study shows that Speedriod will result as a pillar of digital transformation by providing various features like integrity, usability, efficiency and compatible to both iOS and Android, ready to use in regression testing, reporting, logging and minimal tool learning efforts without any knowledge of programming language and contributing scripts.

Keywords Appium · Mobile automation testing · Mobile automation tool
Page object model · Test automation

S. Kapoor (✉)
Department of Computer Science and Engineering, USICT, GGSIPU,
New Delhi, Delhi, India
e-mail: sheetikakapoor@gmail.com

K. Sagar · B. V. R. Reddy
USICT, GGSIPU, New Delhi, Delhi, India
e-mail: sagarkalpna87@gmail.com

B. V. R. Reddy
e-mail: bvrreddy@ipu.ac.in

© Springer Nature Singapore Pte Ltd. 2019
S. Bhattacharyya et al. (eds.), *International Conference on Innovative Computing
and Communications*, Lecture Notes in Networks and Systems 56,
https://doi.org/10.1007/978-981-13-2354-6_29

1 Introduction

Digital transformation is reshaping IT industry by providing various ways of software development. Industries are adopting this transformation to survive in the competitive environment results in the existence of endless applications and platforms to deliver various services [1]. These emerging techniques coexist with older ones like waterfall development model. Different approaches to software development demand an intelligent automation tool which is capable to overcome time-to-market pressure and fulfil business expectations. Automation plays an imperative role in the adoption of digital transformation in the IT industry. Agile and DevOps endorse four weeks sprint [2]. Regression testing is performed at the end of each sprint. This type of testing validates that the changes to the system do not impact existing functionality. Automation plays a crucial role here as it is laborious to perform regression testing manually in the timespan of 4 weeks sprint. Automation testing engages one-time investment but results in great ROI for long-term service projects [3].

Mobile Application Testing is the process of identifying defects in mobile devices across various purposes like functionality, performance, consistency and usability. Mobile applications could be of three types: (1) Native, (2) Web and (3) Hybrid. Native applications are developed for a single platform like Android, iOS which could utilize all mobile device features. Web applications are written in HTML5 and run by the browser. Hybrid applications are multiplatform web apps developed in native wrapper [4]. Automation Testing is a process of verifying Quality with Speed. It provides a smarter way to identify early-stage defects in applications after each change. It could be a part of the accepting build process where automation pack runs against each build. The build would be accepted by QA team for further system testing only when basic automation tests passed [5]. Mobile Automation Testing is the process of identifying defects in mobile applications using real devices or emulators without human intervention. Test Harness depends upon nature of the project. One could have generalized test harness while others could be restricted to use specific to their projects. Test Harness for mobile automation comprises of a mobile app, mobile device or emulators, test cases, test data, common function library, specific function library, test results and end user reports [6]. Nowadays, Page Object Model (POM) technique is used for creating test harness. Mobile Automation Tools currently present in the market includes Appium, Selendriod, Monkey Talk, Robotium, Calabash and Frank. These tools support different platforms still there is a deficiency of mobile automation tool to fulfil the latest industrial trends [7].

Main areas of modern QA embraced intelligent automation testing, smart test environments and Agile organization of QA. 96% organizations have been moved to agile methodology this year [8]. The most important aspects of IT strategy are Customer Satisfaction and Security. The initiative to find the perfect balance between assurance and cost is crucial to IT strategy. Migration to Digital Enterprise results in the creation of hi-tech applications demanding zero tolerance of errors, faster response, highly secure and reliable applications. It results in CIOs investing more on QA and testing. Mobile applications are built to deliver complex functionality with limited resources. Its testing is predominantly impacted by varied OS versions and technological transformations. The amalgamation of elevated complexity and speed of development results in the risk of critical errors and software failures. Furthermore, Mobile applications interact directly with customers. A single critical defect leakage can cause serious damage to brand value and financial losses in terms of penalty and repair costs.

However, due to increase in the volume of mobile testing in the IT industry, organizations are struggling in search of expertise in mobile testing and appropriate test environment. Wherefore, in this paper, we are proposing an intelligent mobile application testing tool called Speedriod. Paradoxically, it is capable of automating test scenarios without any expertise in scripting language along with solutions to other issues addressed worldwide like instability, unavailability of an integrated environment, data management, effective reporting, etc. Furthermore, this paper delineates related work in Sect. 2 along with problem statement in Sect. 3, whereas Sect. 4 embraces proposed methodology. Section 5 describes the comparison of the proposed tool with other existing tools present in the market and Sect. 6 consummate the paper with the conclusion and future scope.

2 Related Work

Amalfitano et al. [9] have proposed various tools and techniques in this province. First, they have proposed MobiGuitar tool for model-based mobile testing applicable for Android apps only. They have specified three steps include abstraction of states, observation, and tracing of GUI of apps for automation testing. Furthermore, a comparison is done between MobiGuitar and other existing tools in the market. Second, GUI Ripping technique is proposed to justify the behaviour of Android apps via its GUI. It is capable of creation and execution of test cases along with identification of crashes at runtime. This technique maintains GUI tree and proved better than Monkey tool [10]. Third, they developed A2T2 tool for Google Android Platform. It is based on three components which involve implementation, crawling and test case generation. It is suitable for regression and crashing testing. Its crawler interprets GUI model automatically that is used for generating test cases [11].

Bo et al. [12] reduced the complexity of test scenarios via their tool MobileTest. It is based on the sensitive-event approach to regression and smoke testing. It could be used in a controlled environment for generation and execution of scripts. Similarly, Vos et al. [13] proposed Testar for GUI-based mobile apps. It is capable of executing fault-finding scenarios rather than contributing to a record and play feature. Hu et al. [14] advised hybrid approach for Android apps capable of finding GUI defects based upon their study upon various application bugs. It includes test case generation and log file analysis.

Nagowah and Sowamber [15] proposed a novel framework called MobTAF which enables testers to perform application testing without any mandatory connection of mobile to computer. This framework is specific to Symbian phones. MobTAF is

Fig. 1 Challenges in mobile testing in 2017 [8]

based upon J2ME and capable to replicate scenarios which are difficult to perform on emulators. Nestinger et al. [16] also proposed a framework called Mobile-C which is based on C/C++. It utilizes a mobile agent to control algorithm and automate other tasks. Similarly, Hao et al. [17] proposed PUMA. PUMA is a programmable framework to automate GUI of apps. Various analyses have been performed on mobile apps using PUMA. Alotaibi and Qureshi [18] proposed a framework which is based on Appium tool. It is capable of speeding up the testing of different mobile browsers and applications. A survey has been conducted to verify the framework. Baride and Kamlesh [19] presented a model to configure mobile devices on the cloud. It enables testers to execute written tests automatically as per given application. Yang et al. [20] proposed an approach for extracting application's model automatically. It is based upon GUI Ripples and reverse engineering performed upon ORBIT tool. It comprises of two modules named as action detection and dynamic crawler. Furthermore, a comparison is done between Guitar and ORBIT.

Choudhary et al. [21] performed a comparison of test data generation tools for Android on the basis of four factors: (1) Code Coverage, (2) Fault Detection Capability, (3) Ease of Use and (4) Compatibility. Further, the pros and cons of different techniques have been discussed. Muccini et al. [22] did analysis in the mobile application domain. Various challenges and role of automation testing in the domain has been addressed. They embraced the difference between desktop and mobile applications. Moreover, testing process, levels and artefacts have been identified as future scope in this domain. A survey has been conducted by Capgemini, Micro Focus and Sogeti to analyse latest testing trends and address various challenges. The survey report is called World Quality Report 2017–18 [8]. In this report, it is mentioned that only 16% of test activities are automated. A No. of organizations had declared that they have been struggling with mobile testing over past 3 years. The shortage of time to test mobile apps has been increased from 36% in 2015 to 52% in 2017. Moreover, 47% organizations are facing issues with testing techniques while 46% do not have the appropriate tool to test mobile applications. Figure 1 describes all such challenges identified in the survey in 2017. Furthermore, in this survey, it is mentioned that 57% testers are lacking in development and coding skills while 55% are not expertise in automation testing [8].

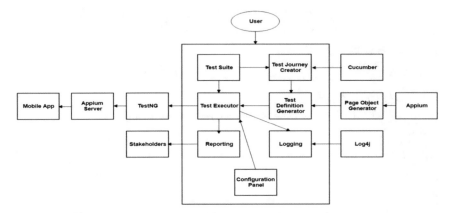

Fig. 2 Architecture of Speedroid

3 Problem Statement

The dynamics of Quality Assurance and Testing is briskly reshaping due to the following trends:

- The rapid growth of smart devices.
- Increased demand for automation.
- Adoption of Agile, DevOps and IoT methodologies.

Although, various mobile automation tools are present in the market still they are failed to achieve the desired level of test automation due to various challenges as described in Figs. 2 and 3. Therefore, in this paper, we are proposing Speedriod which aims to provide solutions of various challenges addressed in worldwide surveys 2017–18 includes:

- Stable and integrated testing environment,
- A suitable framework for functional testing of mobile applications,
- A tool which empowers a manual tester to generate automation scripts which in turn reduces the demand for mobile automation tester experts,
- Generation of test scenarios within less time and efforts without any knowledge of programming language,
- Provide a single platform for common understanding among clients, testers and developers using BDD and Reporting mechanism.

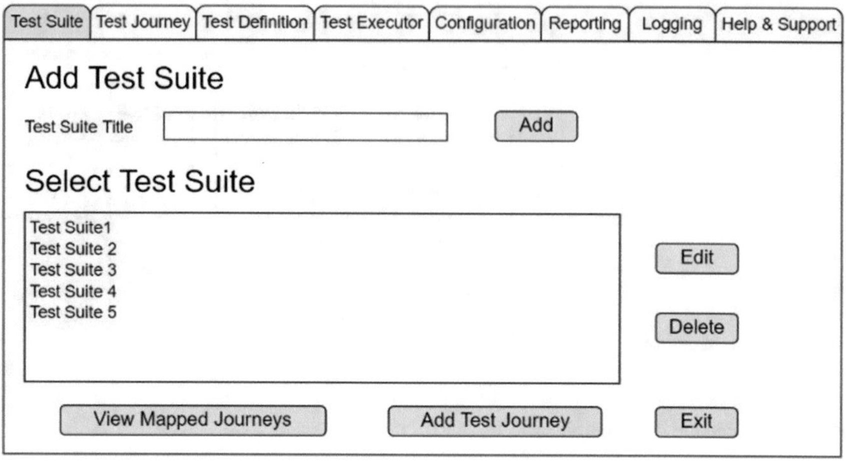

| Test Suite | Test Journey | Test Definition | Test Executor | Configuration | Reporting | Logging | Help & Support |

Add Test Suite

Test Suite Title [   ]   [Add]

Select Test Suite

Test Suite1
Test Suite 2   [Edit]
Test Suite 3
Test Suite 4
Test Suite 5   [Delete]

[View Mapped Journeys]   [Add Test Journey]   [Exit]

Fig. 3 Test suite editor

4 The Proposed Tool

Digital transformation enforces IT industry to speed-up testing practices with a focus on performance and efficiency of mobile applications. Automation testing is a predominating solution of time and efficiency by playing an important role in Regression Testing of each new build. In this paper, we propose a novel tool named as Speedriod. It is cross-platform tool, i.e. works for all platforms includes iOS, Firefox OS and Android applicable for all types of mobile applications, i.e. native, web app and hybrid.

Speedroid is a Java-based mobile automation testing tool. It is an extension of Appium in conjunction with TestNG, Cucumber and Log4j to accomplish various purposes like TestNG works as Unit Test Framework for Speedroid, whereas Cucumber enforces testers to create test scenarios in BDD format along with Log4j for logging purpose. Speedroid provides a platform to testers which empower them to create automation scripts like a 'layman'. It provides various features like Test Case Creation, Traceability, Execution, Validation, Reporting, Error Analysing and Logging. It is comprised of seven components such as (1) Test Suite Editor, (2) Test Journey Creator, (3) Test Definition Generator, (4) Test Executor, (5) Configuration Panel, (6) Reporting and (7) Logging. The architecture of Speedroid is described in Fig. 2.

4.1 Test Suite Editor

A Test Suite is a collection of relevant test cases. Speedroid allows testers to group together related Test Journeys in a Test Suite. Different Test Suites can coexist to accomplish various purposes like one test suit for round 1 of regression testing which includes test journeys of only one application. While another test suite could coexist for round 2 of regression testing comprises of test journeys to validate integration among different applications. A unique ID is assigned to each test suite to distinguish them. The template of Test Suite Editor is picturized in Fig. 3.

4.2 Test Journey Creator

Test Journeys describes the end-to-end flow of a specific scenario of the system. Test Journey Creator enforces testers to create test journeys in BDD format. BDD format is advantageous for building common understanding among all stakeholders, i.e. client, tester, developer and management team. Speedroid provides the sample format of test journeys will be there as default text. Each test journey will be mapped to an existing test suite. Its template is described in Fig. 4.

The Test journeys are also called feature files. One feature file can have multiple scenarios. Test data is also specified in feature file under the Examples section. It allows executing the same script with different test data.

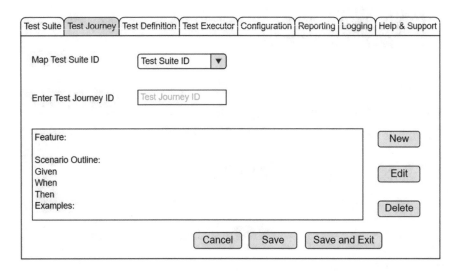

Fig. 4 Test journey creator

4.3   Test Definition Generator

Test Definition Generator creates automation scripts automatically in respect of data provided by the tester on Test Definition screen of Speedroid. First, Page Objects are created by specifying the details of all needed elements (i.e. elements upon which any action or validations to be performed) present on the page of a mobile app. Details include its name, type, identifier and value. Element type could be button, text, dropbox, radio button, hyperlink, etc. Whereas, the various identifier could be used includes XPath, ID, class name, link text, name, CSS selector, tag name etc., along with their respective values need to be provided to create page objects automatically.

Second, actions are selected by the user to be performed on an added element. In Speedroid, various actions are filtered on the basis of element type like click, double-click and get text actions for buttons. Third, validations are specified by users to decide upon the result of test journey. A test journey will be considered as pass only when all validations specified in particular test journey are passed. The template of the Test Definition Generator is displayed in Fig. 5.

The pseudocode of creation of Page Objects is as follows:

Algorithm:

1. Driver=getDriver(Configured Browser)
2. Element.Type ElementName=Driver.Identifier(Value)
3. Return ElementName

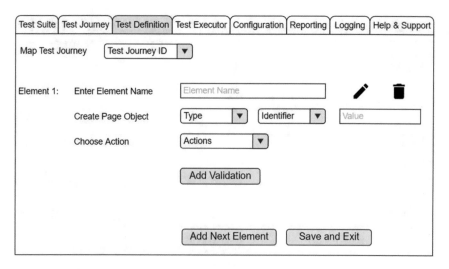

Fig. 5   Test definition generator

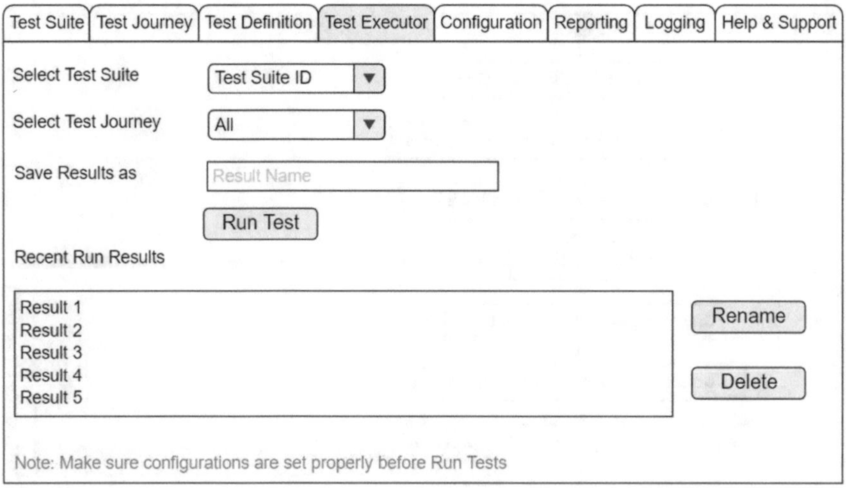

Test Suite	Test Journey	Test Definition	Test Executor	Configuration	Reporting	Logging	Help & Support

Select Test Suite   Test Suite ID ▼

Select Test Journey   All ▼

Save Results as   Result Name

                         [Run Test]

Recent Run Results

Result 1   [Rename]
Result 2
Result 3
Result 4   [Delete]
Result 5

Note: Make sure configurations are set properly before Run Tests

Fig. 6 Test executor

4.4 Test Executor

Test Executor provides a platform to execute the entire test suit as well as a specific journey of a test suite. It asks for the name of the test result to be saved with. All recent test results are being displayed in this tab of Speedroid. A tester can rename and delete recent test results. Its template is described in Fig. 6 along with the functionality in Fig. 7.

The pseudocode of Test Executor is as follows:

Algorithm:

1. TestSuite TS=getSelectedSuite()
2. For (int i=0;TS.length;i++)
   getTestJourneys()
3. BrowserName=getBrowser()
4. For each TestJourney
   Driver=BrowserName.Driver()
   CreatePageObjects(Type,Identifier,Value)
   PerformAction(getAction,getElements)
   if(Assert(Output) then setTestJourney('Pass')
   else { setTestJourney('Fail') and captureScreenshot() }
5. emailReport(TestJourney.getResult(), recipients)

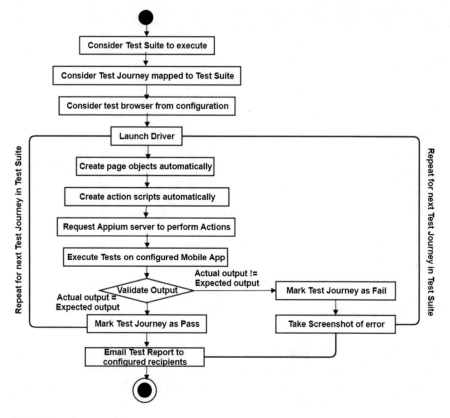

Fig. 7 Flow diagram of test executor

4.5 Configuration Panel

The configuration panel specifies various configurable parameters like mobile app name, test browsers and emulators to be used along with email ID of stakeholders to which the report needs to be sent. The template of this component is picturized in Fig. 8.

4.6 Reporting

Reporting provides the summary of test results graphically. Higher management is generally interested in statistics and figures rather than the details in depth. Reports are eye-catchy and provide a lot of information in less time. Reports contain statistics about Pass/Fail scenarios count, test steps execution count, coverage, execution percentage and other relevant details via graphs and charts.

| Test Suite | Test Journey | Test Definition | Test Executor | Configuration | Reporting | Logging | Help & Support |

Enter Mobile App Name   App Name

Select Test Browser   Chrome ▼

Select Emulator   Emulators ▼

Enter Report Recipients   Email Address

    Save   Set to Default

Fig. 8 Configuration panel

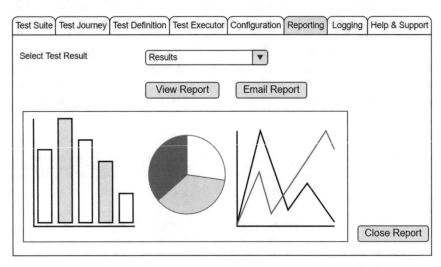

| Test Suite | Test Journey | Test Definition | Test Executor | Configuration | Reporting | Logging | Help & Support |

Select Test Result   Results ▼

    View Report   Email Report

    Close Report

Fig. 9 Reporting

Speedroid provides reporting via two means: (1) View Report and (2) Email Report. View report displays the test result report on the screen under-reporting tab. While Email report takes recipients from configuration tab and email test report to all the stakeholders configured. Once test execution is complete, an automatic email would be sent to all recipients having a message of test execution completion along with test report attached. Whereas, the Email Report feature under-reporting tab allows a user to resend test report to all the stakeholders. It uses Jenkins plugin for reporting. Its template is picturized in Fig. 9.

The pseudocode of email report is as follows:
Algorithm:

1. email = mimeMessage()
2. email.setHost(Host)
3. email.setSubject('TestExecutionReport',Date,Time)
4. email.setTo(getRecipients)
5. email.setFrom(TestingReportAccount)
6. email.setAttachment(ResultReport)
7. Transport.send(email)

4.7 Logging

Logging empowers users to backtrack the issues. A test journey failure does not always indicate code issues, i.e. Defects. Failure can occur due to test data. Let's say login via valid credential throws an error due to a locked user account. So, it does not mean there is an error in login module to the system. It is a test data issue. Such issues are identified via logs.

Speedroid provides logging in two ways: (1) Written Logs and (2) Screenshots of errors. Written logs utilize the features of Log4j for effective logging capability. Moreover, Speedroid takes the screenshot of mobile app page whenever error encounters. It helps testers to find route cause of failed journey easily. Its template is described in Fig. 10.

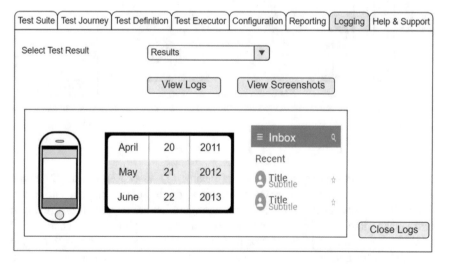

Fig. 10 Logging

Table 1 Comparison of Speedroid with existing mobile automation tools

Features	Appium	Robotium	Selendroid	Keep it functional	Monkey talk	Calabash	Speedroid
Programming language independent	N	N	N	N	N	N	Y
Integrated tool	N	N	N	N	N	N	Y
Cross-platform capability	Y	N	N	N	Y	Y	Y
Auto-screenshot feature	N	N	N	N	N	N	Y
Auto-reporting feature	N	N	N	N	N	N	Y
Various framework support	Y	Y	Y	N	Y	Y	Y
Application support	Y	N	N	N	N	N	Y

5 Speedroid Versus Other Existing Tools

There are various mobile automation testing tools present in the market which accomplish one and more purposes. Famous mobile automation tools are Appium, Robotium, Selendroid, Keep it Functional, Monkey Talk and Calabash. However, the worldwide surveys indicate the issues with the tools present in the market which includes (1) Environmental issues, (2) Skill set challenges, (3) Time Constraint and (4) Lack of right automation tool as described in Fig. 1.

As proposed, Speedroid is a novel tool which is capable to provide the solution of all addressed challenges in the field of mobile application automation testing. A comparison has been done among novel tool Speedroid and the tools which are already present in the market in Table 1.

The comparison reveals that existing tools are dearth in various essentials features like language independent, integrated tool, various logging and reporting capabilities. The deficiency of these features in existing tools results in the birth of challenges described in Fig. 1. The proposed tool Speedroid is an integrated tool which provides all such features which are lacked in existing tools includes language independent, stable and integrated tool. It provides complete end-to-end flow to testers from test case creation to execution and ends up with reporting and logging capabilities. These features of Speedroid enable any tester to create automation scripts without any fulfilment of specific skillset. Moreover, using Speedroid, testers can create automation script in such a less time as it fits best in Agile and DevOps methodologies.

6 Conclusion and Future Scope

Nowadays, mobile phones are the primary source of interaction as well as medium for other critical purposes like financial transactions. Thousands of apps are deployed every day. Diversity in applications challenges tests strategies at each phase. Due to Digital Transformation, organizations are adopting Agile and DevOps methodologies rapidly. In these methodologies, changes are delivered in less time span of generally 2–3 weeks called a sprint. Whereas, with each sprint, No. of test cases to be executed increases exponentially because in such cases, testers need to test new functionality as well as existing functionality. Test Automation provides a mechanism to perform repetitive tests effectively and efficiently. So that, most of the time could spend upon newly created tests rather than upon existing ones.

To accomplish these purposes, organizations hire manual as well as automation testers. However, in the recent worldwide surveys, it has been highlighted that most of the organizations are failed to achieve the required level of automation due to skill deficiency in the creation of automation framework. It has been experienced that the proportion of automation testers are very less as compared to manual testers. Moreover, automation testers also deal with a lot of challenges includes the efforts to learn a new framework, a new programming language, lack of integrated tool and need to write a programming language specific code for supportive functionalities like screenshots handling, error handling, email features, reporting and logging.

Existing tools enables a tester to write simple automation scripts with minimal knowledge of programming language. However, in IT industry, scripts are not really simple. A stable framework needs to be created to fulfil automation purposes. On the other hand, if the framework already exists, a lot of effort is needed to understand that framework. The comparison delineates in Table 1 indicates that these existing tools are deficient in various essential features which plays a significant role in the delivery of quality mobile applications. These features include lack of fast, integrated, language independent and completely featured tool with advanced logging and reporting capabilities.

In this paper, the automation testing tool for mobile apps has been proposed. The tool is named as Speedroid that is proficient to provide the solution of all addressed challenges. Using Speedroid, exceptionally, a manual tester could create automation script like a 'layman' without any knowledge of programming language in significantly less time than other existing tools present in the market. To use this tool, it is not required to understand its framework. It is a well-integrated tool along with other supportive features like auto-screenshot capturing of errors, auto-reporting and logging. In turn, this will increase the level of automation in the industry and contribute to the quality assurance of mobile applications by creating scripts in remarkable less time that makes it fit best in Agile and DevOps methodologies. In this paper, we are proposing Speedroid. In future, its implementation will be done along with other advanced features like parallel execution of automated scripts to reduce test execution time effectively.

References

1. Heilig L, Schwarze S, Voss S (2017) An analysis of digital transformation in the history and future of modern ports
2. Cruzes DS et al (2017) How is security testing done in agile teams? a cross-case analysis of four software teams. In: International conference on agile software development. Springer, Cham
3. Abrahamsson P et al (2017) Agile software development methods: review and analysis. arXiv: 1709.08439
4. Vilkomir S (2017) Multi-device coverage testing of mobile applications. Softw Qual J 1–19
5. Stocco A et al (2017) APOGEN: automatic page object generator for web testing. Softw Q J 25(3):1007–1039
6. Rosenfeld A, Kardashov O, Zang O (2017) ACAT: a novel machine-learning-based tool for automating Android application testing. In: Haifa Verification Conference. Springer, Cham
7. Moran K, Linares Vásquez M, Poshyvanyk D (2017) Automated GUI testing of Android apps: from research to practice. In: 2017 IEEE/ACM 39th international conference on software engineering companion (ICSE-C). IEEE
8. Capgemini, Micro Focus, Sogetti (2017–2018) World quality report 2017–2018. https://www.capgemini.com/thought-leadership/world-quality-report-2017-18
9. Amalfitano D et al (2017) A general framework for comparing automatic testing techniques of Android mobile apps. J Syst Softw 125(2017):322–343
10. Amalfitano D, Fasolino AR, Tramontana P (2011) A GUI crawling-based technique for android mobile application testing. In: The fourth IEEE international conference on software testing, verification and validation, Berlin, Germany, 21–25 March, Workshop Proceedings, 2011, pp 252–261
11. Amalfitano D, Fasolino AR, Tramontana P, Carmine SD, Memon AM (2012) Using GUI ripping for automated testing of android applications. In: IEEE/ACM international conference on automate software engineering, ASE'12, Essen, Germany, 3–7 Sept 2012, pp 258–261
12. Bo J, Xiang L, Xiaopeng G (2007) MobileTest: a tool supporting automatic black box test for software on smart mobile devices. In: Proceedings of the second international workshop on automation of software test. IEEE Computer Society
13. Vos TEJ et al (2015) Testar: tool support for test automation at the user interface level. Int J Inf Syst Model Des (IJISMD) 6(3):46–83 (2015)
14. Hu C, Neamtiu I (2011) Automating GUI testing for Android applications. In: Proceedings of the 6th international workshop on automation of software test. ACM
15. Nagowah L, Sowamber G (2012) A novel approach of automation testing on mobile devices. In: 2012 international conference on computer & information science (ICCIS), vol 2. IEEE
16. Nestinger SS, Chen B, Cheng HH (2010) A mobile agent-based framework for flexible automation systems. IEEE/ASME Trans Mechatron 15(6):942–951
17. Hao S et al (2014) PUMA: programmable UI-automation for large-scale dynamic analysis of mobile apps. In: Proceedings of the 12th annual international conference on mobile systems, applications, and services. ACM
18. Alotaibi AA, Qureshi RJ (2017) Novel framework for automation testing of mobile applications using Appium. Int J Mod Educ Comput Sci 9(2):34
19. Baride S, Dutta K (2011) A cloud based software testing paradigm for mobile applications. ACM SIGSOFT Softw Eng Notes 36(3):1–4
20. Yang W, Prasad MR, Xie T (2013) A grey-box approach for automated GUI-model generation of mobile applications. In: International conference on fundamental approaches to software engineering. Springer, Berlin, Heidelberg
21. Choudhary SR, Gorla A, Orso A (2015) Automated test input generation for Android: are we there yet?(e). In: 2015 30th IEEE/ACM international conference on automated software engineering (ASE). IEEE
22. Muccini H, Di Francesco A, Esposito P (2012) Software testing of mobile applications: challenges and future research directions. In: Proceedings of the 7th international workshop on automation of software test. IEEE Press

Domain-Specific Fuzzy Rule-Based Opinion Mining

Surabhi Thorat and C. Namrata Mahender

Abstract Opinion mining's goal is to recognize the behavior or emotion of people. To encash human thoughts, views in spreading awareness through social media for positive changes in public interest like health, natural disaster, education, and many more. This paper is an attempt for driving social awareness among people relevant to drought. The proposed model considers opinion on drought related issues and predict the sentiment behind each view. The major issue while processing the opinion, view of candidate is about natural fuzziness in-build in words expressed. This fuzziness in words actually helps to understand the depth of the word. So to encash the depth of the word fuzzy rule-based approach is proposed which reduces the computational complexity and increases the interpretability. The patterns allows the system to provide more precise result which is our work has been categorized in 6 major categories that is negative, moderately negative, highly negative, positive, moderately positive and highly positive. After applying this pattern we got more precise results for sentiment analysis.

Keywords Data mining · Sentiment analysis · Social media · Domain-specific Drought · Association rule mining · Fuzzy rule

1 Introduction

In the current scenario, people are generally communicating or express their opinions and views on different issues through the use of social media. A rise in web2.0 applications has resulted in the incremental accumulation of a huge amount of user-generated content via online activities like blogging, social networking, emailing,

S. Thorat (✉) · C. Namrata Mahender
Department of Computer Science & IT, Dr. Babasaheb Ambedkar University,
Aurangabad, Maharashtra, India
e-mail: thorat.surabhi@gmail.com

C. Namrata Mahender
e-mail: nam.mah@gmail.com

© Springer Nature Singapore Pte Ltd. 2019
S. Bhattacharyya et al. (eds.), *International Conference on Innovative Computing and Communications*, Lecture Notes in Networks and Systems 56,
https://doi.org/10.1007/978-981-13-2354-6_30

review posting, tweets and so on. As a result a lot of interest in atomized classification, computational linguistics, and opinion mining arises. Analyzing opinions is a method that forecast opinion sentiment polarity from text like customer review on the product of a movie, blog post, tweets, comments, or product, etc. [1]. The task of data mining as a crucial issue to discover knowledge from database repositories so that meaningful predictions can be done. Opinions are views that are not always based on evidences or facts. To predict opinions from phrases, sentences or in documents we need to handle complex terms or the terms having multiple meanings while expressing the opinions. Opinion words present in text can be predicted in many forms as they compressed of fuzziness. Like, the words "great", "excellent", and "fine" "awesome" has certain boundaries between them which is very difficult to predict. Thus, Fuzzy logic is able to analyze the subjectivity of words, so that word labels can be assigned with approximate degree of membership, which leads to precise interpretation of words.

Machine learning transforms the text data in the form of structured data. That leads learning algorithms for sentiment classification can be utilized directly. Particularly bag-of-words method reflects each word present in the document like feature in a structured data set is a well-known approach used for such kind of classification [2]. According to Cristianini support vector machine is widely used for accurate organization of sentiment data [3]. Rish et al. says that in spite of the impractical unconventional assumptions of Naïve Bayes, its classifier is unexpectedly effective for practical use because its classification judgement frequently lead to correct results even if its approximate probability ratio was inaccurate [4]. Support vector machines trained models have certain boundaries in terms of deep of learning, while the system trained through Naive Bayes does not precisely predict the result because of the necessary hypothesis that all input attributes are completely self-sufficient from each other. To handle this form of data Fuzzy Logic is highly recommended to get the promising result [5].

2 Literature Review

A study says that during the year of 2012, fuzzy approaches for processing the natural language [6] was very less regardless of the appropriateness of fuzzy methods applied for processing and classifying the text. In past years, fuzzy approaches are in full swing in regard to text processing. Previously a fuzzy fingerprint text based classification for organizations had proposed [7], it shows outstanding performance in comparison to generally used non-fuzzy techniques. It can be used to generate corpus in atomized manner which can be utilized for text similarity comparison [8]. The results from this experiment indicate that the fuzzy metrics is highly correlated with people rankings when related with conventional metrics. Unsupervised fuzzy method [9] was used to categorize Twitter users on the basis of masculinity. Fuzzy methodology that is given by Liu and Cocea in 2017 which was shows minimal computational complexity along with sustaining a related impact to other renowned machine learning methods [10].

Jefferson et al. in their work expresses how membership degree inputs were utilized in further superior results in inclusion with various sentiment intensities [11]. They compared the results of the system with machine learning algorithms and found that its results are more promising when we talk about sentiment analysis. In recent years significant work has done in the area of fuzzy login. Ansarul Haque and Rahman assign the weight to the frequent works in the process of sentiment analysis by means of fuzzy logic. In their work POS tagger gets generated first and tags the extracted tweets. Then computes the sentiment values of the tweets [12]. Howells and Ertugan tried to combine public bots along with mining and fuzzy concepts. They propose a system which evaluates text available in social networking sites like social networking site like Twitter and analyze users opinion or views [13].

3 Fuzzy-Based Aspects of Opinion Mining

Research on Fuzzy-based opinion systems is growing gradually. Presently opinion mining techniques generally classify views as positive, negative, and neutral. Sentiment analyzing methods such as holistic lexicon approach will not consent categorization of reviews fineness to identify the precise meaning of each and every sentiment. Major requirement is to grow the views categorization and give weights to several thought representing words. Only Fuzzy logic can predict accurate thought or views and categorized them at various strengths. Less work had done by researches in the area of on using Fuzzy to enhance the thought classification [14, 15]. Fu and Wang [16] also addressed that in Chinese sentence-level sentiment classification most of the available fuzzy-based systems have either not covered several crucial issues of sentiment analysis or may not applied the robust techniques of Fuzzy [16]. In relation with this Animesh and Debael [17] suggested a sentiment mining systems based on fuzzy logic. It is a supervised model that extracts analyses using Fuzzy logic [17]. Now a day's Fuzzy logic is in great need as it works on Natural Language used by humans which compressed of lot of fuzziness. Fuzzy logic implements through proficient knowledge using IF-THEN intellectual rules. They are designed on the basis of having flexible membership functions instead of normal crisp binary logic. Besides, the fuzzy set theory proposes an enhanced and easy ways to present the intrinsic fuzziness in mining the views [18].

After observing the work in regard to fuzzy-based aspects we proposed a system in which we had generated a classification pattern that compressed of POS tagging, domain related words, frequent keywords and polarity labels. This model produces prominent results in domain-specific sentiment classification.

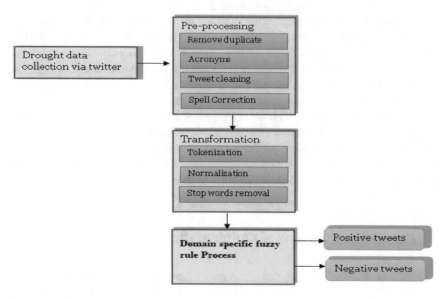

Fig. 1 Proposed model

4 Proposed System

In our model we describe the domain-specific fuzzy rule-based opinion mining system using machine learning. The main objectives of our proposed system are to generate positive and negative opinion from tweets on drought domain (Fig. 1).

4.1 Data Collection

We collected tweets using Twitter Application Program Interface on drought domain. Drought dataset consists of 5000 reviews. As we focused on drought tag only, we initially selected only tweets containing the drought keyword. We had focused on the keyword drought but we found that the word drought is used by the people differently in different context. It is used in many ways that are resemblance is denoting lack of humanity, financial crisis, lack of water. Like, people are tweeting "there is a drought in humanity", "my country is facing drought to fulfilling basic needs", etc. This dataset containing people's unordered or random views on the critical issue drought. In this phase we continue with our preprocessing with all 5000 tweets despite of its relevance with domain. We filtered drought specific tweets associated with water scarcity and its impact using the module domain-specific fuzzy rule of our proposed model which is explained in detail in Sect. 4.3 to enhance the sentiment analysis of our domain.

4.2 Data Preprocessing

Unique tweets selection: Tweets collected from twitter contains duplicate tweets that are coming again and again in the dataset. So we applied the function to eliminate the repetitive tweets and select only the distinct one to reduce the processing task.

Acronym: We had analyzed that people are using the slang expression or non-word which are not found in English dictionary while tweeting the text. Like people are writing "n" instead of "and" or writing "u" instead of "you". To solve this problem we had created a Comma Separated Values (CSV) file which contains the correct word for most of the non-words found in the tweets. After that, we had replaced the non-word with correct words with the help of CSV file created by us. This will become a fruitful corpus for converting the slag or non-words into its correct format. This can be used for any sort of research work where the conversion from slang words to correct words is required.

Tweet cleaning: The collected data generally includes special symbols like @, #, %, emoticons, etc. So we process the data and removed all such contents so that analysis should be smooth out.

Spell correction: After cleaning the non-grammatical words were replaced with correct spell words. This will help to get the accurate meaning of the sentence.

Transformation: In in phase we split sentences into small words or tokens. The segmentation is needed to analyze the sentence. We exclude words like prepositions, numbers, special characters, special symbols, articles, and nouns. By of Parts of speech tagging we extract opinion words from the tagged file.

4.3 Domain-Specific Fuzzy Rule

After transformation, we found that all word token are not useful for predicting sentiments on "Drought". Once we completed preprocessing of tweets we extract tweets relevant to water scarcity only by creating a corpus that include water scarcity terminology. For our system, we had created corpus having all possible word based on "drought" synonym. Based on this corpus out of 5000 tweets we got 3620 relevant tweets. The nouns and noun phrases can be taken as features from it. We had proposed our model in two section. In section A, we focus on opinion words. Like, major adjectives or adverbs that are used to qualify the nouns and express sentiments. Two or three sequential words are mined from the POS tagged tweets if tags found pattern matching based on Table 1. We had generated patterns based on the formula [Pattern + Drought Related Term + Frequent Word + Polarity Classification], i.e., (P + D + F + PScore). Before matching the pattern, we had predicted the polarity of the tweets. These polarity classifications of each label is matched according to the proposed pattern. Like, we get patterns "Drought really too worst farmer", "No

Table 1 Section A and Section B

Section A

Pattern (P)	Drought related term (D)	Frequent word (F)	Polarity score
Unigram pattern			
(Adverb, Adverb, Adjective)	Drought	Farmer	
(Adverb, Adjective)	Dearth	Suicide	
(Adjective)	Shortage	Crop	
Bigram pattern			
(Adverb, Adverb, Adjective)	Water scarcity	Business	
(Adverb, Adjective)	No rainfall	Economy	
(Adjective)	Less water	Drink	

Section B

Proposed fuzzy rule

1. Finding word frequency in all collected tweets
2. Find polarity score of positive and negative words
3. IF Positive score ≥ 1 THEN "Positive"
4. IF Negative score < 1 THEN "Negative"
5. IF (Adverb, Adverb, Adjective) and (D) and (F) Then
6. {IF "Positive" THEN Sentiment = "Highly Positive sentiment" ELSE Sentiment = "Highly Negative sentiment"}
7. IF (Adverb, Adjective) and (D) and (F) Then
8. {IF "Positive" THEN Sentiment = "Moderately Positive sentiment" ELSE Sentiment = "Moderately Negative sentiment"}
9. IF (Adjective) and (D) and (F) Then
10. {IF "Positive" THEN Sentiment = "Positive sentiment" ELSE Sentiment = "Negative sentiment"}

rainfall horrifying poor economy", "Dearth extremely scare suicide", etc. This pattern matching shows better accuracy in sentiment prediction results on "Drought".

5 Results

Based on the proposed fuzzy rule we are able to analyze the sentiment more precisely in comparison of the simple polarity evaluation based on SentiWordNet. We had considered all cases individual word polarity, word frequency, pattern matching and drought synonym. The results shown in Fig. 2. Indicates sample result that we are getting and its more accurate on "Drought" domain as we considered the factors associated with water shortage.

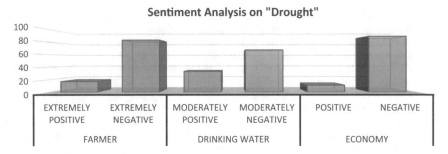

Fig. 2 Sentiment result on drought

6 Conclusion

We tried to proposed system to accomplish sentiment classification of tweets on the "drought" domain in combination with the fuzzy rule-based to predict the people opinion on drought. We have collected 5000 tweets out of which only 3620 tweets had extracted based on our domain relevance. Our domain-specific opinion mining system involves various phases like (1) Preprocessing phase, (2) Transformation (3) Fuzzy Rule. This model proposes a fuzzy logic based sentiment analysis framework on a challenging domain "Drought". Social media can be considered for raising a critical issue related to water. We had tried to find the internet users sentiment on it. The results based on our model say that people are tweeting more on negative aspects of "Drought" in comparison to "Positive". The fuzzy rule in combination with polarity score and drought terminology improves our result.

References

1. Ohana B, Tierney B, Delany S, (2011) Domain independent sentiment classification with many lexicons. In: 2011 IEEE workshops of international conference on advanced information networking and applications (WAINA). IEEE, pp 632–637
2. Sivic J (2009) Efficient visual search of videos cast as text retrieval. IEEE Trans Pattern Anal Mach Intell 31(4):591–605
3. Cristianini N (2000) An introduction to support vector machines and other kernel-based learning methods. Cambridge University Press, Cambridge
4. Rish I (2011) An empirical study of the Naive Bayes classifier. In: IJCAI 2001 workshop on empirical methods in artificial intelligence, vol 3, no 22, pp 41–46
5. Clerk Maxwell J, Zadeh LA (1965) Fuzzy sets. Inf Control 8(3):338–53
6. Caravalho J, Batista F, Coheur L (2012) A critical survey on the use of fuzzy sets in speech and natural language processing. In: 2012 IEEE international conference on fuzzy systems, pp 1–8
7. Batista F, Carvalho JP (2015) Text based classification of companies in CrunchBase. In: 2015 IEEE international conference on fuzzy systems, pp 1–7
8. Chandran D, Crockett KA, Mclean D, Crispin A (2015) An automatic corpus based method for a building multiple fuzzy word dataset. In: 2015 IEEE international conference on fuzzy systems, pp 1–8

9. Vicente M, Batista F, Carvalho JP (2015) Twitter gender classification using user unstructured information. In: 2015 IEEE international conference on fuzzy systems, pp 1–7
10. Liu H, Cocea M (2017) Fuzzy rule based systems for interpretable sentiment analysis. The international conference on advanced computational intelligence, pp 129–136
11. Jefferson C, Liu H, Cocea M (2017) Fuzzy approach for sentiment analysis. In: 2017 IEEE international conference fuzzy systems (FUZZ-IEEE), pp 9–12. ISSN 1558-4739
12. Ansarul Haque Md, Rahman T (2014) Sentiment analysis by using fuzzy logic. Int J Comput Sci Eng Inf Technol (IJCSEIT) 4(1)
13. Howells K, Ertugan A (2017) Applying fuzzy logic for sentiment analysis of social media network data in marketing. In: 9th International conference on theory and application of soft computing, computing with words and perception, ICSCCW 2017, 24–25, Budapest, Hungary. Procedia Comput Sci 120:664–670
14. Samaneh N, Murad MA, Kadir RA (2010) Sentiment classification of customer reviews based on fuzzy logic. In: International symposium on information technology (ITSim), pp 1037–1040
15. Huhn J, Hullermeier E (2009) FURIA: an algorithm for unordered fuzzy rule induction. Data Min Knowl Disc 19(3):293–319
16. Fu G, Wang X (2010) Chinese sentence-level sentiment classification based on fuzzy sets. In: COLING 2010: poster volume, Beijing, Aug 2010, pp 312–319
17. Kar A, Mandal DP (2011) Finding opinion strength using fuzzy logic on web reviews. Int J Eng Ind 2(1):37–43
18. Subasic P, Huettner A (2001) Affect analysis of text using fuzzy semantic typing. IEEE-FS 9:483–496

A Shallow Parsing Model for Hindi Using Conditional Random Field

Sneha Asopa, Pooja Asopa, Iti Mathur and Nisheeth Joshi

Abstract In Natural Language Parsing, in order to perform sequential labeling and segmenting tasks, a probabilistic framework named Conditional Random Field (CRF) have an advantage over Hidden Markov Models (HMMs) and Maximum Entropy Markov Models (MEMMs). This research work is an attempt to develop an efficient model for shallow parsing which is based on CRF. For training the model, around 1,000 handcrafted chunked sentences of Hindi language were used. The developed model is tested on 864 sentences and evaluation is done by comparing the results with gold data. The accuracy is measured by precision, recall, and F-measure and is found to be 98.04, 98.04, and 98.04, respectively.

Keywords Conditional random fields · CRF · Parsing · Shallow parsing · Hindi Chunking

1 Introduction

Natural language is a collection of infinite set of strings called sentences. These sentences are made up of smaller units called tokens. Processing of natural language itself is a challenging task as it includes some phases to be completed in order to get the desired output. Considering the example of an infant, who tries to learn and understand the language. When a sentence is uttered in front of a child, she splits the sentence into smaller units and then tries to grasp the utterance and meaning of

S. Asopa (✉) · P. Asopa · I. Mathur · N. Joshi
Banasthali Vidyapith, P.O. Banasthali Vidyapith, Newai 304022, Rajasthan, India
e-mail: asopasneha@gmail.com

P. Asopa
e-mail: pooja.asopa@gmail.com

I. Mathur
e-mail: mathur_iti@rediffmail.com

N. Joshi
e-mail: nisheeth.joshi@rediffmail.com

© Springer Nature Singapore Pte Ltd. 2019
S. Bhattacharyya et al. (eds.), *International Conference on Innovative Computing and Communications*, Lecture Notes in Networks and Systems 56,
https://doi.org/10.1007/978-981-13-2354-6_31

295

(a)   **(b)**

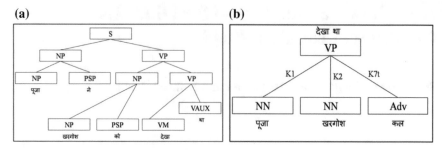

Fig. 1 **a** Example of phrase structure. **b** Example of dependency grammar

each word. Similarly, the first phase of natural language processing (NLP) divides a
sentence into tokens and then with the help of morphological analysis, we find the
morphemes which are the smallest unit in a word, which further helps us in understanding the meaning of each word. Now, based upon this, categorization of words
can be formulated. For example, in Hindi, these categories can be Noun, Adverb,
Verb, and many more. Based upon these words and their categorization, grouping
of words is done which in terms of NLP is known as chunking and in more formal
terms, it is also known as shallow parsing. For the generation of complete parse tree,
shallow parsing is required. Syntactic analysis or parsing is an important intermediate phase of representation for semantic analysis, and thus, plays a significant role
in application areas of NLP. In natural languages, every token is ordered according
to some structure in a sentence and this structure is assigned by the grammar.

The need of parsing in natural languages can be understood more clearly by
understanding the structure of a sentence which is hierarchical in nature and consists
of words which on combining form phrases. Parsing can be defined as the process
to analyze and determine the grammatical structure of sentences when the formal
grammar of a sentence is given. These grammar formalisms are of two types, phrase
structure and dependency grammar. In phrase structure, basically, we identify phrases
and recursive structures in a sentence with the help of grammar rules which are
broadly divided into different categories like noun phrases, verb phrases, and adverb
phrases Asopa et al. [1]. Figure 1a depicts the phrase structure of a given sentence
"पूजा ने खरगोश को देखा था" by considering the following grammar rules:

$$S \rightarrow NP\,VP \quad NP \rightarrow NP\,PSP$$
$$VP \rightarrow NP\,VP \quad VP \rightarrow VM\,VAUX$$

In dependency grammar, a dependency graph is created which consists of lexical nodes known as dependencies. Here, the lexical nodes are linked with the help
of binary relations. Words are connected by head (regent/governor) and dependent
(modifier) relations. For example, the sentence "पूजा ने खरगोश को कल देखा था"
will have dependency structure as shown in Fig. 1b.

Along with grammar formalism, some techniques like rule-based, data-driven, and
generalized approaches are used to generate the parse tree. Hindi is highly aggluti-

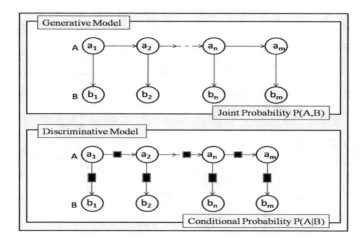

Fig. 2 Generative and discriminative model

native in nature. In comparison to phrase structure, dependency grammar formalism is efficient in handling free word order languages. Here, in this paper, a model for phrase-based shallow parsing has been developed using CRF.

2 Probabilistic Models

Generative models use joint probability distribution and based upon this distribution, the feature vectors are generated when label vectors are given, whereas conditional probability distribution is used in discriminative model for assigning label vectors directly to the feature vector. This is illustrated in Fig. 2.

Naïve Bayes classifier is based on joint probability distribution P(A, B). The assumption taken in Naïve Bayes classifier is that all the features of B are conditionally independent, if once the label class A is known. The extended form of Naïve Bayes is HMM. HMM is used for structured data which is sequential and also show the dependencies of A and B variables as joint probabilities. The computation of joint probability is complex. HMM is a generative model. The maximum entropy model (ME) uses conditional probability and is a discriminative model like Naïve Bayes, and it also classifies a class A when several features of B are given. If a classifier learns from the unknown values which are considered to be in a Markov chain, they are classified as MEMM which is the extension of ME. CRF model is also the sequential extension of ME. CRF was introduced by Lafferty et al. [2]. CRF is a probabilistic framework for probability computation p (A|B) of output $A = (a_1, a_2, \ldots, a_n) \in A_n$ with the given observation $B = (b_1, b_2, \ldots, b_n)$. CRF is used for labeling and segmenting. With the help of CRF, many tasks of NLP, for example, chunking and part of speech (POS) tagging can be done. Shallow parsing based on CRF has

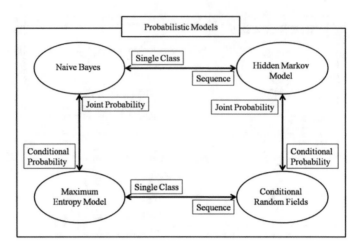

Fig. 3 Probabilistic models

an advantage of utilizing and giving best of generative and discriminative models. Figure 3 shows the overview of probabilistic models.

3 Related Works

The inference problem of HMM and label bias problem of MEMM have given motivation to Lafferty et al. [1] to develop a probabilistic framework named as CRF. In Rao and Yarowsky [3], enhanced the features by using CRF and have shown that how enhancing features can improve machine learning. In the work, shallow parsing is treated as a knowledge discovery process. Also, Gahlot et al. [4] compared in terms of accuracy with support vector machine and maximum entropy model, CRF performed better for Hindi language. An approach for performing shallow parsing by using CRF is proposed by Sha and Pereira [5]. CRF were used to do the chunking of Manipuri in Nongmeikapam et al. [6, 7] presented a deterministic dependency parser which is based on memory-based classifier and makes use of linear time algorithm for parsing to parse English text by using Penn Treebank Data. Ghosh et al. [8] have discussed about their work which was part of the NLP Tool contest at ICON 2009. They have used statistical CRF-based model and rule-based post-processing. The accuracy obtained is 69% during CRF run. Bharati and Sangal [9] have built a parser for Hindi using A Karaka-based approach.

Fig. 4 Sample of the input
file used for model training

पंद्रह	QTC	B−NP	
मिनट	NN	I−NP	
में	PSP	I−NP	
हम	PRP	B−NP	
रौस	NNP	B−NP	
पहुँच	VM	B−VP	
गए	VAUX		I−VP
।	PUNC	O	
लोहे	NN	B−NP	

4 Development of Proposed Model

The work is done by using CRF++ package.[1] The developed system is divided into several phases. First, around 1,000 raw sentences having approximately 13000 words were considered. For creating an input file, these sentences were chunked by the help of rule-based chunker developed by Asopa et al. [2]. The generated chunks were analyzed and the chunks which were found incorrect were corrected with the help of gold annotated data. The generated input file with a template file was then given as an input to the CRF++ package to train and develop a model. The sample of the input file is used to train the model as shown in Fig. 4.

The model is trained using IOB tagging for giving chunk tags. Table 1 shows the description of tags.

After development, the model is evaluated by giving a test file in the format accepted by the model. Table 2 gives the detail of the total number of chunks and the correct chunks generated by the model.

A template file is used to generate a set of feature functions and consists of unigram template for defining unigram features and bigram template for defining bigram features. Unigram template generates the feature functions by using Eq. 1.

$$\text{Number of feature functions} = R * Q. \tag{1}$$

Here, the total output classes are represented by R and total unique string which are expanded by unigram are Q. Bigram template generates the distinct features by using Eq. 2.

---

[1] https://taku910.github.io/crfpp/.

Table 1 Description of tags in IOB tagging

Phrases	Chunk tags
Noun phrase	Beginning chunk tag B-NP
	Intermediate chunk tag I-NP
Verb phrase	Beginning chunk tag B-VP
	Intermediate chunk tag I-VP
Conjunction phrase	Beginning chunk tag B-CCP
	Intermediate chunk tag I-CCP
Adjective phrase	Beginning chunk tag B-JJP
	Intermediate chunk tag I-JJP
Adverb phrase	Beginning chunk tag B-RBP
	Intermediate chunk tag I-RBP
Outside	Outside chunk tag O

Table 2 Analysis of total and correct chunks

Chunk names	Total chunks	Correct chunks
Noun phrases	5523	5390
Verb phrases	1393	1383
Adverb phrases	13	13
Adjective phrases	59	59
Conjunction phrases	331	326

$$\text{Total distinct features} = R * R * Q. \qquad (2)$$

Here, the total output classes are represented by R and total unique string which are expanded by bigram are Q.

The model is trained by giving the input file of around 1000 sentences and number of features. Figure 5 illustrates the complete process.

5   Evaluation

The developed model was evaluated by taking 864 sentences. The total chunked tags generated by the model were 7314 and total tags which should be given according to gold standard data were 7314. The result was compared with gold standard data and this was found that correct tags generated by the model were 7171. For calculating accuracy, three major parameters which are considered as standard indicators namely precision, recall and F-measure were calculated by using the Eqs. 3, 4, and 5, respectively.

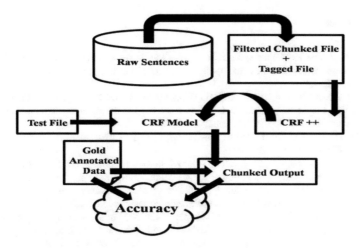

Fig. 5 Steps in the development and evaluation of CRF model

Table 3 **a** Precision. **b** Recall. **c** F-measure

(a)

Precision (%)	Correct matches	System generated
98.04	7171	7314

(b)

Recall (%)	Correct matches	Total outputs
98.04	7171	7314

(c)

F-Measure (%)	2 * P * R	P + R
98.04	19223.68	196.08

$$\text{Precision (P)} = \text{Correct Matches/System Generated.} \quad (3)$$

$$\text{Recall (R)} = \text{Correct Matches/Total Output.} \quad (4)$$

$$\text{F-Measure} = (2 * P * R)/P + R. \quad (5)$$

All the three measures were found to have the same score, i.e., 98.04 (Table 3).

6 Conclusion

This work is an attempt to perform shallow parsing by using a probabilistic framework known as conditional random fields and have obtained precision, recall, and F-measure as 98.04, 98.04, and 98.04, respectively. Although the system has shown good accuracy, it can further be enhanced by introducing more features like tense, aspect, and modality.

References

1. Asopa S, Asopa P, Mathur I, Joshi N (2016) Rule based chunker for Hindi. In: 2016 2nd international conference on contemporary computing and informatics (IC3I), 14 Dec 2006. IEEE, pp 442–445
2. Lafferty J, McCallum A, Pereira F (2011) Conditional random fields: probabilistic models for segmenting and labeling sequence data. In: Proceedings of the eighteenth international conference on machine learning, ICML, vol 1, pp 282–289
3. Rao D, Yarowsky D (2007) Part of speech tagging and shallow parsing of Indian languages. Shallow Parsing South Asian Lang 8:17
4. Gahlot H, Krishnarao AA, Kushwaha DS (2009) Shallow parsing for Hindi—an extensive analysis of sequential learning algorithms using a large annotated corpus. In: IEEE international advance computing conference, 2009, IACC 2009, 6 Mar 2009. IEEE, pp 1158–1163
5. Sha F, Pereira F (2003) Shallow parsing with conditional random fields. In: Proceedings of the 2003 conference of the North American chapter of the association for computational linguistics on human language technology, vol 1. Association for Computational Linguistics, pp 134–141
6. Nongmeikapam K, Chingangbam C, Keisham N, Varte B, Bandopadhyay S (2014) Chunking in Manipuri using CRF. Int J Nat Lang Comput (IJNLC) 3(3)
7. Nivre J, Scholz M (2004) Deterministic dependency parsing of english text. In: Proceedings of the 20th international conference on computational linguistics, 23 Aug 2004. Association for Computational Linguistics, p 64
8. Ghosh A, Das A, Bhaskar P, Bandyopadhyay S (2009) Dependency parser for Bengali: the JU system at ICON 2009. NLP tool contest ICON
9. Bharati A, Sangal R (1990) A karaka based approach to parsing of Indian languages. In: Proceedings of the 13th conference on computational linguistics, vol 3, 20 Aug 1990. Association for Computational Linguistics, pp 25–29

A Model of Fuzzy Intelligent Tutoring System

Pooja Asopa, Sneha Asopa, Iti Mathur and Nisheeth Joshi

Abstract Learning is a process which is actively constructed by the learner itself. With the advent of technology in the field of Artificial Intelligence, learning is no more restricted to traditional classroom teaching. Cognitive Systems can be used to impart pedagogical education to learners. Cognitive systems are those intelligent systems which can think, decide, act, and analyze learner's learning accordingly and can help them in generating significant learning. Intelligent Tutoring System (ITS) provides its learners with a one-to-one tutoring environment where intelligent agents can act as tutors. In this paper, a model of Fuzzy-ITS is proposed and evaluated using fuzzy rule set which has capabilities to enhance the skills of learners by providing instructions and feedback.

Keywords Intelligent tutoring systems · Fuzzy-ITS · Intelligent agents
Cognitive systems · Fuzzy inference system · ITS

1 Introduction

The construction of knowledge is a recursive process that is based on our previous experiences and reflecting on those experiences, the reconstruction of knowledge occurs. Pedagogical education can be imparted to learners related to various complex topics using cognitive systems. The idea is to generate significant learning through which different mental models are constructed in different contexts. Cog-

P. Asopa (✉) · S. Asopa · I. Mathur · N. Joshi
Banasthali Vidyapith, P.O. Banasthali Vidyapith, Newai 304022, Rajasthan, India
e-mail: pooja.asopa@gmail.com

S. Asopa
e-mail: asopasneha@gmail.com

I. Mathur
e-mail: mathur_iti@rediffmail.com

N. Joshi
e-mail: nisheeth.joshi@rediffmail.com

© Springer Nature Singapore Pte Ltd. 2019
S. Bhattacharyya et al. (eds.), *International Conference on Innovative Computing and Communications*, Lecture Notes in Networks and Systems 56,
https://doi.org/10.1007/978-981-13-2354-6_32

Fig. 1 Components of fuzzy intelligent tutoring system

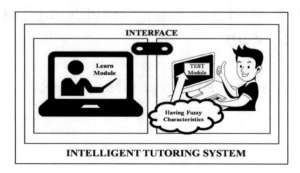

nitive systems are those systems which have immense capabilities to think, decide, act and analyze, which are similar to humans. These kinds of intelligent systems based on cognitive knowledge and realizations are modeled in the field of Artificial Intelligence. From the early 1970s, various researchers have been working on intelligent agents. These intelligent agents can act as tutors. The creation of intelligent agents leads to various challenges as agents need to be autonomous, flexible, and more adaptive as per the requirements for the better understanding of students.

ITS consists of a tutor/agent, learner, and the environment where a tutor provides a learning environment to its learners. The generation of ITS is a complex task but it can be generated with the help of authoring tools having an agent-based programmable environment where one can generate their own ITS model and can provide its learners with a one-to-one learning environment. This research paper describes the design and development of the proposed Fuzzy-ITS model. This paper is an extension of Asopa et al. [1] where Fuzzy Inference System (FIS) was proposed for a Fuzzy-ITS. The proposed model consists of learn module and a fuzzy test module as depicted in Fig. 1. The fuzzy test module consists of the fuzzy rule set which is described in Asopa et al. [1].

2   Related Works

Conati [2], conducted a survey on various challenges of Intelligent Tutoring Systems using intelligent agents. The concept of intelligent agents has emerged from artificial intelligence and cognitive science. The idea is to support students for problem-solving in various domains. These intelligent agents can act as tutors. According to the paper, the intelligent agents need to have knowledge about the domain area and should have the capability to make a student understand the complex problem by using pedagogical strategies. For imparting the knowledge to students, the intelligent agents need to have communication knowledge in order to present the information to the students. This paper has described different characteristics of intelligent agents which consist of a pedagogical model, domain model, and student model. The paper has also discussed the differences between Computer-Aided Instruction (CAI) and

Intelligent Tutoring Systems (ITS). The author claims that ITS has great potential to improve higher order teaching skills.

Yang [3], discussed the phases through which ITS has emerged. According to the author, intelligent tutoring system constitutes, five modules termed as Students module, Instruction module, Domain module, Adaptive curriculum planning module and user interface module. According to various researches on cognitive studies, students learn four times faster when they are tutored individually in comparison with the learning in traditional classrooms.

Abu-Naser et al. [4], introduced a web-based ITS for learning Java Objects (JO) known as JO-Tutor. The system includes learning material of Java Objects and has the power to generate problems automatically in order to judge the understanding of learners. The experiments were conducted on the students of CS at Al-Azhar University, Gaza and the concluded result showed a positive impact on learner's understanding. The primary objectives of the research work were to generate ITS which produced unlimited number of problems automatically and to build a system which was required to have a friendly interface with the abilities to produce learner's individual progress information.

Marcia and Mitchell [5], described the designed framework of an ITS called as CHARLIE. The idea was to support distance learning using CHARLIE. It is high-level software-based tutorial having four components, two databases, and an explanation module. The ITS was implemented for providing a distance learning course for C++ programming using six agents. The architecture of CHARLIE has four major components named as control component, instructional component, text analysis component and student modeling component. The interaction between the software agents and the operating system are handled by the control component. The aspects related to instructions per session are dealt by instructional component. The evaluation and analysis of partial syntax and semantics per session are handled by the text analysis component. The role of student modeling component is to determine the best model for learning and also to evaluate the level of learner's progress during the session.

Wang and Mitrovic [6], presented the framework for generating suitable problems for the learners using the techniques of Artificial Neural Networks for SQL-tutor. While learning in the environment provided by SQL-tutor the next relevant problem as per the learner would be generated using neural networks. This generation of the problem for the learner is selected by the agent who is modeled using ANN. The idea is to check the correctness and errors made by the learners in solving a specific problem and based upon this, the next problem would be generated. Thus, with the help of NN, performance of the students can be evaluated.

Koedinger et al. [7], discussed the outcomes of an experiment conducted on 500 high school students using Intelligent Tutoring Systems on problem-solving related to Algebra. The intelligent agent created was named as PAT. The outcomes of the experiment have shown a remarkable difference in traditional classroom learning and the learning of students by intelligent agent (PAT).

Reiser et al. [8], described the intelligent tutoring system for LISP programming. According to the paper, the intelligent agent helps students to learn LISP programming using pedagogical principles. The intelligent tutor has capabilities to help students built programs in LISP by providing instructions and by providing feedback to students. The paper presents the methodology based on three modules called as Ideal student module, Bug catalog knowledge module [9], and Tutoring control module. The ideal student module consists of the knowledge related to the domain for problem-solving and the bug catalog module handles the common mistakes made by the students. The tutoring control module consists of pedagogical strategies through which the tutor generates the interaction with the student.

3   Design of Fuzzy-ITS Model

The proposed model consists of a learner, a learn module, and a fuzzy test module. For the development of the proposed model, 20 agents were created [10]. The below sections of this research paper discuss the proposed model, learn module, and the fuzzy test module.

3.1   Proposed Model

The proposed ITS is incorporated in fuzzy inference system developed by Asopa et al. [1]. The proposed model presently consists of two levels each for the learn module and the fuzzy test module. The learn module helps the learner to understand the topic thoroughly. After learning, the learners can also test their skills by clicking on test button. The fuzzy test module counts on each and every step of the learner whether it was a correct or wrong response. The fuzziness is categorized with the help of fuzzy rules and based upon their assessment, the graph plot of a respective learner is generated.

3.2   Learn Module

The learn module consists of tutorials which helps the learners in understanding various concepts in C language. Presently, the developed model focuses on the basic concepts of C; like data types, logic building, operators, and decision-making. Here, understanding the basics of data types and some level of logic building like swapping has been categorized in the first level and the concepts related to operators and decision-making are categorized in the second level. Figure 2 illustrates the learn module of the developed model which consist of a tutor window, various buttons, the display window, and the rules window.

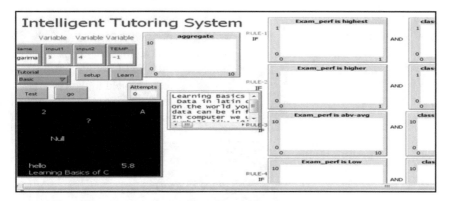

Fig. 2 The learn module of the developed intelligent tutoring system

3.3 Fuzzy Test Module

The fuzzy test module is designed to evaluate the skills acquired by learners during their learning phase. Here, if the learner qualifies in the basic level, only then, the learner is promoted to next higher level. The fuzzy test module consists of three phases, the first two phases belong to the basic module of learning and the third phase belongs to the next higher level of learn module. In the first phase, we have evaluated our learner by generating basic questions on data types. If the learner answers them correctly, then the learner is promoted to the second phase of the fuzzy test module. However, if the learner finds difficulty in answering the questions correctly that means the student has a certain amount of fuzziness in her understanding level. Here, the system can interact with the learner and help her in understanding the concept more clearly. On each and every step of a learner, agents of the proposed model motivate the learners by their encouraging words. Figure 3 depicts the first phase of the fuzzy test module.

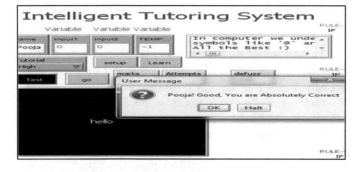

Fig. 3 The first phase of fuzzy test module in the developed Fuzzy-ITS

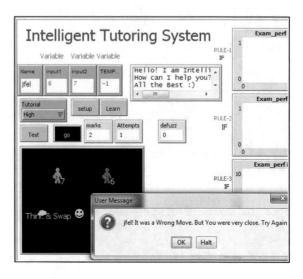

The second phase deals with the logic building part of the learner. Here, the learner is assigned a task to input two numbers and swap them by using the third variable. If the learner has taken the right steps for swapping, then the agents would change their positions accordingly, and the learner would be promoted to the next higher level. However, if the steps taken by the learner are not appropriate that means the learner is not clear with the concepts thus, the learner has to repeat this level again. Figures 4 and 5 illustrate the steps of the learner in the second phase. Each time with the assigned question at every level, the learner has three attempts to improve her skills. For the evaluation of learner, each and every step of the learner is counted upon and the value associated with it is stored in the variable named as "marks". In the third phase, the learner is assigned a question on nested conditional operator and based upon this, the learner is evaluated.

4   Results and Discussions

The developed model is evaluated on 50 students and the results have shown satisfactory improvements in the performance of learners. The evaluation of the proposed system is done by considering the class record and exam performance as inputs and based on this student performance is estimated. For the evaluation, we took a small sample of 50 students and assigned class record as input from their previous class assessment. The exam performance is estimated by the current assessment of the learner while she is performing on the proposed model. Considering these inputs, the student performance is estimated using the fuzzy rules. Some sample fuzzy rules for the "Less Fair", "Poor", and "VPoor" learner's are shown in Table 1.

Fig. 5 The test module in the system indicating the wrong move of learner

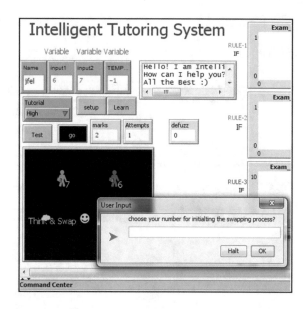

Table 1 Sample fuzzy rules

Input 1	Operator	Input 2	Result
EP is Abv Avg	AND	CR is unsatisfactory	SP is less fair
EP is low	AND	CR is satisfactory	SP is poor
EP is low	AND	CR is unsatisfactory	SP is VPoor

EP Exam performance, *CR* Class record, *SP* Student performance

Considering these rules, the proposed system has generated the outputs and those students who were categorized as "poor" learners in their previous class assessment have raised their performance and were categorized as "less fair" learners when they were evaluated on the proposed system. This is due to the fact that the learners have a significant number of chances to improve their performance and with each and every chance the learners create a better mental model as compared to the previous one. Another important aspect for the improved performance of learners can be the motivating words of the agents which can encourage them to try and try again. In Fig. 6, rule 3 illustrates a learner having poor performance in the class assessment but the performance of the same student is improved in the proposed Fuzzy-ITS system and is now categorized as "less fair" learner. Figure 7 depicts the performance of an excellent learner.

The estimated results of the system were further evaluated by comparing them with the human assessments. Among them, 40 assessments of both matched. The accuracy of the system is found to be 80% using the given Eq. 1.

$$Accuracy = Total\ Matches/Total\ Performance\ Scores. \qquad (1)$$

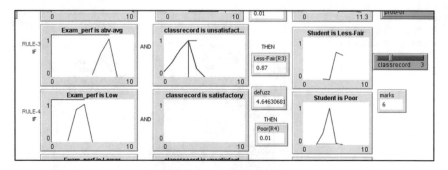

Fig. 6 The improved performance of a poor learner in the proposed system

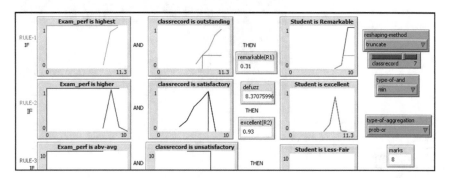

Fig. 7 The performance of an excellent learner

5   Conclusion

Understanding C language is a complex task for learners who are not from the programming background. In order to assist learners and make them understand their specific learning goals, ITS can help them by giving a dynamic learning environment. Visualization can play a significant role in constructing a better mental model of various concepts. In the proposed Fuzzy-ITS model, various agents are designed to visualize the concepts related to C Language. The fuzzy nature of the proposed system helps the learner in providing step-by-step learning as per the uncertainty levels of the learners. The proposed system is further evaluated with human assessments which give 80% of accuracy. However, by adding more rules, the accuracy of the system can be further improved. Thus, it can be concluded that both teaching and learning can be done effectively with the help of Fuzzy-ITS.

Acknowledgements We would like to acknowledge undergraduate students of BCA I year of Banasthali Vidyapith for their contribution in experimental analysis. The age group of the students was between 18 and 20, except one student having age 17.

References

1. Asopa P, Asopa S, Joshi N, Mathur I (2016) Evaluating student performance using fuzzy inference system in fuzzy ITS. In: 2016 International conference on advances in computing, communications and informatics (ICACCI). IEEE, pp 1847–1851
2. Conati C (2009) Intelligent tutoring systems: new challenges and directions. In: Proceedings of the 14th international conference on artificial intelligence in education (AIED). Brighton, England, pp 2–7
3. Yang FJ (2010) The ideology of intelligent tutoring systems. ACM Inroads 1(4):63–65. https://doi.org/10.1145/1869746.1869765
4. Abu-Naser S, Ahmed A, Al-Masri N, Deeb A, Moshtaha E, Abu Lamdy M (2011) An intelligent tutoring system for learning Java objects. Int J Artif Intell Appl (IJAIA) 2(2). https://doi.org/10.5121/ijaia.2011.2205
5. Mitchell MT (2007) An architecture of an intelligent tutoring system to support distance learning. Comput Inf 26:565–576
6. Wang T, Mitrovic A (2002) Using neural networks to predict student's behavior. In: Kinshuk, Lewis R, Akahori K, Kemp R, Okamoto T, Henderson L, Lee C-H (eds) Proceeding international conference on computers in education ICCE 2002. IEEE Computer Society, Los Alamitos, CA, pp 969–973
7. Koedinger KR, Anderson JR, Hadley WH, Mark MA (1997) Intelligent tutoring goes to school in the big city. Int J Artif Intell Educ 8:30–43
8. Reiser BJ, Anderson JR, Farrell RG (1985) Dynamic student modelling in an intelligent tutor for LISP programming. In: Proceedings of IJCAI-85, Los Angeles, CA, pp 8–14
9. Gupta D (2018) Taxonomy of GUM and usability prediction using GUM multistage fuzzy expert system. Int Arab J Inf Technol
10. Wilensky U (1999) NetLogo. Center for Connected Learning and Computer-Based Modeling, Evanston, IL

Improving Accuracy of IDS Using Genetic Algorithm and Multilayer Perceptron Network

Thet Thet Htwe and Nang Saing Moon Kham

Abstract Adoption of the use of Internet and increased in reliance on technology, the security of computer network and information system become more and more important for all types of organizations. Since the security is all-important for any types of organizations, we make an empirical study on network intrusion detection system which is one of the essential layers of organization security. In this study, genetic algorithm and multilayer perceptron network are used as methodologies. We divide the dataset into three parts according to the protocol and then applied a genetic algorithm for attribute selection. Multilayer perceptron network is used to train for each classifier. We examine performance differences between some recent research work and our system. The result shows that our protocol-based Genetic Algorithm for Multilayer Perceptron Network (GA-MLP) model has slightly increased in detection rate.

Keywords Intrusion detection system · NSL-KDD dataset · Genetic algorithm Multilayer perceptron network

1 Introduction

The Internet has completely changed the world in a way leading people to ease of access and communicate in everyday life. An increased usage of Internet results in the great concern for organizations to protect their network and private information from unauthorized and malicious users. Networks are the most favorable medium for attackers to compromise and gain access to perform malicious activities. Many organizations deploy firewalls, intrusion detection, and prevention systems, honeypots,

T. T. Htwe (✉)
Cyber Security Research Lab, University of Computer Studies, Yangon, Myanmar
e-mail: thetthethtwe@ucsy.edu.mm

N. S. M. Kham
Faculty of Information Science, University of Computer Studies, Yangon, Myanmar
e-mail: moonkhamucsy@gmail.com

© Springer Nature Singapore Pte Ltd. 2019
S. Bhattacharyya et al. (eds.), *International Conference on Innovative Computing and Communications*, Lecture Notes in Networks and Systems 56,
https://doi.org/10.1007/978-981-13-2354-6_33

as a defensive mechanism as they do know that they are expecting to be attacked in the near future. Moreover, we cannot deny that researching in the field of security especially in the intrusion detection system (IDS) is not the new one but surely necessary. An intrusion detection system (IDS) is a security mechanism that monitors the unauthorized activities and policy violations against information system. Depending on their placement and data they used for detection of attacks, IDS can be divided into Network-based IDS (NIDS) or Host-based IDS (HIDS). This proposed work will discover more about Network-based IDS (NIDS).

The placement of NIDS may be behind or in front of the firewall and can also be placed in the demilitarized zone (DMZ). In Network-based IDS, data collected from the network traffic are examined in detail in order to discover whether the network is intruded or not. NIDS can be classified into anomaly or misuse IDS depending on different analysis method that they used to detect the attacks. In anomaly-based IDS, the normal behavior or characteristics of the network are first recognized and then make analysis whether any deviation occurs. Any deviation from this normal behavior can be considered as an attack. The attractive characteristic of anomaly-based IDS is that it can detect novel attacks. As a downside, the anomaly-based IDS suffer from false alarms and difficulty in defining what is normal. In misuse-based IDS, user activities are compared with known patterns of attack which is already stored in the signature database. Upon performing a matching process, an intrusive activity can be determined. The weak point of this type of IDS is they can only recognize attacks that have a previously defined signature. However, the fast detection rate can be achieved by the use of this type of IDS.

This paper will intend to examine a part of the research work that will use to solve the security issue that we faced today. In this empirical work, the NSL-KDD dataset will be used as the training dataset for the proposed system and multilayer perceptron network will be applied as a classifier. For an attribute selection method, genetic algorithm will be deployed on the dataset which is divided based on protocols. The remains of paper are arranged as follows.

Section 2 will be related work. The study of the intrusion dataset is discussed in Sect. 3. The consecutive section, Sect. 4 is a talk on methodologies used in this study. In Sect. 5 comparative result will be mention in term of accuracy. Section 6 will point out the conclusion and also information resources linked to this paper are added as a reference.

2 Related Work

Some of the research work on intrusion detection system with NSL-KDD is mentioned in this section. An intrusion system that combines a swarm intelligence and Random forest data mining technique is proposed by S. Revathi and Dr. A. Malathi in 2014. In their proposed work Simplified Swarm Optimization (SSO) is used to find the more appropriate set of attributes for classifying network intrusions and Random Forest (RF) is used as a classifier. In the preprocessing step, dimensionality reduction is come up with the proposed SSO-RF approach and an optimal set of features is identified. The overall accuracy is 98.72% and compared with other methods: Particle Swarm Optimization (PSO)-RF and Simplified Swarm Optimization (SSO). The empirical results show that their research work gives a better performance than other approaches for the detection of all kinds of attacks present in the dataset [1].

In 2015, Y. P. Raiwani, and S. S. Panwar make an experimental work that uses four different neural networking algorithms with Correlation-based feature selection (Cfs) Subset data reduction algorithm using Waikato Environment for Knowledge Analysis (WEKA) data mining tools to evaluate the performance of each algorithm. The Cfs Subset reduced 42 attributes of the NSL-KDD full dataset into six selected attributes and the reduced dataset is applied to each neural network algorithms: Logistic regression, Voted perceptron, Multilayer perceptron, and Stochastic gradient descent. The various performance measure results are presented and all classifiers have the favorable accuracy of nearly 90 and over 90%. According to the Kapa Statistic, multilayer perceptron with Cfs Subset evaluator (Cfs-MLP) has an accuracy of 98% [2]. The difference with our proposed work is that we make use of protocol-based and we use the genetic algorithm as attribute selection method.

Another research work on intrusion detection system is the enhancement of network attack classification using particle swarm optimization (PSO) and multilayer perceptron (MLP). It has been initiated by I. M. Ahamed in 2016. In his research work, PSO has been used to improve the learning capability of the MLP by setting up the linkage weight in an attempt to enhance classification accuracy of the multilayer perceptron network. The percentage of classification reached is 98.9% with 1.1 false alarm rate [3].

3   Intrusion Dataset

In this paper, we use NSL-KDD dataset for the training of our proposed system. NSL-KDD is a refined version of KDD99 which is artificially created at MIT Lincoln Labs by using the tcpdump for evaluation of intrusion detection system. The most attractive fact for the researcher on NSL-KDD dataset is that it does not contain redundant and duplicate records which can lead to negative effect on training process of a classifier.

3.1   Dataset Description

In NSL-KDD, both train set and test set are provided. It consists of 41 features and one class attribute which defines the connection attacks or normal. The overall true classes of the NSL-KDD data set, both training, and testing are shown in Table 1.

Attacks in NSL-KDD fall into four main categories: Denial of Service (DoS) attack, User to Root (U2R) attack, Remote to Local (R2L) attack, and Probing attack. In DoS attack, the attacker makes some computing or memory resource too busy or

Table 1 NSL-KDD dataset

NSL-KDD dataset	Normal	Attack	Total_Occurance
KDD_Train+	67343	58630	125973
KDD_Test+	9711	12833	22544

Table 2 List of attacks in each protocol-dataset of KDD Train+

Protocol	Attack types	Total number of attack	Attack categories
TCP	neptune, warezclient, potrsweep, satan, nmap, ipsweep, back, guess_passwd, ftp_write, multihop, rootkit, buffer_overflow, imap, warezmaster, phf, land, loadmodule, spy, perl	19	DoS, Probe, U2R, R2L
UDP	teardrop, satan, nmap, rootkit	4	DoS, Probe, U2R
ICMP	ipsweep, nmap, smurf, pod, satan, portsweep	6	DoS, Probe

too full to handle legitimate requests or denies legitimate users access to a machine. In U2R attack, the attacker gains access to normal user and then make use of the vulnerability in order to get the root access to the system. In R2L attack, the attacker has the ability to send packets to a machine over a network but does not have an account on that machine and exploits some vulnerability to gain local access as a user of that machine [4]. In probing, the attacker collects information about the targeted network by sending various packets so that the attacker can make a decision to defeat the security system of the victim network.

3.2 Attack Types Based on Protocols

The network traffic collected in NSL-KDD dataset includes three different protocols: transmission control protocol (TCP), user datagram protocol (UDP) and Internet control message protocol (ICMP). According to the observation of NSL-KDD train set and test set based on categorized protocol, there are 19 different attacks types in TCP and four different types of attacks in UDP, and six different attacks types in ICMP. The list of attack types that contain in NSL-KDD listed according to the protocols they used is represented in Table 2.

The most susceptible protocol that attackers used to launch an attack is TCP and it is apparent that all four types of attack categories occur on this protocol. Although we need strong and robust communication protocols for transmission packets from one system to another on the network, the ease of use and transparency the existing

Table 3 Subsets of NSL-KDD dataset

Protocol	Total records	Normal (%)	DoS (%)	Probe (%)	U2R (%)	R2L (%)
TCP	102689	52.1964	41.5351	5.2343	0.0068	0.9329
UDP	14993	82.9320	5.9494	11.0985	0.0200	0
ICMP	8291	52.7882	34.3384	49.8734	0	0

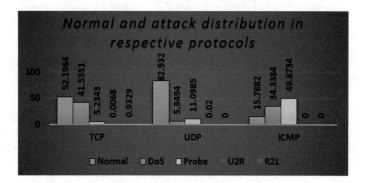

Fig. 1 Normal and attacks distribution in each protocol-dataset

protocols make vulnerable to the network. This vulnerability becomes the entry point for the attacker to launch attacks. For example, SYN flooding attacks take the advantages of a flaw in TCP handshake. Since attackers may exploit the vulnerability of the interested network by using protocols-weakness, we make a consideration of protocols and its related attributes.

As each protocol could have its own vulnerability our idea is to build a learning model based on separated protocols datasets. We separate the NSL-KDD dataset into three subset-datasets. By applying Genetic-based attribute selection algorithm to each subset of the dataset, we can get strong and more related attributes for detection. Total records of each subset-dataset are shown in Table 3.

The normal and attack distribution of each protocol-dataset are shown in Fig. 1. The dataset separation is done with the help of Linux console terminal. Majority of attack patterns in TCP dataset is DoS attack and very few U2R attacks contain in this dataset. Probing is the distinct attack in ICMP dataset. We should also notice that test set and train set does not have the same probability distribution. We can also clearly see that normal and attack records in each protocol-dataset are not fairly distributed.

Table 4 Features selected by GA

Protocol-dataset	Selected features	Total number of features including class label
TCP	4, 5, 6, 7, 12, 13, 22, 24, 25, 29, 30, 31, 33, 35, 36, 38, 39	18
UDP	3, 5, 6, 8, 14, 23, 24, 29, 31, 34, 40	12
ICMP	3, 5, 7, 23, 24, 29, 30, 31, 32, 34, 35, 36, 37	14

4   Methodology

In this paper, we used two algorithms: genetic algorithm and multilayer perceptron network to construct our proposed work. The Genetic algorithm is used in feature selection process and multilayer perceptron network is used for classification purpose.

4.1   Feature Selection Using Genetic Algorithm

The feature selection is an imperative machine learning technique that is significant and considerable in building a better classification system. By reducing features, we can get more desirable results with lower computation costs and improved classification performance [5]. Here we use Genetic Algorithm (GA) as feature selection method. GA is a part of an evolutionary algorithm that mimics the concept of natural selection in which select the best and discard the rest principle is practiced.

GA solve the optimization problems of the NSL-KDD dataset in the following stages.

1. Randomly initialized the population of interested search space.
2. Solutions with the best fitness values are selected.
3. Selected solution is remixed using mutation and crossover operators.
4. Insert new generation into a population.
5. If the stopping criterion is met, then return the solution with their best fitness values. Else go to stage 2.

Table 4 shows the most relevant features that are intelligent exploitation of random search by GA for each protocol-dataset. GA possess some interesting properties. One of the attraction points of GA is robustness that is the balance between the time and an ability to produce the intended result. Moreover, it is useful when the search space is very large with the number of parameters. No gradient information is required to find a global optimal or sub-optimal solution, and it has self-learning capabilities.

4.2 Multilayer Perceptron Network

For this study, we train the system with multilayer perceptron (MLP) network which actually used a supervised learning method called backpropagation algorithm. It consists of multiple layers of nodes in a directed graph and each layer is fully connected to the next layer via weights. The artificial neurons in the network are actually processing elements at which activation function is applied for mapping some weighted input to output. Since MLP is a supervised learning method, a calculated output is compared with the desired output using error function and if an error exists then this error is backpropagated from output layer to the input layer via hidden layers. The calculation of the input layer to output layer is called forward sweep and error correction from the output layer to input layer is called a backward sweep. The weight adjusting is done during the backward sweep.

The multilayer perceptron has main advantages of that they are easy to use, adaptable and good generalization capability. Intrusion detection system needs to be intelligent in order to act in a particular way like a human being. The reason why we need to get security system intelligence is that the variation in attacks from time to time and the need to lessen the administrative work. Moreover, we need to detect intrusion activities without replacing new security mechanism in spite of the advancement in attacks methods. In other words, one of the ways to achieve such kind of intelligence is the use of neural networks which have a good learning and generalization ability.

5 Result and Discussion

In this empirical work, we use Ubuntu 14.04 LTS and Waikato Environment for Knowledge Analysis (WEKA) tool. First, we separate the NSL-KDD dataset into three subset-datasets. Then apply the genetic algorithm to each dataset to get strong and more related attributes. Finally, we use the multilayer perceptron network for classification. The dataset segregation is done in Linux terminal by using the script. The construction of classifier is done in WEKA environment. WEKA, a collection of machine learning algorithms is implemented by the University of Waikato. It contains tools for data preprocessing, classification, regression, clustering, association rules, and visualization. It is an open source software that is well-suited for developing new machine learning schemes [6].

Weka supports different kinds of evaluation metrics such as Receiver Operating Characteristics (ROC), Accuracy, Kappa, Mean Absolute Error (MAE) Sensitivity, Specificity, etc. The performance measure in this study is carried out by matric called accuracy which is the number of correct predictions when expressed in percentage terms. The performance of the classifier can be visualized by confusion matrix. Accuracy can be calculated by the formula:

$$\text{Accuracy} = (TP + TN)/(TP + TN + FP + FN) \tag{1}$$

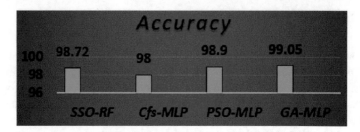

Fig. 2 Performance comparison of classifiers

TP = True_Positive;   TN = True_Negative;   FP = False_Positve;   FN = False_Negative.

TP indicates the instances which are predicted as normal correctly. FN means the wrong prediction since the classifier predicts the attack as a normal connection. FP gives a false alarm, i.e., the classifier detects attacks which are normal in reality. TN indicates instances which are correctly detected as an attack [7]. The performance comparison of the proposed work and other research works previously discussed in Sect. 2 are shown in Fig. 2. Our GA-MLP system with protocol analysis has a slight improvement in detection rate.

6 Conclusion

Some intrusion experts believe that most novel attacks are variants of known attacks and the signature of known attacks can be sufficient to catch novel variants [8]. The research area for intrusion detection continues to an active area because it is difficult to detect unknown attacks make the system to generate so many false positive and great amount of alert volume that causes a decrease in performance and accuracy of the entire system. In this paper, we present an empirical study of our system compared with other systems as part of the research work. We try to keep separate NSL-KDD dataset into three distinct protocol-datasets and then apply the genetic algorithm for a better attribute. It is because most of the attacks are tried to make use of the vulnerability in the protocol. Multilayer perceptron network is used to build our protocol-based GA-MLP intrusion detection system. The performance comparison shows that part of our system gives a slight increase in accuracy compared with other systems.

References

1. Revathi S, Malathi A (2014) IJECS 3:3873
2. Raiwani Y, Panwar SS (2015) Int J Emerg Trends Technol Comput Sci (IJETTCS) 4(1):219
3. Ahmed IM (2016) Int J Comput Appl 137(12)
4. Tavallaee M, Bagheri E, Lu W, Ghorbani AA (2009) In: IEEE symposium on computational intelligence for security and defence applications, 2009. CISDA 2009. IEEE, pp 1–6
5. Aziz ASA, Azar AT, Salama MA, Hassanien AE, Hanafy SEO (2013) In: 2013 Federated Conference on Computer Science and Information Systems (FedCSIS). IEEE, pp 769–774
6. Witten IH, Frank E, Hall MA, Pal CJ (2016) Data mining: practical machine learning tools and techniques. Morgan Kaufmann
7. Choudhury S, Bhowal A (2015) In: 2015 International conference on smart technologies and management for computing, communication, controls, energy and materials (ICSTM). IEEE, pp 89–95
8. Stolfo J, Fan W, Lee W, Prodromidis A, Chan PK (2000) Results from the JAM project by Salvatore, pp 1–15
9. Dua S, Du X (2016) Data mining and machine learning in cybersecurity. CRC Press
10. Simon D (2013) Evolutionary optimization algorithms. Wiley
11. Bassis S, Esposito A, Morabito FC et al (2016) Advances in neural networks: computational and theoretical issues. Springer
12. Bahl S, Sharma SK (2015) In: 2015 Fifth international conference on advanced computing & communication technologies (ACCT). IEEE, pp 431–436
13. Dhangar K, Kulhare D, Khan A (2013) Int J Comput Appl 65(23) (2013)
14. Bujas G, Vukovic M, Vasic V, Mikuc M (2015) Smart CR 5(6):510
15. Dhanabal L, Shantharajah S (2015) Int J Adv Res Comput Commun Eng 4(6):446
16. Low CH (2015) Github—defcom17/nsl kdd: Nsl-kdd dataset. https://github.com/defcom17/NSL_KDD. Accessed 23 Mar 2018

Analysis of Ensemble Learners for Change Prediction in an Open Source Software

Ankita Bansal

Abstract As a software system evolves, the changes are introduced at every stage of its development. The timely identification of change prone classes is very important to reduce the costs associated with the maintenance phase. Thus, the author has developed various models which can be used in the design phase (one of the early phases) to identify the parts of software which are more change prone than others. Software metrics along with the change data can be used for developing the models. In this study, the author has investigated the performance of various ensemble learners and a statistical model for identifying the classes which are change prone. The use of ensemble learners gives the researchers an opportunity to analyze and investigate them in the area of change prediction. The empirical validation is carried on five official releases of Android operating system. The overall results of the study indicate that the ensemble learners are capable of effective prediction.

Keywords Ensemble learners · Software quality · Change prediction
Validation, metrics component

1 Introduction

Introduction of changes in software can be due to various reasons such as to increase the functionality, to cater the demands of the customer, to fix defects, etc. As software progresses through its software development life cycle, it becomes more and more difficult to incorporate these changes. Thus, it becomes very costly and time consuming to incorporate a change in the maintenance phase as compared to incorporating a change during the initial phases. In this paper, the author has developed models to identify the classes which are more change prone than others so that focused attention can be given to such classes leading to saving of resources. Also,

A. Bansal (✉)
Information Technology Division, Netaji Subhas Institute of Technology, Dwarka,
New Delhi, India
e-mail: ankita.bansal06@gmail.com

© Springer Nature Singapore Pte Ltd. 2019
S. Bhattacharyya et al. (eds.), *International Conference on Innovative Computing and Communications*, Lecture Notes in Networks and Systems 56,
https://doi.org/10.1007/978-981-13-2354-6_34

correct prediction gives the alternatives for the design of the classes, which may be incorporated by the designers during early phases leading to reduction of errors in the later phases. Software metrics are widely used to measure various aspects of software such as coupling, cohesion, inheritance, etc. In literature, there are several studies which have established relationship between software metrics and change proneness, then constructed change prediction models using machine learning and statistical methods.

In this work, the author has developed various machine learning models which are primarily Ensemble Learners (EL) and compared their performance with the traditional statistical method, logistic regression. Ensemble learners combine multiple other learners which are primarily called as base learners or weak learners to solve a particular problem [1]. One of the important characteristics/advantages of ensemble learners is that they have the capability of boosting the weak learners (whose prediction capability is close to or slightly better than random guess) to strong learners [2]. There are various empirical studies of EL in different fields such as bioinformatics, medical sciences, etc. [3–6]. Due to the wide applications of EL in different fields, the author is keen to know their performance in the prediction of change prone classes. In this study, five EL are used, bagging, LogitBoost (LB), AdaBoost (AB), Random Forest (RF), and Logistic Model Trees (LMT). The dataset used for validation is Android Operating System. The author has analyzed six releases of Android and models have been validated on each release. In addition to this, inter-release validation has also been conducted.

The following Research Questions (RQ) are addressed in this study:

RQ1: How is the performance of EL in prediction of change prone classes?
RQ2: Among multiple EL discussed, which learner is suitable for change prediction of classes of Android?
RQ3: Compare the results of both the types of validation used in this study (inter-release and 10-cross validation).

This organization of the paper is as follows: In Sect. 2, the basics of the work are explained where the focus is on the variables and the dataset used. In the next section, the author has explained the data analysis methods and the performance measures used to evaluate the models. The results of validation are discussed in Sect. 4. The work is concluded in Sect. 5.

2   Basics of the Work

This section gives the brief description of the independent and the dependent variables used. The data used for empirical validation is also described.

Table 1 Univariate results

Metrics	Android 2.3	Android 4.0	Android 4.1	Android 4.2	Android 4.3	OO paradigm
CBO	++	++	++	++	++	Coupling
RFC	0	++	0	0	0	
LCOM	++	++	++	++	++	Cohesion
DIT	–	–	–	–	–	Inheritance
NOC	++	0	++	0	0	
NOA	++	++	++	++	++	Size
NPRM	++	++	++	++	++	
NPROM	++	++	0	++	++	
NPM	++	++	++	++	++	
LOC	++	++	++	++	++	
WMC	++	++	++	++	++	
NOM	++	++	0	++	++	
NIM	++	++	++	++	++	
NIV	++	++	++	++	++	
NLM	++	++	++	++	++	

2.1 The Variables

The independent variables used in this study are various metrics which represent different characteristics of OO paradigm. In this study, the famous CK metrics [7] are used along with some additional metrics which measure size in different ways. They are specified in Table 1.

The dependent variable used is change proneness. Change proneness is defined as the prediction of change in the successive or future versions of software based on the changes in the current or previous version.

2.2 Data Used for Empirical Validation

For empirical validation, Linux Operating System, Android, written in Java programming Language is used. Nowadays, Android is being widely used in mobile devices as well as computers. There are a number of versions of Android released in market, the latest being Android 8.0. In this study, the first six stable and popularly used versions, i.e. 2.3, 4.0, 4.1, 4.2, 4.3 and 4.4 have been used.

This work may be extended to incorporate the other versions as well. Data collection is done with the help of a tool developed by the author which is known as CRG tool (Change Report Generator). The detailed description of the tool can be obtained from Malhotra et al. [8]. The source codes of all these versions are available at

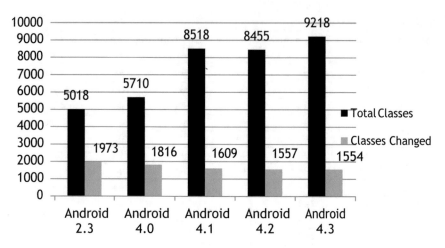

Fig. 1 Details of software used

http://source.android.com/source/initializing.html. Figure 1 shows the detail of each version. In the figure, "Total Classes" demonstrate the number of common classes between the successive versions.

3   Research Methodology

In this section, various data analysis methods and the performance evaluation measures used in this study are discussed.

3.1   Data Analysis Methods

Statistical Model

The statistical model used in Logistic Regression (LR). In this study, two types of LR, univariate and multivariate LR are used [9]. Univariate LR is used to find the association between each individual independent variable with the dependent variable. Multivariate LR is used for model prediction.

Ensemble Learning Techniques

The author has used 5 EL in this study for model construction. Their brief description is provided in this section. The default settings of Weka 3.6 software have been used to construct the prediction models.

1. Bagging: In Bagging, random samples with replacement are selected from a given dataset. These samples constitute a new dataset. Training is conducted on this new dataset multiple number of times. The result of classification is achieved from all the classifiers. Then, a majority vote is taken as the final result of classification.
2. Adaboost (AB): In Boosting, each instance in the training set is assigned identical weight which is equal to $1\backslash S$, where S is the size of the training set. Then, the classifier is trained and the weights of the misclassified instances in increased. This ensures that the classifier focuses on hard examples. This process is repeated n number of times and the majority vote is taken as the final result of classification.
3. Logitboost (LB): LB is a variant of boosting which uses regression function as the base classifier.
4. Random Forest (RF): Random forest creates a number of decision trees which is known as a forest. For creating each decision tree, it uses a subset of training data which is randomly chosen. When any unknown input is given to the random forest, each tree produces the result (known as vote) or the output class for that unknown input. Then, the votes of all the trees are aggregated to produce the final class of the unknown input.
5. Logistic Model Trees (LMT): LMT is a combination of decision tree which implements logistic regression to perform classification.

3.2 Performance Measure Used

The author has used area under the Receiver Operating Characteristics Curve (AUC) to evaluate the results. The ROC curve is plot of sensitivity and (1-specificity) on y and x-axis respectively. The area under the curve is the measure of accuracy of prediction. Thus, higher area is desirable for high accuracy of correct prediction.

To avoid the problem of overfitting, testing is not performed on the same dataset on which training has been done. To achieve this, k-fold cross validation [10] and inter-release validation have been used. The value of k is taken as 10. In inter-release validation, training and testing are performed on the different releases of the same dataset.

4 Result Analysis

In this section, the results of univariate analysis, hypothesis testing and the validation results are discussed.

4.1 Univariate Result Analysis

In this subsection, univariate analysis is conducted to analyze the relationship between each independent variable and the dependent variable. For each metric, the statistical significance (Sig.) and the coefficient (β) have been found. If the "Sig." value of the metric is below 0.01 (significance level), then the relationship between the metric and change proneness is considered to be strong. The coefficient (β) depicts the impact of the independent variable and its sign indicates whether the metric is associated with the dependent variable directly or in an inverse manner. It is found that all the metrics except DIT have shown positive impact. For Android 2.3, all the metrics except RFC are significant at 0.01 significance level. For Android 4.0, NOC is found to be insignificant. For Android 4.1, NOM, RFC, and NPROM are insignificant at 0.01. For Android 4.2 and 4.3, NOC and RFC are found to be insignificant. Only the significant metrics are used to construct the model.

After conducting univariate analysis; the author has shown the degree of the significance in Table 1 using the following notations. ++ shows the significance at 0.01, + shows the significance at 0.05, − shows the significance at 0.01 but in an inverse manner, − shows the significance at 0.05 but in an inverse manner, and 0 shows that the metric is not significant.

4.2 Discussion of Validation Results

This section discusses the validation results. Before model construction, feature selection has been done using Correlation based Feature Selection (CFS) and univariate logistic regression. CFS selects those independent variables which are correlated with the dependent variable but not amongst themselves. For all the models constructed using EL, CFS is used, whereas for the statistical model, univariate LR is used. The results of evaluation using 10-cross validation are shown in Table 2. The evaluation measure used is AUC and the highest AUC is shown in bold. It can be observed from the table that among the models of Android 2.3, Bagging has shown the best performance with the highest AUC as 0.77 followed by RF. Bagging has also shown highest performance for all the other releases of Android. RF follows bagging with the second highest AUC for all the releases except Android 4.0. For release 4.0, the AUC of bagging and RF are same. Performance accuracy obtained by the statistical model (LR) is comparable to the accuracies of other ELs. The author proposes the use of bagging and RF to the practitioners and academicians for conducting change prediction on Android or similar dataset.

Statistical Tests

The results are also statistically evaluated using a non-parametric test known as Friedman test. The result of Friedman test has given the p-value of 0.000 which is less than the chosen level of significance 0.05. This concludes the acceptance of

Table 2 Results of 10-Cross validation

Technique	2.3	4.0	4.1	4.2	4.3	Average
Bagging	**0.77**	**0.75**	**0.77**	**0.76**	**0.73**	0.756
LB	0.72	0.70	0.69	0.69	0.68	0.696
AB	0.70	0.67	0.69	0.69	0.67	0.684
RF	0.76	**0.75**	0.76	0.75	0.72	0.748
LMT	0.74	0.69	0.73	0.68	0.68	0.704
LR	0.71	0.66	0.66	0.69	0.69	0.682

Table 3 Ranks given by friedman test

Technique	Ranks
Bagging	1.1
LB	4
AB	5.1
RF	1.9
LMT	4.1
LR	4.8

Table 4 Inter-release validation results

Technique	2.3 on 4.0	2.3 on 4.1	2.3 on 4.2	2.3 on 4.3	Average
Bagging	**0.72**	**0.71**	**0.71**	**0.72**	0.715
LB	0.68	0.69	0.69	0.7	0.690
AB	0.67	0.67	0.68	0.69	0.678
RF	**0.72**	**0.71**	**0.71**	**0.72**	0.713
LMT	0.68	0.68	0.65	0.67	0.670
LR	0.68	0.66	0.68	0.69	0.678

alternate hypothesis and rejection of null hypothesis. In addition to p-value, Friedman test also gives rank (lowest rank indicates the best performance) which indicates the performance of the techniques. The rank of the technique is shown in Table 3. It can be observed that the Friedman test also shows the best performance by Bagging (lowest rank) and then, RF.

Inter-Release Validation

The classification accuracy of the models is also determined using inter-release validation. In inter-release validation, the models are first trained using Android 2.3 and then these models are used to identify change prone classes of the other releases (4.0, 4.1, 4.2, and 4.3). Table 4 shows the results of validation. It can be observed that the results are comparable to the results of 10-cross validation with bagging and RF outperforming the other models. Thus, to identify the change prone classes of any future release of Android, the researchers may use bagging and RF with the expectation of achieving approximately the comparable accuracy.

5 Conclusion

The importance of predicting change prone parts of software has been well stated. It allows the managers to allocate the resources judiciously among their team members. In other words, more number of resources in terms of time, money, and manpower can be allocated to the change prone classes as compared to the other classes. In addition to this, accurate prediction helps designers to suggest alternate designs, if required. With this motivation, the author has developed several ensemble learners to predict change prone classes. A statistical model is also developed using logistic regression and the performance of ensemble learners is compared with logistic regression. The author has used 10-cross validation as well as inter-release validation to bring out the results. The area under the curve obtained using 10-cross validation is highest for bagging followed by random forest among all the other ensemble learners. The area under the curve of different models obtained using inter-release validation is comparable with the area under the curve obtained using 10-cross validation. This proves the effectiveness of bagging and RF to identify the change prone classes of any version of android which may be released in future. The results are also statistically evaluated using the Friedman Test. The results of Friedman test also showed the superior performance of bagging among the other ensemble learners.

This work can be extended by incorporating the newer releases of Android. This will lead to generalized and more conclusive results. In addition to inter-release validation, cross-project validation may also be conducted by including additional datasets.

References

1. Zhou ZH (2012) Ensemble methods: foundations and algorithms. Chapman and Hall/CRC
2. Schapire RE (1990) The strength of weak learnability. Mach Learn 5(2):197–227
3. Dietterich T (2000) Ensemble methods in machine learning. In: Multiple classifier systems, pp 1–15
4. Nagi S, Bhattacharyya D (2013) Classification of microarray cancer data using ensemble approach. Netw Modeling Anal Health Inf Bioinform 2(3):159–173
5. Chen T (2014) A selective ensemble classification method on microarray data. J Chem Pharm Res 6(6):2860–2866
6. Dittman DJ, Khoshgoftaar TM, Napolitano A, Fazelpour A (2014) Select-bagging: effectively combining gene selection and bagging for balanced bioinformatics data. In: IEEE international conference on bioinformatics and bioengineering, pp 413–419
7. Chidamber SR, Kemerer CF (1994) A metrics suite for object-oriented design. IEEE Trans Softw Eng 20:476–493
8. Malhotra R, Bansal A, Jajoria S (2016) An automated tool for generating change report from open-source software. In: IEEE international conference on advances in computing, communications and informatics, pp 1576–1582
9. Hosmer D, Lemeshow S (1989) Applied logistic regression. Wiley, New York
10. Stone M (1974) Cross-validatory choice and assessment of statistical predictions. J R Soc Ser A 36:111–114

A Robust Framework for Effective Human Activity Analysis

Awadhesh Kr Srivastava, K. K. Biswas and Vikas Tripathi

Abstract Human activity analysis is an interesting and challenging problem among the researchers of computer vision area. The applications of human activity analysis are monitoring and surveillance. There are various surveillance approaches available in the literature for witnessing activities, events, or persons. In this paper, we present a robust framework for human action analysis. In the proposed framework, we extract the features named as generate motion image from frames deviation. Random forest is used as a feature classifier. To show the robustness of the proposed framework, we analyze and classify the publicly available HMDB dataset. The average accuracy of classification is 46.83% achieved.

Keywords Computer vision · Human activity recognition · Random forest

1 Introduction

Computer vision is concerned with the analysis and understanding of images in terms of the properties that are present in the image frames. The major promising utilization of the computer vision includes the detection and identification of human activities. Human activity recognition is an efficient approach to perceive and examine the activities of individuals in camera encapsulated matter. While recognizing the

A. K. Srivastava (✉)
CSE Department, Uttarakhand Technical University, Dehradun, India
e-mail: srivastava_awadhesh@yahoo.co.in

A. K. Srivastava
IT Department, KIET, Ghaziabad, India

K. K. Biswas
CSE Department, Bennett University, Greater Noida, India
e-mail: kanad.biswas@bennett.edu.in

V. Tripathi
CSE Department, GEU, Dehradun, India
e-mail: vikastripathi.be@gmail.com

© Springer Nature Singapore Pte Ltd. 2019
S. Bhattacharyya et al. (eds.), *International Conference on Innovative Computing and Communications*, Lecture Notes in Networks and Systems 56,
https://doi.org/10.1007/978-981-13-2354-6_35

human activity, motion in the body is more important key compared to the structure of the human body. Decision rule can be used to extract features from the images sequences [1]. By taking the difference between two-pixel values in consecutive frames, human activities can be determined. There are numerous types of methodologies for the recognition of the different levels of activities. Some techniques used for vision-based activity recognition are based on background subtraction [2]. The ability to recognize the simple and complex human activities allows us to use activity recognition feature in various fields. The applications of the human activity recognition include animation and synthesis. Some other examples which are closely related to the action and activity recognition are classification of human actions. Even though a substantial advancement has been made in vision-based human activity recognition but still, it is far away from reaching on top in its league. We described in this paper of our efforts to present a novel method for the generation of a motion image using deviation frames with less computational power consumption. The paper is structured as follows: we first provide the previous work performed on the vision-based action recognition in Sect. 2. In Sect. 3, methodology of the proposed work has been explained. Section 4 contains the analysis of the result and Sect. 5 gives conclusion of the paper.

2 Literature Review

In the domain of computer vision, video surveillance has been one of the most vigorous research areas. Security and protection in various aspects have been enhanced by the video-based surveillance systems. The human motion analysis using computer vision is evolving actively over the ages due to the availability of the more revealing vision-based algorithms and the advancement of video camera technologies. There are various approaches proposed by researchers for human activity analysis. Motion History Image (MHI) is adept for motion-based analysis as it describes the changes of some moving objects over the image sequence. The analysis of human activities using MHI and its variants by using intensity pixel is presented in their survey by various researchers. In survey, the researchers have also provided information regarding various descriptors which can encode human activities by its motion, shape, and orientation information such as Histograms of Oriented Flow (HOF), Histograms of Oriented Gradients (HOG), Spatio-Temporal Interest feature Points (STIP), etc. The descriptors like HOF, HOG provides effective representation and computation of human activities. To determine the interest points, a window-based descriptor is computed [3]. HOG and HOF are window-based descriptors, and the window is focused on the extracted features and after that, it is sliced into a frequency histogram which is generated from each cell of the grid [n x n], is used to show the edge orientation in the cell. Motion using Optical flow is analyzed by the HOF and it quantizes orientation of flow vectors [4]. Yussiff et al. [5] have applied HOG approach to locate people in each video frames, which is captured for surveillance. A paper by Zicheng et al. [6] in which they used HOG to detect and analysis unstructured human activity in unstructured environment. Huang [7] uses lookup table along with the method of integral image to speedup HOG. Some researchers combined two or three different-different descriptors to enhance the accuracy of their results. Optical flow model with new robust data obtained from HOG. In terms of optical flow method and frame sampling rate, the trade-off between computational efficiency and accuracy of descriptors such as HOF, HOG, and MBH are used by some researchers. Srivastava et al. [8] used local motion histograms for human action detection. The authors used random forest as their classification model. Another method to encode valuable information based on shape analysis known as Hu moments was proposed by Hu which is immune towards scaling, translation, and rotation. Tripathi et al. [9] used Motion History Image and Hu moments to extract features from the video. An algorithm in which a fusion of HOG descriptor, Hu moments, and Zernike Moments are used to detect human activity from a single point of view is proposed by Sanserwal et al. [10].

3 Methodology

The proposed methodology makes use of some computer vision-based methods for the recognition of human activity. In this method, the video is taken as an input and then converted into frames. Further, we take three consecutive frames and calculate pixel-wise Standard Deviation of all the three frames. The resultant values are provided to our new framework to extract features from the image sequences. Further, to classify the obtained features, the Random Forest classifier is used. The framework is clearly depicted in the Fig. 1. The algorithmic representation of our framework is represented in Fig. 2.

3.1 Motion Generation Image

In the presented work, for the generation of the motion image, we have used standard deviation method. First, we calculate the mean of all the values of the three consec-

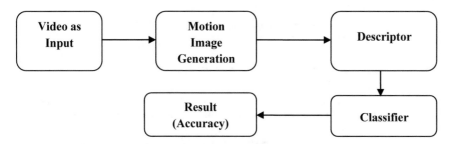

Fig. 1 Architecture of the proposed action recognition framework

Input: Video	

1. Compute frames
2. Initialize x=0
3. Initialize n=3
4. While x < frames do
    a. Initialize y=0
    b. While y < n
        i. Buffer frame
        ii. y= y + 1
    c. Calculate SD frame
    d. x= x + 1
5. Compute Descriptor
	1. Computing the frames of the video.
	2. Initialize x by 0.
	3. Initialize n by 3.
	4. Frames = Total number of frames computed.
	    a. Initialize y by 0.
	    b. Taking the Standard deviation of 3 frames only
	        i. Buffer $(x+y)^{th}$ frames
	        ii. Increment y by 1.
	    c. Increment x by 1.
	5. Computing Descriptor.

Fig. 2 Algorithm for generation of descriptor

utive frames and after that, we take the standard deviation of the three frames using Eq. (1).

$$SD = \sqrt{\sum (a - u)^2 / n} \tag{1}$$

where

a = pixel value in the frames.
u = mean of the corresponding pixel values of three frames.
n = 3 (three continuous frames each time).

3.2 Descriptor

In this approach, a new feature extraction-based framework is used to obtain motion features. The descriptor calculates x and y derivatives of image using convolution operation as shown in Eqs. 2 and 3.

$$A_x = A * p_x \text{ where, } p_x = [-1\,0\,1] \tag{2}$$

$$A_y = A * p_y \text{ where, } p_y = \begin{bmatrix} 1 \\ 0 \\ -1 \end{bmatrix} \tag{3}$$

Now, gradient and magnitude of image can be computed using Eqs. 4 and 5:

$$|M| = \sqrt{I_x^2 + I_y^2} \tag{4}$$

$$\theta = arctan\frac{I_x}{I_y}. \tag{5}$$

We have used Random Forest Classifier that works by creating multiple decision trees during training. In our case, the model has been trained using Random Forest Classifier which creates 100 trees.

4 Result and Discussion

The framework has been trained and tested using Python 2.7 and OpenCV at computer system having Intel i5, 2.4 GHZ processor with 32 GB RAM on the videos for recognizing the human activities performed in the image sequences. We have trained the framework for mining the valuable information from the sequence of images on the HMBD dataset which contains a total of 51 action classes and each action class has 100 videos. Testing is done on different videos from the one used for the training purpose. For validating our result, we have tested our system with three different sets of videos in test and train. Thereafter, we consider the accuracy of our proposed framework 46.83% by taking the average of accuracy, which can be analyzed by the Table 1. Figure 3 shows a pictorial representation of the values of instances in test and train dataset of all the three train sets in terms of the frames. We have taken approximately 1,000,00 instances for the training purpose and 40,000 for the testing purpose which shows the ratio of 60–40% as depicted in the Fig. 3.

Table 1 Accuracy of descriptor

	Accuracy (%)
Train 1	45.8501
Train 2	46.8987
Train 3	47.7030
Average	46.8272

Fig. 3 Graph between instances of test and train in terms of frames

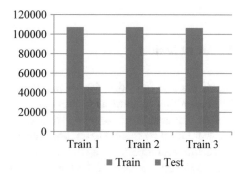

Table 2 Result analysis

	TP-rate	FP-rate	Precision	Recall	F-measure	ROC
Train 1	0.459	0.015	0.44	0.459	0.442	0.853
Train 2	0.469	0.014	0.446	0.469	0.448	0.859
Train 3	0.477	0.014	0.454	0.477	0.457	0.865

Table 3 Comparison with other algorithms (in percentage)

Algorithm used	Result (%)
Improved dense Trajectory [4]	57.2
WFDT [11]	52.1
DT [12]	46.6

Table 2 depicts the True Positive rate (TP) which shows the number of instances prophesied positive that are actually positive, False Positive (FP) means it is predicted positive but actually it was negative, precision which is predicted positive, Recall means the fraction of the predicted positive class and total positive class, F-measure consists of the combination of recall and precision to provide a single value of measurement, and Receiver Operating Characteristic (ROC) values of all 51 classes for 3 independent testing Datasets.

To observe the potential of our framework, we have performed a comparative analysis with other HMDB dataset-based algorithms used to detect human activities, as shown in Table 3. It has been observed that the Dense Trajectory method makes use of two descriptors, i.e., HOG and HOF along with the trajectory and they got an accuracy of 46.6% which is 0.2% lesser compared to the accuracy of our proposed method. Other algorithms like Improved Dense Trajectory [4] method gives 57.2% accuracy and W-Flow Dense Trajectories (WFDT) [11] gives 52.1% accuracy, which outperforms with more accurate analysis than our method but uses more number of attributes and, however, these approaches are complex and usually requires human supervision as it works on the selection of interest points. Hence, the proposed framework significantly detects the human activities by using less computation in comparison to other frameworks.

5 Conclusion

In this paper, a deviation frames based and computationally efficient human activity recognition method were introduced. The deviation-based frames capture the motion characteristics from an action sequence. Further, these motion characteristics were used by feature extraction framework and average recognition of 46.83% was achieved, outperforming the existing HMDB dataset based methods. To classify the actions on the basis of the features obtained, the Random Forest classifier was used. The real-time computation can also be implemented for more proficiency.

References

1. Veenendaal A, Daly E, Jones E, Gang Z Vartak S, Patwardhan RS (2015) Decision rule driven human activity recognition. J Comput Sci and Emerg Res 3
2. Garrido-Jurado S, Salinas R, Madrid-Cuevas FJ, Marin-jimenez MJ (2016) Automatic generation and detection of highly reliable fiducial markers under occlusion. J Pattern Recogn 47(6):2280–2292
3. Chakraborty B, Holte M, Moeslund T, Gonzalez J (2012) Selective spatiotemporal interest points. J Comput Vis Image Underst 116(3):396–410
4. Wang H, Schmid C (2013) Action recognition with improved trajectories. In: IEEE International conference on computer vision, pp 3551–3558
5. Yussiff A, Yong S, Baharudi BB (2014) Detecting people using histogram of oriented gradients: a step towards abnormal human activity detection. Adv Comput Sci Appl 1145–1150
6. Oreifej O, Liu Z (2013) Histogram of oriented 4d normals for activity recognition from depth sequences. In: IEEE conference on computer vision and pattern recognition, pp 716–723
7. Huang C, Huang J (2017) A fast HOG descriptor using lookup table and integral image. arXiv: 1703.06256
8. Srivastavac A, Biswas KK (2017) Human activity recognition using local motion histogram. In: Springer international conference on next generation computing technologies. (accepted)
9. Tripathi V, Gangodkar D, Latta V, Mittal A (2015) Robust abnormal event recognition via motion and shape analysis at ATM installations. J Electr Comput Eng
10. Sanserwal V, Pandey M, Tripathi V, Chan Z (2017) Comparative analysis of various feature descriptors for efficient ATM surveillance framework. IEEE ICCCA
11. Jain M, Jegou H, Bouthemy P (2013) Better exploiting motion for better action recognition. CVPR
12. Wang H, Klaeser A, Schmid C, Liu C (2013) Dense trajectories and motion boundary descriptors for action recognition. Int J Comput Vis

A Trust Rating Model Using Fuzzy Logic in Cloud

Vidhika Vasani and Vipul Chudasama

Abstract Cloud computing provides services from the available pool of resources. Even with the available condition, cloud computing can reach the peak of success amongst cloud user. The issue they face is the barrier of trust between the end users for using the given services. Conventional security and protection controls keeps on being executed on cloud; however, because of its liquid and dynamic nature, a testable trust estimate of the cloud is required. This research paper exhibits an analysis of the present trust administration strategies for cloud operations. In this research paper, we proposed a model for trust administration using Fuzzy Logic, which can be beneficial for cloud service providers to select trusted datacenter for consumers.

1 Introduction

Cloud computing has been widely accepted by the organization due to its multi-function's software, computation, and storage services. It is utilized to address the shortage issues of its customers by giving them on-request pay-per-use administrations. The cloud platform is usually suitable for high-performance server machines, high-speed storage devices, and an organized network. The cloud computing has been used by many leading companies like Amazon, Google, and Microsoft as they require big servers in order to keep adding the data and new customers. In Cloud, shared capacity and computing resources are provided by Infrastructure as a service (IaaS) which includes memory, hard disk space, and CPU. Evaluation for any framework is provided by Platform as a Service (PaaS). A client who requires programming and frameworks without disturbing its licenses and establishment is given

V. Vasani (✉) · V. Chudasama
Department of Computer Science Engineering, Institute of Technology, Nirma University, Ahmedabad 382481, Gujarat, India
e-mail: 16mcei26@nirmauni.ac.in

V. Chudasama
e-mail: vipul.chudasama@nirmauni.ac.in

© Springer Nature Singapore Pte Ltd. 2019
S. Bhattacharyya et al. (eds.), *International Conference on Innovative Computing and Communications*, Lecture Notes in Networks and Systems 56,
https://doi.org/10.1007/978-981-13-2354-6_36

a Software as a Service (SaaS). All these services are run on the framework of cloud providers.

In the present scenario of cloud computing, trust is considered as the main factor because it largely depends on the insight of influences and value judgment by the providers of the cloud services. Due to its computing structure, which collects huge amount of data in an isolated framework can increase the risk of data being hacked or corrupted due to any misconfigured malicious servers. Trust plays a big role in cloud as it is almost impossible to know the outcome of what could go wrong. However, trust management framework should have the capacity to deal with and maintain the trust connections between cloud specialist organizations and their consumers, which will help secure the customer's data.

2 Literature Survey

Table 1 gives literature surveys of different research papers of how they cope with cloud computing security issues and their trust management methodologies and different techniques for secure cloud.

3 Problem Definition

Trust is the most significant relationship between an organization and its client. When the customer shares their confidential data with a company, they accept all the data to be secure. Customer needs to be sure that their data will be secured with the organization they share with. This is where the trust management framework takes a big part. Trust models are created to help the customer better realize the process of how their data is being secured by the CSPs or the third-party providers that are providing the cloud services. The cloud service users (CSUs) always must keep trust on the cloud service providers (CSPs), and the CSPs must keep trust on the CSUs for a strong creation of cloud services. In the cloud computing scenarios, the CSUs give all their digital resources in the hands of the CSPs, and the CSPs hold direct control over nearly all the security factors. This is the reason why it is necessary to have proper trust management between CSP and CSUs. Trust management is an important factor to use cloud services without being concerned about data being compromised. The relationship of trust is established between the two parties by stating the process of how the data will be secured and being transparent about how their data will be used. However, there are still some concerns of trust between CSP and datacenter which are categorized as follows:

- Is it possible for CSP to evaluate trusted resources of datacenter?
- Does CSP provide trust interaction, which is useful for cloud services in datacenter?

Table 1 Literature survey

Sr. no.	Author	Implementation	Advantage/disadvantage	Future scope
1	Singh and Chatterjee [1]	Overview on fundamental highlights of the cloud computing, security issues, threats, and their answers. Furthermore, the paper portrays a few, key subjects identified with the cloud, to be specific cloud design system, administration and sending model, cloud security ideas, threats, and assaults	An advantage of brisk organization, cost proficiency, extensive storage room and simple access to the framework whenever and anyplace	The discourse of some open issues in the cloud
2	Qu and Buyya [2]	Using different Fuzzy Inference System for IaaS service selection	It enhances cost proficiency and administration dependability when utilizing mists by considering execution varieties in the choice stage	Create choice arrangements for cloud clients and intermediaries in view of our trust assessment framework to additionally mechanize cloud sending
3	Pearson and Benameur [3]	Overview on how safety, trust then protection problems happen popular the specific situation of cloud subtracting then talk about courses popular which they might remain tended to	Lessening cost by sharing, figuring, and capacity assets, consolidated with an on-request provisioning system depending on a compensation for every utilization plan of action	Adjusting to legitimate security necessities and meeting customer protection what is more, security desires
4	Supriya [4]	CSPs are rated using fuzzy logic system	Enable customers to settle on an educated decision towards choosing the suitable CSP according to their prerequisite	Expansion to rating distinctive CSPs this model can be utilized to choose an arrangement in view of the clients need
5	Merrihan et al. [5]	Review about the various trust management techniques and present trust management strategies with respect to the execution of cloud specialist organizations study over different parts of protection, security, reliability, client criticism, and so on	A trust administration framework used to assign individual weights for criticisms as indicated by the fame of their senders and as for the exchange popularity	Future scope is to implement an effective trust management framework to enable cloud to benefit for customers in distinguishing their reasonable distributed computing specialist organizations

(continued)

Table 1 (continued)

Sr. no.	Author	Implementation	Advantage/disadvantage	Future scope
6	Hwang and Li [6]	Utilizing Data coloring and programming watermarking systems to assemble a trust in cloud computing	Utilize data coloring at different security levels in light of the variable cost work connected	Development of security as an administration and information assurance as an administration for cloud
7	Huang and Mnicol [7]	Study of available instruments for launching trust, and comment on their boundaries	The integrated view of trust based on evidence and chain help to make a various decision for cloud stack holders	A framework having reasoning about trust with models, languages for calculation of trust
8	Chiregi and Navimipour [8]	Trust value is calculated using five attributes; availability, loyalty, information integrity, identity and capability	Propose a strategy for assessment pioneers and troll substance recognizable proof utilizing three geographical measurements, including response amount, yield amount and notoriety measures	plan to investigate the effect of transformative calculations on trust and notoriety assessment
9	Noor and Sheng [9]	The plan and execution of CloudArmor, a notoriety based put stock in administration system	Gives an arrangement of functionalists to guide trust as an administration (TaaS), which combine (i) the novel convention to demonstrate the validity of trust criticisms and protect clients privacy	Plan to join diverse trust administration procedures, for example, notoriety and proposal to expand the trust comes about exactness
10	Sule [10]	Displays some security components that empower cloud benefit end clients to assess the trust level of different cloud administrations and assets	Security systems are assessed in view of the fuzzy logic on a Eucalyptus cloud stage	Incorporate executing and assessing the procedures. The assessment would incorporate utilizing fuzzy rationale calculations to choose reliable cloud stage

- Due to uncertainty in trust for various administration operations, how CSP will provide trust rating to datacenters?

In response to the above concerns, a model of trust was created which requires the help of CSP to explain the relationship between trust interaction and trust management service. Figure 1 shows the model of the trust administration framework where services layer and service request layer are shown. In service layer cloud, services and trust management services are available which are connected through trust interaction.

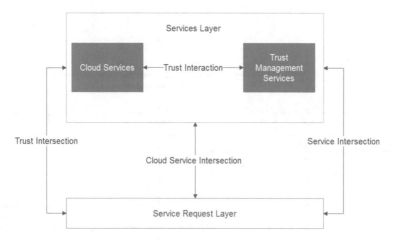

Fig. 1 Model of the trust administration framework

4 Proposed Solution

In the proposed model, it shows that CSP collects the data from the datacenter and calculate the trust value and provide fuzzy rating to datacenter. We experimented on datasets for performance analysis of our method. We practically implemented our method for trust evaluation using fuzzy logic. In Sect. 4.1, we introduce the characteristics of resources such as Availability, Reliability, Data Integrity, Identity and Capability to calculate trust value of the datacenter. Then, in Sect. 4.2, we introduced our proposed algorithm. Then, in Sect. 4.3, we discussed the result using MATLAB tool. The dataset for simulation remains available on www.dataset12.expertCloud. ir. 2000 resources were taken as a dataset with different features.

4.1 Characteristics of Resources

1. **Availability (AVL)**: The data storage and provision resources are defined by AVL, and it remains made of serviceable and reachable request by approved object. It means that the services are presented even when many nodes undergo failure (Fig. 2).
   It is the related to time of a system or component for the usefulness of total time which is mainly essential to purpose. The AVL of resources r is estimated via Eq. (1).

$$AVLr = Ar/Nr \tag{1}$$

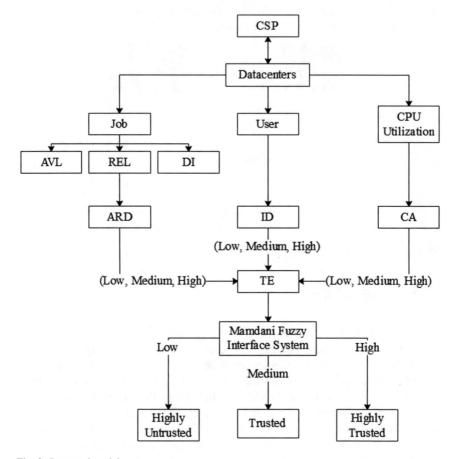

Fig. 2 Proposed model

For Nr means, the total number of jobs is given to the resource r, Ar means the total number of jobs is taken by the resource r.

2. **Reliability (REL)**: For trust, REL is an important factor. It denotes to the capacity of a software element to dependably make affording to its provisions. The REL of resources r is estimated via Eq. (2).

$$RELr = Cr/Ar \qquad (2)$$

For Cr means, the total number of jobs is completed by the resource r, Ar means the total number of jobs is taken by the resource r.

3. **Data integrity (DI)**: Security is a main problem that desires different considera-tion in the clouds. The term Data Integrity stands for confidentiality and accuracy of the information. The trust or assurance about executor's behavior is given by the honesty of the executor. Data protection and data precision are included by

security. Information cost capacity occurs due to poor network inactivity. Accuracy damage capacity occurs due to superseded calculating infrastructure. The DI of resources r is estimated via Eq. (3).

$$DIr = Dr/Cr \tag{3}$$

For Cr means, the total number of jobs iscompleted by the resource r, Dr means the total number of jobs is data integrity preserved by the resource r.

4. **Identity (ID)**: Identity of user needs to be checked through different levels of security. The level of the cloud service user has been confidential into the subsequent levels: Entity Protection level, Authorization level, and Security level. This level differs since 0–1. The ID of resources r is estimated via Eq. (4).

$$IDr = ALr + ELr + SLr \tag{4}$$

where ALr equates the authorization level of resource k, ELr equates the entity protection level of resource r, and SLr equates the security level of resource r.

5. **Capability (CA)**: The term CA of the cloud resources gives a valuable performance of data or file transfer and application performance. It is calculated through CPU utilization parameters such as processor speed of resource r (Pr) and memory speed of resource r (Mr) and network parameters such as bandwidth (Br) and latency (Lr) of resource r. The CA of resources r is estimated via Eq. (5).

$$CAr = (Br * (Pr + Mr + (1/Lr)))/(Br * (PrMax + MrMax + (1/Lrmin))) \tag{5}$$

Br equates the quantity of data moved during the duration of the rth resource, Lr equates the delay to reach rth resource, MrMax equates the maximum speed of memory that exists in the system, Lrmin equates the minimum delay that exists in any link of the system, and PrMax equates the maximum speed of processes that exist in the system.

6. **Trust Evaluation (TE)**: In the earlier section, AVLr, RELr, DIr, IDr, and CAr are found via the relationships presented. Using these values, we can find the value of TEr. The TE of resources r is estimated via Eq. (6).

$$TEr = ARDr + IDr + CAr \tag{6}$$

where ARDr represents the trust value of resource(r) based on the AVLr, RELr, and DIr.

4.2   Proposed Algorithm

Algorithm 1 Calculate Trust

Step 1: Collect the data from data centre.{ A , N , C , D , AL , SL , EL , B , L , P , M}
(where A = No. of accepted jobs, N = No. of submitted jobs, C = No. of completed jobs, D
= No. of data integrity preserved jobs, AL = Authorization Level, SL = Security Level, EL
= Entity Protection Level, B = Bandwidth, L = Latency, P = Processor Speed, M = Memory
Speed.)

Step 2: $ARD \leftarrow ((N*C) + (A*C) + (N*D))/N*A*C$ (where ARD = Availability +
Reliability + Data Integrity)

Step 3: $ID \leftarrow AL + SL + EL$ (where ID = Identity)

Step 4: $CA \leftarrow (B*P + M + (1/L)) / (B * Max(P) + Max(M) + (1/Min(L)))$ (where CA =
Capability)

Step 5: $TE \leftarrow ARD+ID+CA$ (where TE = Trust Evalution)

Step 6: *Fuzzy Rules ← TE(Low,Medium,High)*

Step 7:

if*T E > High* **then return** Highly Trusted(DC)

else

if *Low* <=TE<= *High* **then return** Trusted(DC)

else

if*T E < Low* **then return** Highly Untrusted(DC)

close;

Step 8:   **Output** ← *Highly_Trusted(DC), Trusted(DC), Highly Untrusted(DC)* (where
DC=Datacenter)

4.3   Result

The proposed model has three inputs specifically of ARD, ID, and CA, as shown
in Table 2. Each input has three-member functions. The output TE has also three-
member functions low, medium, and high. The outputs found from the FIS are Highly
Untrusted, Trusted, and Highly Trusted. Trust rating output is calculated from three
blocks ARD, ID, and CA using Mamdani FIS. However, the fuzzy model selected
here is Mamdani FIS, so the output is a crisp value, i.e., low-, medium-, and high
performance of FIS file are shown in two different figures. One figure shows the rules
of fuzzy logic for trusted value and another figure shows the surface with respect to
ARD, ID, CA inputs, and TE output. In Table 2 it shows the result of FIS file. In that
file, it shows the range of value for inputs and outputs and also shows membership
functions.

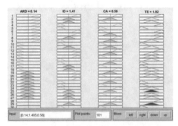

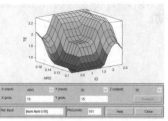

Rules   Surface

Table 2 Result

Input	Range of value	Membership functions	Output	Range of value	Membership functions and rating of DC
ARD	[0.0501, 0.1000, 0.1251]	Low	TE	[−0.0680, 0.8748, 1.6000]	Low and (Highly Untrusted (DC))
	[0.1130, 0.1370, 0.1602]	Medium			
	[0.1486, 0.1800, 0.2120]	High			
ID	[−0.1050, 0.5655, 1.0600]	Low		[1.0391, 1.9400, 2.7700]	Medium and (Trusted (DC))
	[0.7711, 1.4100, 2.0000]	Medium			
	[1.6954, 2.2935, 2.8654]	High			
CA	[0.0339, 0.2204, 0.4397]	Low		[2.1550, 3.0100, 3.7300]	High and (Highly Trusted (DC))
	[0.2820, 0.5600, 0.7994]	Medium			
	[0.6407, 0.8208, 1.1800]	High			

5   Conclusion and Future Work

In this paper, we discussed the survey of different trust management techniques for cloud computing. We have also discussed primary field related trust value for the datacenters. Trust value is estimated by applying the fuzzy logic module which can provide a guide for the CSPs to prefer a datacenter as per their requirement. However, this estimator cannot satisfy all constraints in a cloud to evaluate the datacenter. In the future, we will incorporate our algorithm in open-source cloud simulator for trust monitoring to provide the rating to the datacenter.

References

1. Singh A, Chatterjee K (2017) Cloud security issues and challenges: a survey. J Netw Comput Appl 79:88–115
2. Qu C, Buyya R (2014) A cloud trust evaluation system using hierarchical fuzzy inference system for service selection. In: 2014 IEEE 28th international conference on advanced information networking and applications (aina). IEEE
3. Pearson S, Benameur A (2010) Privacy, security and trust issues arising from cloud computing. In: 2010 IEEE second international conference on cloud computing technology and science (CloudCom). IEEE
4. Supriya M (2012) Estimating trust value for cloud service providers using fuzzy logic

5. Monir, MB et al (2015) Trust management in cloud computing: a survey. In: 2015 IEEE Seventh International Conference on Intelligent Computing and Information Systems (ICICIS). IEEE
6. Hwang K, Li D (2010) Trusted cloud computing with secure resources and data coloring. IEEE Internet Comput 14(5):14–22
7. Huang J, Nicol DM (2013) Trust mechanisms for cloud computing. J Cloud Comput: Adv Syst Appl 2.1:9
8. Chiregi M, Navimipour NJ (2016) A new method for trust and reputation evaluation in the cloud environments using the recommendations of opinion leaders' entities and removing the effect of troll entities. Comput Hum Behav 60:280–292
9. Noor TH, Sheng QZ (2011) Credibility-based trust management for services in cloud environments. In: International conference on service-oriented computing. Springer, Berlin, Heidelberg
10. Sule M-J (2016) Trusted cloud computing modelling with distributed end-user attestable multilayer security. Brunel University London, Diss

Neural Networks for Mobile Data Usage Prediction in Singapore

Saksham Jain, Shreya Agrawal, Arpit Paruthi, Ayush Trivedi
and Umang Soni

Abstract Pricing strategy is driven by prediction. Forecasting using neural networks has indicated the rise in use of artificial intelligence. The manuscript presents the prediction of mobile data usage with the use of Artificial Neural Networks. A dataset from Singapore was used for generating results for reporting the potential use of artificial intelligence in forecasting. Recently, pricing war in the mobile service industries has been contagious and has affected the decision-making of leading enterprises. The manuscript aims to report the use of neural networks for prediction and insight development. The results are generated using MATLAB, and neural network toolbox have been discussed with inferences in the manuscript.

Keywords Neural network · Artificial intelligence · Mobile data · Drill rate

1 Introduction

Mobile devices are the most vital communication instrument for this century. As its peripherals overcome the complex challenges of interconnectivity, its adaptation prediction remains a challenge. A study conducted by C.C. Yang shows that some of the factors on which smartphone acceptance is dependent for a certain population in

S. Jain · S. Agrawal · A. Paruthi · A. Trivedi (✉) · U. Soni
Netaji Subhas Institute of Technology, Dwarka, New Delhi, Delhi, India
e-mail: ayush_trvd@yahoo.com; ayushtrvd28@gmail.com

S. Jain
e-mail: jain.saksham01@gmail.com

S. Agrawal
e-mail: shreya.31ag@gmail.com

A. Paruthi
e-mail: paruthiarpit@gmail.com

U. Soni
e-mail: umangsoni.iitd@gmail.com

© Springer Nature Singapore Pte Ltd. 2019
S. Bhattacharyya et al. (eds.), *International Conference on Innovative Computing
and Communications*, Lecture Notes in Networks and Systems 56,
https://doi.org/10.1007/978-981-13-2354-6_37

Singapore are age, past adoption behavior, and gender [1]. Mobile commerce being a central region of investment for innovation is now a life-changing tool for any country [2]. A study suggests that in terms of usage and exchange, printed material seems to remain the most preferred format for sharing information and electronic devices seem to be quite low in terms of preference [3]. The inclusion of such details in order to predict drill rate from primary acquired data becomes harder, and it is also indicative of modeling human choices and decision-making. Determinants for drill rate include how its use is seen and its rate of use [4]. A Bloomberg report generated in September this year reported that Singtel is creating competition in Singapore's already crowded market [5]. With relevant competition now escalating pricing war in Singapore, relying on insight from data could now ensure success. The factors that motivated mobile data usage are modeled as satisfier, whereas the factors that demotivate data usage are labeled as dissatisfier [6].

2 Artificial Intelligence

Reducing problems that can be solved with available tools in artificial intelligence is an alternative which has recently gained significant attention. With growing data, there is a growing need to determine useful inferences. Tools are now becoming more available. Communities have developed new firmware and fixed previous bugs, and now these soft tools are compatible with more operating systems. MATLAB has recently added an interface with AlexNet.

3 Methodology

3.1 Insight from Previous Reports

According to a survey conducted by Circles. Life, a mobile operator in Singapore, 67% of the surveyed Singaporeans in the age group of 16–54 felt that they were deprived of the required amount of mobile data. 60% people claimed that they required more than 6 GB of data on an average in order to be satisfied [7]. The mobile penetration rate in Singapore according to latest data is about 149.8%, which is more than double the global mobile penetration rate [8, 9].

3.2 Data

The dataset of quarterly recorded mobile data usage, from July 1, 2004 to December 31, 2016 was considered for prediction. The dataset of monthly recorded mobile penetration rate for the same time period was averaged over every 3 months to obtain a quarterly record. The source of this data is the Infocomm Media Development Authority (IMDA), under the Singapore Open Data License [10, 11]. Mobile Penetration Rate was chosen as an exogenous variable for the target variable, volume of Mobile Data Usage is in petabytes, because of a regression value of 0.99 which indicates high correlation between the two variables.

3.3 Neural Network

Remus and O'Connor [12] wrote that Neural Networks are highly effective for time series forecasting of quarterly data, more than other statistical methods. The nonlinear autoregressive neural network was proven to perform better than persistence models for multi-step-ahead forecasting by Ahmed and Khalid in their work. Differencing data were substantiated to be the best method for building an efficient NN model. Previously, this toolbox was used by Qi and Zhang for their time series forecasting model in 2008 and its effectiveness was confirmed by Narayanan Manikandan and Srinivasan Subha in their work in 2016. Manikandan and Subha used the NARX (nonlinear autoregressive network with exogenous inputs) model for a robust model. Therefore, for this time series prediction, the NARX library was later used due to the presence of an exogenous (independent) variable. In their model, they chose 25 neurons (hidden layer) further high accuracy was achieved with a delay of 20.

4 Single-Step and Multi-step Prediction

4.1 Nonlinear Autoregressive (NAR) Neural Network Model

4.1.1 Single-Step-Ahead Mobile Data Usage Forecasting Over 50 Quarters

The NAR architecture was used. A statistically significant feedback delay, as indicated by the target autocorrelation function, is added to the network. The Levenberg–Marquardt (LM) training method is chosen as the optimization algorithm. For the purpose of forecasting, the historical data is divided randomly into three parts: 65% for training, 15% for validation, and 20% for testing. It was observed that a variation of 5–10% in the division of data did not have a significant effect (Fig. 1).

The single-step time series response observed after inputting the data into the NAR network is shown in Fig. 2.

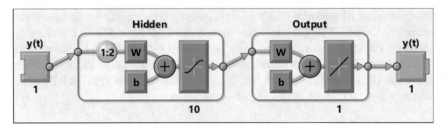

Fig. 1 Architecture for the selected NAR neural network

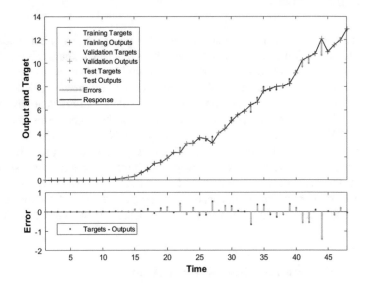

Fig. 2 Time series prediction response for mobile data usage over 50 quarters

4.1.2 Multi-step-Ahead Mobile Data Usage Forecasting Over 50 Quarters

Multi-step prediction is a domain that demands contribution, especially since not much work has been done for a longer term time series predictions for telecommunications planning and management. Multi-step-ahead prediction of mobile data usage with a NAR model is carried out by utilizing a closed-loop form. Figure 3 shows the closed-loop model. For this model, the first 45 quarters were used for training the model and is used to predict the values for the 5 quarters ahead.

The closed-loop NAR model did not perform well. The final time series response obtained has been shown in Fig. 4. It can be clearly observed that the model was unable to learn the trend accurately.

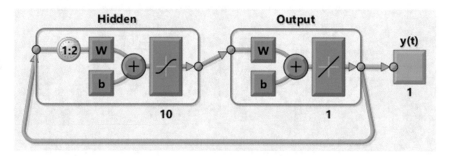

Fig. 3 Closed-loop model for multi-step-ahead prediction using NAR model

Fig. 4 Five-step-ahead
prediction of mobile data
usage time series using NAR
model

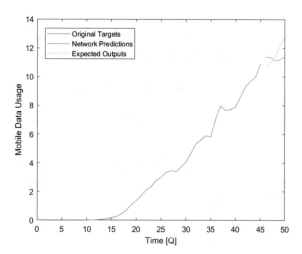

4.2 Nonlinear Autoregressive Model with Exogenous Input (NARX) Neural Network

4.2.1 Single-Step-Ahead Mobile Data Usage Forecasting Over 50 Quarters

NARX (Fig. 5) architecture was used. Statistically significant input delays indicated by the input/target cross-correlation function, are added to the network. Here, an exogenous input is also involved, for a better result. In this model, the mobile data usage was the time series to be predicted with mobile penetration rate being the exogenous input time series.

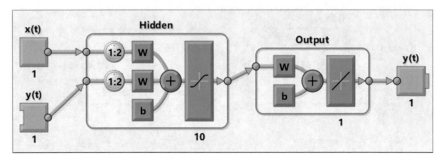

Fig. 5 Architecture of the default NARX network

4.2.2 Multi-step-Ahead Mobile Data Usage Forecasting Over 50 Quarters Using NARX Model

Multi-step-ahead prediction for a NARX model is carried out by utilizing a closed-loop form [13]. The one-step-ahead forecast is fed back into the model to obtain the results for the next step ahead. Figure 6 shows the closed-loop model. For this model, the first 45 quarters were used for training the model and used to predict the values for the 5 quarters ahead.

The resulting time series response obtained has been shown in Fig. 7. As can be observed, the model learned the input-target relation well, and predicted the trend accurately (Table 1).

It was observed that by varying the parameters and settings of the model, i.e., number of neurons to 25 and the delay to 20, the accuracy of the model shows a considerable increase (Fig. 8).

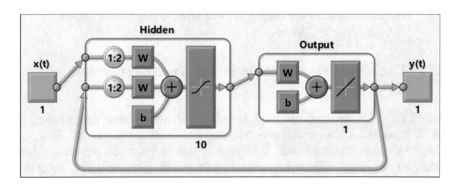

Fig. 6 Closed-loop model for multi-step-ahead prediction

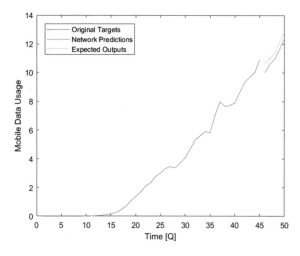

Fig. 7 Five-step-ahead prediction of mobile data usage time series using NARX model

Table 1 Difference between NAR and NARX models for multi-step prediction

	NAR model	NARX model
Target time series	Mobile data usage	Mobile data usage
Exogenous input series	–	Mobile penetration rate
Transfer functions	tansig/purelin	tansig/purelin
Epoch	8	9
Performance	0.7327	0.0442
Gradient	0.0652	0.0301
Mu	0.01	0.01

4.3 Insight for Strategic Decision-Making

Both Mobile Data Usage and Mobile Penetration Rate can be presented as time series with each step corresponding to a single quarter [14, 15]. The rationale behind time series forecasting is that the output values can be represented as a function of past values. Pricing war in Singapore is dependent on planning, optimizing risk strategies which could be addressed using such competitive prediction methods.

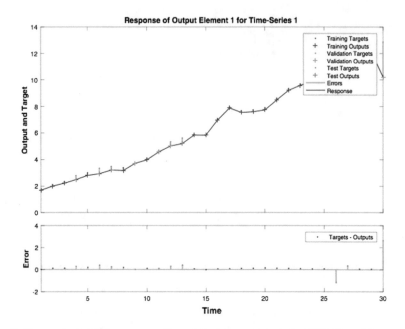

Fig. 8 Time series prediction response for mobile data usage, optimized NARX model

5 Conclusion

The results obtained from the study show a slight deviation for the last quarter value. A more effective approach to solve this problem is supplementing this with fuzzy logic which can generate a more effective forecast. The premium outcome of this effort is also to present insight from the forecast result. The insight about penetration rate rising suggest innovations increasing more adaptation to the use of digital utilities.

References

1. Yang KC (2005) Exploring factors affecting the adoption of mobile commerce in Singapore. Telematics Inform 22(3):257–277
2. Anil S, Ting LT, Moe LH, Jonathan GPG (2003) Overcoming barriers to the successful adoption of mobile commerce in Singapore. Int J Mob Commun 1(1–2):194–231
3. Majid S, Tee Tan A (2002) Usage of information resources by computer engineering students: a case study of Nanyang Technological University, Singapore. Online Inf Rev 26(5):318–325
4. Pagani M (2004) Determinants of adoption of third generation mobile multimedia services. J Interact Mark 18(3):46–59
5. Wong S (2017) Singapore's biggest phone company jumps into mobile pricing war. Bloomberg, Singapore. Web, Sep 2017
6. Lee S, Shin B, Lee HG (2009) Understanding post-adoption usage of mobile data services: the role of supplier-side variables. J Assoc Inf Syst 10(12):2

7. Circles. Life | Data Deprivation Survey (2017). https://pages.circles.life/data-deprivation-survey/
8. Statistics Singapore | Latest Data (2017). http://www.singstat.gov.sg/statistics/latest-data
9. Global mobile phone internet user penetration 2019 (2017). https://www.statista.com/statistics/284202/mobile-phone-internet-user-penetration-worldwide
10. Mobile Data Usage (2017). https://data.gov.sg/dataset/mobile-data-usage
11. Mobile Penetration Rate (2017). https://data.gov.sg/dataset/mobile-penetration-rate
12. Remus W, O'Connor M (2001) Neural networks for time-series forecasting. In: Principles of forecasting. Springer, Boston, MA, pp 245–256
13. Benmouiza K, Cheknane A (2013) Forecasting hourly global solar radiation using hybrid k-means and nonlinear autoregressive neural network models. Energy Convers Manag 75:561–569
14. Verkasalo H (2009) Contextual patterns in mobile service usage. Pers Ubiquit Comput 13(5):331–342
15. Lee Y (2017) Grab starts mobile wallet service at Hawker stalls in Singapore. Bloomberg, Singapore. Web, Nov 2017

Candidates Selection Using Artificial Neural Network Technique in a Pharmaceutical Industry

Abhishek Kumar Singh, S. K. Jha and A. V. Muley

Abstract Artificial neural network (ANN) is a useful technique in decision-making which can replicate the biological thinking pattern of the human decision maker. By providing the learning way of a supervised and unsupervised process, we can train the ANN to give the output as accurate to the human judgment. This technique has been used in solving multiple problems including forecasting and to predict the solution. In this work, ANN has been used for the candidates' selection in the pharmaceutical company. To mimic the human judgment in the selection process of human resource management by supervised learning. So that we can eliminate the human judgment with ANN.

Keywords Artificial neural networks (ANN) · Decision-making (DM)
Prediction · Forecast

1 Introduction

The decision-making in any industry is a task of thinking and analyzing. To think and analyze are the tasks of the biological neural network to give the solutions of the problem, but now with the help of higher processing capability of computers, finding the solution using ANN is within reach [1]. And, if we considered the medicine manufacturing industries which deals with hazardous and potentially harmful chemicals for the environment. It is a risky task, if due precaution is not taken during its operation. As considering the pharmaceutical company wants to recruit the candidates for their industry, then HRM department has to take the decision for selecting

A. K. Singh (✉) · S. K. Jha · A. V. Muley
Division of MPAE, NSIT, Dwarka 110078, New Delhi, India
e-mail: Abiksingh7@gmail.com

S. K. Jha
e-mail: skjha63@rediffmail.com

A. V. Muley
e-mail: avmuley2000@gmail.com

© Springer Nature Singapore Pte Ltd. 2019   359
S. Bhattacharyya et al. (eds.), *International Conference on Innovative Computing and Communications*, Lecture Notes in Networks and Systems 56,
https://doi.org/10.1007/978-981-13-2354-6_38

Fig. 1 Neuron model

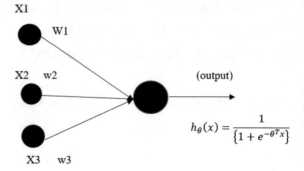

the candidate with the help of experts or decision maker for recruitment. Because safety is a major concern in the pharmaceutical industry and the candidate should be wisely selected with good knowledge to avoid accidents. These accidents can be so severe that it can lead the company towards downfall and chemical hazard may lead to atmospheric or environmental degradation, and the chemical waste produced also wants proper waste management and decomposition system [2, 3]. For the selection of suitable medical scientist for such a crucial industry is a primal task for human resource department of the company. In this case, the prediction of selecting the candidate is done using ANN tool.

2 Artificial Neural Network

ANN is based on biological functioning of neurons, as now computers are capable of processing complex data thereby making ANN an important algorithm. This algorithm is capable of learning highly nonlinear functions and infers knowledge from the data by adjusting synaptic weights using backpropagation algorithm. The model consists of an input layer, a hidden layer, and an output layer. During training, the training data is used to train the model and validation data is used to check how well the model is generalizing on the untrained data. Then, a test data is used to test the accuracy of the model to know if it performs accurately on the unseen data [4].

2.1 Neurons in ANN

They replicate biological neurons. In the artificial neuron, the input vector $x = (x_1, x_2 \ldots x_n)$. is given which is multiplied with the connection parameters $w^T = (w_1, w_2 \ldots w_n)$. then, it is added before passing through the transfer function $\sigma\left(\sum w^T x\right)$ to get the output for other neurons. Where σ is an activation function (Fig. 1).

Neural Networks were used in the early 1980s and late 1990s but due to the incapability of computing resources for them, the algorithm was difficult to use, but now the computers are more capable, so the neural network can process the complex data by using the mathematical models and algorithms. The research in the past is now led to developments of prediction method [5].

The prediction of ANN is automated [6], statistical models must be improved with time, hence using past data supervised learning is introduced using ANN [7]. The supervised learning is based on the theory which can self-organizing and evaluate the input patterns [8].

2.2 Learning Process of ANN

ANN is based on the classification-based technique and can work with qualitative, quantitative data that can easily provide results [9]. Supervised learning process takes place by using examples and by guidance's unsupervised learning process, as the name suggests there are no existing output labels for the given data in it.

ANN technique replicates the biological neurons, and these neurons are adaptive to the character of the assigned problem. And for that particular network, it can be designed and trained according to the problem by altering the values. And by observing the data, the network can predict the best functioning path of the network [10]. It is been differentiated that backpropagation networks of ANN predicted better results than logistic regression models while predicting the commercial bank failures. It consists of layers of input, output, consisting of number of neutrons in each layer, and transfer function, and the layers are densely connected to each other.

2.3 Using Feedforward Backpropagation Algorithm Network Type

It consists of interconnected rows so that the weights can be adjusted accordingly. Today, it is a mainstay of neuron computing. In backpropagation, the error moves backward from the output node towards the input nodes. In which the error is calculated after which the weight is modified [11] (Fig. 2).

3 Experimentation and Methodology

Table 1 shows the data of selected candidates from the industry, this data is used to train the network as an input. This can be inferred from experimentation on the dataset that the ANN based model can be used to recruit medical scientists in pharmaceutical

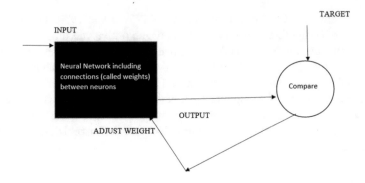

Fig. 2 Block diagram of backpropagation

Table 1 Input instance of dataset

Language test	Professional test	Safety rule test	Professional skills	Computer skills	One on one interview	Group discussion
85	85	85	85	85	85	85
60	60	60	60	60	60	60
67	67	67	67	67	67	67

industry. This Table 1 of the input data is collected by the pharmaceutical HRM department. And, the Table 2 is the results given by the decision maker (DM) of the HRM department. By using ANN, the new results are found which is validated by comparing it with the results of decision maker of HRM.

Table 2 is fed as an output to make the supervised learning. Using these data, the network is created using ANN toolbox in MATLAB. The purpose of this data is to provide past data with input and output so that NN can generate the neurons interconnected with the network [12]. In this network, there are a total of seven inputs which consist of criteria as mentioned above (Group discussion, Language test, Professional test, Safety rule test, Computer skills, one-on-one interview, and Professional skills). The total number of hidden layers is 10 which consist of 10 sigmoidal units. One output node, using tan-sigmoid transfer function is used in the above neural network.

These are used for pattern recognition problems. It is used in learning algorithms

$$logsig(n) = \frac{1}{1 + e^{-n}}$$

Table 2 Data by DM of HRM of 17 candidates

Candidates	Outputs
1	74.02
2	81.19
3	80.84
4	73.23
5	77.18
6	78.41
7	75.52
8	76.41
9	80.09
10	75.29
11	71.04
12	77.39
13	77.95
14	77.40
15	85.95
16	77.93
17	83.81

Table 3 Key features of the ANN model used

Hidden layer activation function	Tan-sigmoid transfer function
Output layer activation function	Linear function
Training algorithm	Levenberg–marquardt backpropagation
Epoch	1000
Goal error	$1E^{-7}$
Mu (Starting)	0.001
Mu (decrease factor)	0.1
Mu (increase factor)	10
Mu (max)	10^{10}
Data division	Random
Performance estimation	Mean square error

As the problem is of the nonlinear least square form, then the appropriate algorithm used is TRAINLM. The data provided is to start the training of the data even if it is far away from the guess. By repetitive training of the model, network will learn the prediction of data by backpropagation method (Table 3).

4 Results and Discussion

There are two samples considered for validation of the model, one is candidate 6 and other is candidate 8, and the results are compared with the output by ANN. Output by ANN for first sample is −78.4125, output by ANN for second sample is −76.0538. This above prediction by the neural network is almost similar/close to the results given by decision maker as shown in (Table 2). The first sample is matched with the candidate number 6 and the second sample is close to the value of candidate number 8.

The two samples predicted using ANN have miniscule error between the predicted value and that is given by the decision maker (DM), which validates that the model and training of data are almost accurate. This confirms that the backpropagation algorithm used in the neural network realizes learning with least error. Thus, this model can supplant the role of the decision maker or human judgment by using ANN, so that reliable decisions for employee evaluation can be predicted by ANN. The results outcome is almost similar to the DM by comparing. ANN results to the decision maker results. Now, ranking of the candidate can be done accordingly without the help of the decision makers (Figs. 3 and 4).

In the present work, ANN is using one hidden layer with seven neurons. For that training, the algorithm has to be changed. And, increase in neurons leads to the better approximation of results. After using the ANN technique, and checking it under various conditions, by observation, we can say that model can be used for selection purpose in any industry and can be used in other selection processes and would give the appropriate results. Thus, the model developed using ANN can be used to replace a human expert for the purpose of employee selection.

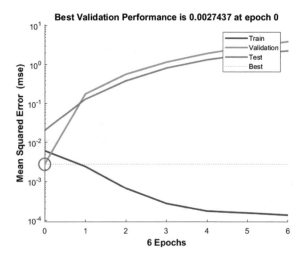

Fig. 3 Neural network training performance

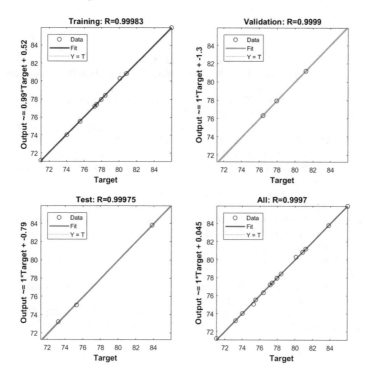

Fig. 4 Training regression graph

5 Conclusion and Future Scope

This study is done in the requirement process of the candidates for the pharmaceutical company for the recruitment of the medical scientist. Because employing a human resource department in a nascent company is cost intensive, therefore ANN based technique introduced in this study provides alternative to it. Furthermore, ANN technique is used in this study for the selection of candidates in the company to replace recruitment expert decision maker in the selection process. In this study, the outcome of ANN is highly accurate and is similar to the results given by the expert. Which shows that the ANN model can be used as the decision maker for the selection process and eliminate the decision maker.

Furthermore, other machine learning techniques can be used to compare and improve over our Neural Networks.

References

1. Heckerman DE, Shortliffe EH (1992) Artif Intell Med 4:35–52
2. Gathuru IM, Buchanich JM, Marsh GM, David DG (2015) Health hazards in the pharmaceutical industry
3. Sebastine IM, Wakeman RJ (2003) Consumption and environmental hazards of pharmaceutical substances in the UK. Process Saf Environ Prot
4. Agatonovic-Kustrin S, Beresford R (2000) Basic concepts of artificial neural network (ANN) modeling and its application in pharmaceutical research. J Pharm Biomed Anal
5. Hiew M, Green G (1992) Beyond statistics. A forecasting system that learns
6. Hoptroff RG (1993) The principles and practice of time series forecasting and business modeling using neural nets. Neural Comput Appl
7. Widrow B, Sterns SD (1985) Adaptive signal processing. Prentice-Hall, Englewood Cliffs, NJ
8. Carpenter GA, Grossberg S, Reynolds JH (1991) ARTMAP: Supervised real-time learning and classification of nonstationary data by a self-organizing neural network. Neural Netw
9. Partovi FY, Anandarajan M (2002) Classifying inventory using an artificial neural network approach. Comput Ind Eng 41(4)
10. Gluck MA, Bower GH (1991) 1Y88, Evaluating an adaptive model of human learning. J Mem Lang 27:166–195
11. Artificial neural network in applied sciences with engineering applications international peer-reviewed academic. J Call Pap 3 1(2013)
12. Cyril V, Gilles N, Soteris K, Marie-Laure N, Christophe P, Fabrice M, Alexis F (2017) Machine learning methods for solar radiation forecasting: a review. Renew Energy

Distant Supervision for Large-Scale Extraction of Gene–Disease Associations from Literature Using DeepDive

Balu Bhasuran and Jeyakumar Natarajan

Abstract Understanding the genetic mechanism of a disease can solve a variety of problems such as personalized precision medicine, new drug development or repurposing them. Technological advancement in biology leads to methods like next-generation sequencing (NGS) which produced large sets of genes associated with diseases in the form of variants and biomarkers. All these findings are reported as a huge collection of scientific literature, which is rapidly growing every day. In this study, we present a distant supervision methodology for finding the gene–disease associations from literature in large scale. We used DeepDive system, which is highly successful in relation extraction from a wide variety of sources such as image and text. In this study, we build a gene–disease relation extractor by feeding a highly sophisticated feature set to DeepDive and extracted associations from 879585 PubMed articles. Our system identified candidate gene–disease associations from abstracts and calculated a probability for each association. Overall, our system produced a set of 75595 associations using a domain-specific distinct feature set from over 879585 abstracts.

Keywords Text mining · BioNLP · Gene–disease relation extraction
Distant supervision · DeepDive

B. Bhasuran · J. Natarajan (✉)
DRDO-BU Center for Life Sciences, Bharathiar University, Coimbatore 641046,
Tamil Nadu, India
e-mail: n.jeyakumar@yahoo.co.in

B. Bhasuran
e-mail: balubhasuran08@gmail.com

J. Natarajan
Data Mining and Text Mining Laboratory, Department of BioInformatics,
Bharathiar University, Coimbatore 641046, Tamil Nadu, India

© Springer Nature Singapore Pte Ltd. 2019
S. Bhattacharyya et al. (eds.), *International Conference on Innovative Computing
and Communications*, Lecture Notes in Networks and Systems 56,
https://doi.org/10.1007/978-981-13-2354-6_39

367

1 Introduction

One of the major issue and challenging problem in the field of biology and medicine is determining the genetic mechanism causing diseases. It has a wide variety of applications such as early diagnosis, personalized precision medicine, DNA screening and drug repurposing [1]. Advanced biomedical techniques such as genome-wide association studies (GWAS) and next-generation sequencing (NGS) identified a large number of potential genes associated with diseases and reported as research outputs through scientific articles [2]. Both academia and industry successfully accepted text mining (TM) as the highly reliable method for knowledge discovery from this rapidly growing literature.

Relation extraction in biomedical text mining is a well-studied area and it contributed in large to knowledge discovery [3]. Using relation extraction associations among entities such as protein\gene, miRNA, and disease was identified as protein–protein interactions, disease-gene, and miRNA-disease associations [4]. These extracted relations have applications ranging from personalized precision medicine to drug repurposing [1]. Developing a supervised relation extraction model poses a greater challenge that a silver standard or a gold standard corpus is a necessity. In this scenario, distant supervised learning emerged as a successful solution [5]. Using a knowledge base distant supervision heuristically generates a labeled corpus which is otherwise performed manually by domain experts. The labeling is performed by using the source corpus in which a given pair is listed as a known association in the knowledge bases. DeepDive is a distant supervision-based framework which allows the user to perform relation extraction with probability [6].

In general, the major goal of any text mining system is information extraction from the unstructured text in a context-specific nature. DeepDive is released as a modeled system which performs extraction of information from various sources including text [6]. The DeepDive system comes as a framework with two main parts application code and interface engine. Users are expected to write the application codes for their relation extraction task (in this study gene–disease association), and interface engine provides the DeepDive system. It is also worthy to be mention that applying distant supervision in biomedical context is a challenging task due to shared context, ambiguity, and naming conventions among concepts like genes and diseases. Due to its wide range of applications, text mining researchers applied distant supervision methods for pathway extraction and gene–gene interaction extraction, but it has not been applied for a large-scale gene–disease extraction task [7, 8]. By considering all these factors, we are proposing a distant supervision-based methodology for large-scale extraction of gene–disease associations from the literature using DeepDive.

2 Materials and Methods

In this study, we created a relation extraction methodology using DeepDive in which we input a literature corpus to the extractor and created a knowledge base from it. The gene–disease relation extraction methodology pipeline is depicted in Fig. 1.

The current methodology comprises of four phases namely text preprocessing, gene–disease extractor, deep dive, and system tuning. First, the downloaded PubMed gene–disease relation corpus is parsed for sentences, and the corresponding token was generated using OpenNLP [9] in the text preprocessing step. Next, the sentences were fed to the gene–disease extractor resulted in a set of candidate relations. The gene–disease co-occurring sentences along with relationship describing features are the main components in the generated candidate relations. Finally, a tuning process was carried out for performance improvement. A full schematic architecture used in this work is depicted in Fig. 2.

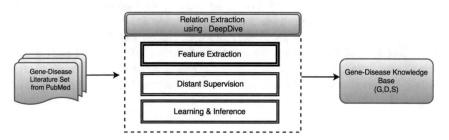

Fig. 1 Workflow of the large-scale gene–disease relation extraction methodology. A literature corpus is used as the input and a gene–disease association knowledge base has been generated as the output

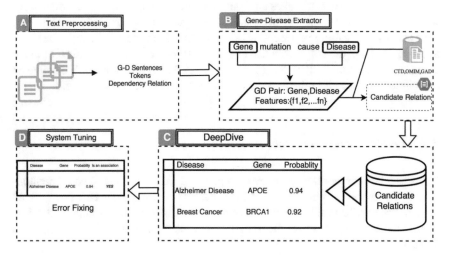

Fig. 2 Schematic architecture of the gene–disease relation extractor

2.1 Biomedical Literature Corpus

We used MEDLINE literature set as the biomedical literature data set for extracting large scale of gene–disease association using distant supervision. We searched in PubMed for gene–disease associated literature by giving the following query:

("Diseases Category" [Mesh] AND "genetics" [Subheading]) AND has abstract [text] AND ("1980" [PDAT]: "2017" [PDAT]) AND "humans" [MeSH Terms] AND English [lang].

The search resulted in 879585 articles in XML format from upon which we applied manual curation and preprocessing to prepare the data set for distant supervision.

2.2 Ground Truth

In order to perform distant supervision, we need to provide a ground truth data set from which the algorithm learns the true positive and true negative associations. For validated positive associations, we used Online Mendelian Inheritance in Man (OMIM) [10], Comparative Toxicogenomics Database (CTD) [11], and Genetic Association Database (GAD) [12] databases. For negative associations, we used negatively annotated associations from gold standard corpora PolySearch2 [13], GAD [12] and EU-ADR [14].

2.3 Gene–Disease Extractor

The gene–disease extractor takes an input sentence and makes possible associations with it. These associations have two major parts: a gene–disease mention in a sentence level and a set of features representing that relation pair. In order to identify the genes and diseases from the corpus, we used in-house developed high performance named entity recognition tools BCC-NER [15] and D-NER [16] respectively. We define an association sentence if both gene and disease both co-occur in that sentence.

2.4 Feature Engineering

Features play an important role in every supervised machine learning based methodologies. Since the current study is on the same line of work, we used a fine-grained domain-specific feature set to represent the gene–disease associations. A detailed representation of the features incorporated into the methodology is represented in Table 1.

Table 1 Feature engineering patterns with system representation

Feature type	System representation
Word sequence window	Word_Seq (mutation_cause)
1-word window	GID_w1 (mutation)
2-word window	GID_w2 (mutation_causes)
3-word window	GID_w3 (mutation_causes_the)
Begin index	Gen|Dis_begin_index
End index	Gen|Dis_end_index
Tokens	Tokens
Lemma	Lemma
POS	Pos_tags
NER	Ner-tags
Regex pattern	Regex_pattern (mutation_causes)
Dependency path	nsub (mutation_gene) vb (cause) pobj (causes_disease)

2.5 Distant Supervised Learning

In supervised machine learning approaches the state-of-the-art methodology is to use a gold standard annotated corpus to train and test the model for the corresponding task such as named entity recognition (NER) and relation extraction (RLE). The provided sentences in the corpus need to be labeled with entities and relations between them by a set of qualified annotators with a high annotation agreement score. This process is time consuming, tedious, costly and the resulting classifier tends to bias toward the domain of interest. In order to overcome the pitfalls like low precision, complexity in mapping and semantic drift the widely accepted alternative paradigm called distant supervision [5–7].

Technically distant supervision can be defined as a learning scheme in which a set of the weakly labeled training set is used to train a classifier for a large-scale extraction task. The basic assumption about distant supervised method is that a given entity pair that co-occurs in a known relational database is most likely to exhibit a relation in some way. There seem to be no compelling reasons to argue that distant supervision algorithm successfully eliminates the problem of domain dependence and over fitting by using databases instead of labeled text.

The basic steps involved in implementing a distant supervised algorithm for gene–disease relation extraction are given below.

(1) Set of labeled training data from gene–disease corpus.
(2) Pool of literature evidence of gene–disease association from OMIM [10], CTD [11], GAD [12], PolySearch2 [13], and EU-ADR [14].
(3) A relational operator that samples the data from the unlabeled data and labels them.

(4) The algorithm that combines the original labeled data and the noisy output of relational operator to produce the final output.

We employed DeepDive, the most widely accepted and successfully employed open-source implementation of distant supervision [6].

2.6 *DeepDive*

DeepDive is an open-source engine which utilizes probabilistic inference on random variables to construct a knowledge base [6]. There are some notable works that can be advanced to support the claim that DeepDive can be used as a single system for relation extraction, integration, and prediction in biological and other fields of research [7, 8]. DeepDive provides an end-to-end data pipeline for the user so that they can improve the quality of the system by simply focusing on crucial points like feature engineering and classification [8]. DeepDive is a trained machine learning system that expects the developer to input fine-grained features for better performance whereas it can significantly reduce the imprecision and various forms of noises from its learning experience.

To feed the DeepDive with the input gene–disease associations, a random variable is generated as is_a_relation for every pair. We input all the gene–disease associations with an is_a_relation label for learning in a way such that for feature weight generation and testing 50% of input samples is held out.

3 Results and Discussions

In order to perform the large-scale extraction of gene–disease relations, we obtained a query-driven literature corpus from Medline as discussed in Sect. 2.1 of biomedical literature set. As a result of the given query, we obtained a set of 879585 kinds of literature, which is used as the input corpus for the distant supervision method. We used the known positive associations as ground truth from well-known databases like CTD, GHR, and GAD.

Given the result of gene–disease extractor, a probability is assigned by the Deep-Dive to indicate it as a true relation. From the identified 75595 relations, a total of 8993 relations were given a 0.9 probability score followed by 15362 relations with 0.8 score. From the feature engineering perspective, our system computed a total of 16452 features for the sentences from with a total of 856 features assigned a weightage greater than 0.1. The sentences described with top features come with interaction words like 'autosomal', 'biomarker', 'mutation', 'polymorphisms' connecting the concepts' gene and disease. A detailed representation of the results is given in Table 2.

Table 2 Detailed representation of the results obtained

Result type	Count
Literature corpus	879585
Features generated	16452
Features with weightage >0.1	853
Identified gene–disease relations	75595
Probability of associations >0.9	8993
Probability of associations >0.8	15362

The distant supervision method resulted in a set of 75595 genes associated diseases in sentence level. Among the resulted sentences' diseases like cancer, hepatitis, heart, and Alzheimer diseases are plenty, shows a large-scale study already has been done in these disorders. A closer look into the data indicates that more studies need to be carried out in diseases like cerebral and pulmonary edema, infectious and perinatal diseases.

4 Conclusions

There is overwhelming evidence corroborating the notion that a large-scale extraction of relations of genes associated to diseases from literature can support drug development and precision medicine. In this present study, we used DeepDive to extract large-scale gene–disease associations from Medline literature using fine-grained domain-specific feature set. As a future enhancement, we are planning to combine these features into an ensemble model and apply to the entire MEDLINE database for a full-scale extraction of all reported gene–disease associations. We strongly believe that the resulted output can give a new dimension in knowledge discovery from the text and can support investigations on the genetic mechanism behind diseases.

Acknowledgements This work received funding from DRDO-BU Centre for Life Sciences, Bharathiar University, Tamil Nadu, India.

References

1. Luo Y, Uzuner O, Szolovits P (2016) Bridging semantics and syntax with graph algorithms-state-of-the-art of extracting biomedical relations. Brief Bioinform
2. Metzker ML (2017) Sequencing technologies the next generation

3. Kilicoglu H (2017) Biomedical text mining for research rigor and integrity: tasks, challenges, directions. Brief Bioinform
4. Holzinger A, Dehmer M, Jurisica I (2014) Knowledge discovery and interactive data mining in bioinformatics—state-of-the-art, future challenges and research directions
5. Zeng D, Liu K, Chen Y, Zhao J (2015) Distant supervision for relation extraction via piecewise convolutional neural networks. In: Proceedings of the 2015 conference on empirical methods in natural language processing, 1753–1762
6. Niu F, Zhang C, Ré C, Shavlik J (2012) DeepDive: web-scale knowledge-base construction using statistical learning and inference. VLDS 2012. i
7. Poon H, Toutanova K, Quirk C (2015) Distant supervision for cancer pathway extraction from text. Pac Symp Biocomput 20:120–131
8. Mallory EK, Zhang C, Ré C, Altman RB (2015) Large-scale extraction of gene interactions from full-text literature using DeepDive. Bioinformatics 32:106–113
9. Morton T, Kottmann J, Baldridge J, Bierner G (2005) Opennlp: a java-based nlp toolkit. EACL
10. Amberger JS, Bocchini CA, Schiettecatte F, Scott AF, Hamosh A (2015) OMIM.org: online mendelian inheritance in man (OMIM®), an online catalog of human genes and genetic disorders. Nucleic Acids Res 43:D789–D798
11. Davis AP, Grondin CJ, Johnson RJ, Sciaky D, King BL, McMorran R, Wiegers J, Wiegers TC, Mattingly CJ (2017) The comparative toxicogenomics database: update 2017. Nucleic Acids Res 45:D972–D978
12. Becker KG, Barnes KC, Bright TJ, Wang SA (2004) The genetic association database. Nat Genet 36:431–432
13. Liu Y, Liang Y, Wishart D (2015) PolySearch2: a significantly improved text-mining system for discovering associations between human diseases, genes, drugs, metabolites, toxins and more. Nucleic Acids Res 43:W535–W542
14. van Mulligen EM, Fourrier-Reglat A, Gurwitz D, Molokhia M, Nieto A, Trifiro G, Kors JA, Furlong LI (2012) The EU-ADR corpus: annotated drugs, diseases, targets, and their relationships. J Biomed Inf 45:879–884
15. Murugesan G, Abdulkadhar S, Bhasuran B, Natarajan J (2017) BCC-NER: bidirectional, contextual clues named entity tagger for gene/protein mention recognition. Eurasip J Bioinform Syst Biol 2017
16. Bhasuran B, Murugesan G, Abdulkadhar S, Natarajan J (2016) Stacked ensemble combined with fuzzy matching for biomedical named entity recognition of diseases. J Biomed Inf 64:1–9

Opinion Mining of Saubhagya Yojna for Digital India

Akshi Kumar and Abhilasha Sharma

Abstract A government policy is an ideology of principles, ethics, regulations, and set of rules introduced or constituted by government in interest of normal citizens across country. The purpose of making a policy or scheme is to explore and strengthen the favourable outcomes or profits and to minimize the adverse effects noticed by government. On the flip side, all the government policies are destined to be efficacious but on the contrary may affect the routine life of a common man. An efficient and endless process of policy evaluation may leads to accurate assessment of progress graph or outcomes of policy in order to improve the social and economic conditions of different stakeholders. The aim of this paper is to evaluate one of the recent government policy, Pradhan Mantri Sahaj Bijli Har Ghar Yojana, or Saubhagya initiated by Indian government to accomplish global electrification across country. In this paper, an attempt has been made to analyse the public perception of this scheme by using opinion mining techniques to understand the positive and negative impact of this policy over Indian citizen. Twitter as a social media tool has been used for collecting and extracting public opinion or sentiments over this scheme.

Keywords Opinion mining · Government policy · Saubhagya · Twitter

1 Introduction

Transforming India into digital empowered society and knowledge economy was the proclamation of Bhartiya Janta Party (BJP) in Indian General Elections, 2014. The party aims to digitally literate every household and individual with the help of information technology in order to make India universal nerve centre of knowledge. The automation of various government departments with the use of Information and

A. Kumar · A. Sharma (✉)
Delhi Technological University, New Delhi, Delhi, India
e-mail: abhilasha_sharma@dce.ac.in

A. Kumar
e-mail: akshikumar@dce.ac.in

© Springer Nature Singapore Pte Ltd. 2019
S. Bhattacharyya et al. (eds.), *International Conference on Innovative Computing and Communications*, Lecture Notes in Networks and Systems 56,
https://doi.org/10.1007/978-981-13-2354-6_40

Communication Technologies (ITC) has spring up e-Governance, which was firmly emphasized in manifesto. e-Governance provides a digital platform to broadcast information for execution of government administration activities [1]. The miscellaneous initiatives of e-Governance are playing a vital role in shaping up the progress graph of government strategies for the nation. To speed up the nation-wide implementation of e-Governance, a programme approach needs to be adopted, guided by common vision and strategy [2].

Thus, Digital India, campaign launched on 2 July, 2015 by Honourable Prime Minister of India Narendra Modi, is a programme to digitally empower the citizen which in turn prepare the nation for knowledge based transformation [3]. The major projection areas of this programme are [3]:

- broadband rural area (highways and villages connectivity)
- collaborative and participative governance (e-Governance, e-Kranti)
- digital learning (public internet access, global information)
- electronics manufacturing (tele-medicine, mobile healthcare) and others.

For the sake of delivering good governance to citizens and promoting e-Governance, the National e-Governance Plan (NeGP), comprising of 27 Mission Mode Projects and 8 components (initially) has been defined and formulated [4]. The Prime Minister gave its approval on 18 May, 2006 as a smart way forward for Digital India [4]. The plan is conceptualized with an integration of e-Governance initiatives across nation into a collective vision. The main objective of NeGP is to provide electronic delivery of government services to citizens by creating a nation-wide network. Various citizen centric projects meeting the target of NeGP are made operational and some of them are under process. Numerous policies and schemes have been launched as a part of these projects.

The process of policy making is based on the conversion of a thought into action where citizens are the generators of these ideas. It is highly beneficial to openly express public concern or share their views about how these policies are affecting the lives and making impact over the community. Thus, the development of a government policy highly relies over the opinion of citizens.

As a step forward to this, social web [5] has made it possible to effectively share the opinion and experiences of people. Different social media channels or networking sites are high in use to get the reaction of public and to vocalize their sentiment over any government scheme/policy. They provides a huge amount of data which can be processed and analysed for better decision making and further strategic movement of the policy. Opinion mining of this enormous amount of data, big data, can be done for the extraction of their viewpoint from web [6] in order to make the policy a success.

This paper uses the concept of opinion mining for a recent government scheme "Pradhan Mantri Sahaj Bijli Har Ghar Yojana, or Saubhagya" launched by Prime Minister of India, Narendra Modi to make "electricity for all" a verity. This scheme ensured electrification of all willing households and around 4 crore families in rural and urban areas are in vision to provide electricity connections by December 2018 [7].

Twitter, being a popular social networking service has been used as a source of information for this conduct as it has various diversified topics [8], variety and volume of people associated with, creating a vast scope for interest of researchers. Various messages or tweets posted by massive amount of users can be seen by millions of people in order to get their common concern over any specific topic. Government bodies and officials are exploiting this tool in order to increase their reach toward citizens. Twitter is quick in nature and its speed make it a valuable tool to act as an early warning system which help government organizations in better decision making for policy evaluation. Among all the existing social network services, Twitter is the one which provides a worldwide forum for sharing individual thoughts or reactions over governmental proceedings and practices.

Data acquisition has been done from Twitter for approximate one month after the launch of the scheme which is divided into four phases: 9 October, 2017 to 15 October, 2017; 16 October, 2017 to 22 October, 2017; 23 October, 2017 to 29 October, 2017 and 30 October, 2017 to 8 November, 2017. This splitting of data collection period has been done in order to get the opinion of people about this scheme for over a period of time. A comparative analysis for assessing the performance of sentiment classification using standard measure of evaluation (Precision, Recall and Accuracy) has been done by applying five famous techniques of machine learning (ML), i.e. Naive Bayes (NB), Support Vector Machine (SVM), Multilayer Perceptron (MLP), K-nearest neighbour (kNN) and Decision Trees (DT).

The remaining paper is structured as follows: Sect. 2 elaborates about the background of *Saubhagya* scheme as a socio centric service. Section 3 discusses the evaluation of this scheme which includes compilation of data collection, statistics representation of data in different forms and executing opinion mining process for calculating precision, recall and efficiency. Section 4 tabulates the results and findings with their graphical representations. At the end, Sect. 5 concludes the paper.

2 Saubhagya Scheme—A Socio-Centric Service

A socio centric service is an integration of necessities and utilities provided by government required for a stable society. The whole idea to project such type of services is to resolve the social, environmental, and economic problems for the sake of public welfare and to provide better quality of life.

The Pradhan Mantri Sahaj Bijli Har Ghar Yojana, or Saubhagya, one such socio-centric service, has launched by Indian Prime Minister Narendra Modi on September 25, 2017. The scheme funds Rs 16,230 crore to provide power access and electricity connections to all those households which are still out of electric power by last mile connectivity. About 4 crore un-electrified households are estimated for electrification out of which 300 lakhs are expected to be covered under this scheme [9]. The statistics of present status of electrification in India is represented in Fig. 1.

As per the government vision about the scheme, it will be a global household electrification across the country that helps in improving energy consumption which

Status Count of Electrification under Saubhagya(in lakhs)

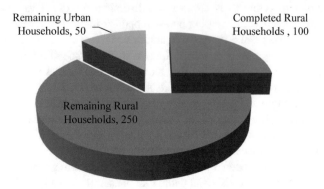

Fig. 1 Nation-wide statistics of households electrification under Saubhagya [9]

leads to better quality life of people. The scheme will be extremely useful for people in their daily life as it will have a positive impact over various sectors like health; education; agriculture; entrepreneurship (agro-based industry setup). The scheme will also facilitate economic growth of the country as well as the employment creation. The scheme provides free of cost electricity connections to all the poor households which are un-electrified. Although, the electricity consumption bill has to be paid by consumer. The baseline is free connection and not free power.

Saubhagya scheme appends the series of similar type of projects in government's village electrification programme. The two major services already there under execution prior to this scheme are as follows:

  i. Deen Dayal Upadhyay Gram Jyoti Yojana (DDUGJY) [10] for rural areas launched in July 2015.
 ii. Integrated Power Development Scheme (IPDS) [11] for urban areas launched in December 2014.

Both the schemes are flagship programmes of Ministry of Power and aims to ensure 24 × 7 Power for All. Following are the major objectives of both the schemes [10–13]:

- Village electrification across India
- Ensure sufficient power supply to consumers, especially farmers
- To empower the distribution and sub transmission network
- Reduction of losses by metering
- Strengthening customer care services by use of IT applications.

Although, both the schemes were utilized for electrification of rural as well as urban areas across country, but still a huge number of households are still there that do not have electricity connections for certain reasons like absence of electricity pole, poor financial condition for paying initial connection charges, lack of awareness as how to get connection and so on. Therefore, Saubhagya has been casted to fill such voids and overcome the issues of last mile connectivity and providing electricity connections to all un-electrified households in country. The government envisage several benefits from the scheme which are as follows [14, 15]:

- Electrification of rural and urban areas across country
- Various pathways of employment by establishing new trades and business enterprises
- Enrichment in availability of internet and communication technologies
- Enhancement of technology in agriculture, education, banking and health industry
- Improvement in societal security
- Open the gateways for extensive development of rural sector.

Along with all these multiplier effects, the scheme will help in boosting the socioeconomic sphere of the country.

3   Saubhagya Yojna Evaluation Using Twitter

Electricity plays a significant role in the economic development of a country and the most essential part is its systematic and efficacious administration of distribution. Therefore, a government policy is required to develop the infrastructure and perpetual enhancement for electricity distribution that makes usage of laws, order, standards, regulations and principles. Although, policies are made to improve the process of decision-making, to ensures positive end results that elevates a community or society and are driven impartial, the reaction may contrasting over disparate group of people or society. Therefore, bringing and pinpointing attention over individual's opinion in response to governmental decisions is a critical step towards finalizing the consequent government course of actions.

Saubhagya, is undeniably a favourable step to shape up the brighter and better prospects for India. However, the facts and records reflect the mixed reactions from indian citizens for this policy. Thus, concerning on the subject, the paper reflects the inclination of citizens about this government policy by implementing the concept of opinion mining. Figure 2 represents system architecture for sequential execution of modules. The process of collecting the relevant data and the mechanism of opinion mining by applying five machine learning techniques using three performance measures to this data has been discussed in subsequent sections.

3.1   Data Aggregation and Preprocessing

Twitter, being a popular social networking service is used as a tool for the collection of data. It provides a multiplex platform by posting various messages or tweets and by sharing the public concern over a specific topic with millions of people across

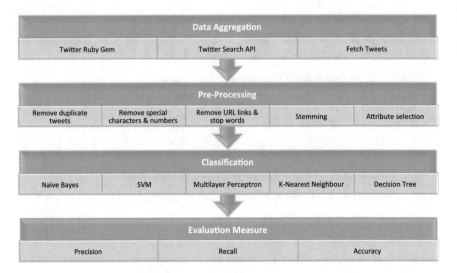

Fig. 2 System architecture for opinion mining of tweets

Table 1 Daily status of tweet collection

Phase I		Phase II		Phase III		Phase IV	
Date	Number of tweets	Date	Number of tweets	Date	Number of tweets	Date	Number of tweets
9/10/2017	12	16/10/2017	59	23/10/2017	56	30/10/2017	65
10/10/2017	19	17/10/2017	177	24/10/2017	282	31/10/2017	112
11/10/2017	4	18/10/2017	33	25/10/2017	75	1/11/2017	11
12/10/2017	4	19/10/2017	17	26/10/2017	36	2/11/2017	9
13/10/2017	37	20/10/2017	17	27/10/2017	3	3/11/2017	2
14/10/2017	135	21/10/2017	31	28/10/2017	9	4/11/2017	0
15/10/2017	76	22/10/2017	59	29/10/2017	16	5/11/2017	27
						6/11/2017	38
						7/11/2017	50
						8/11/2017	4

world. Therefore, to extract the tweets related to Saubhagya scheme, Twitter search
API has been used.

An application has been developed using Ruby on Rails which utilise the gem
"twitter". The twitter ruby gem is a ruby interface which is used to extract tweets
from the twitter search API. The developed application has been registered with
Twitter and received the access token, i.e. o auth credentials for code integration.

#Saubhagyascheme, #Saubhagyayojna, #Saubhagya are some of the few hashtags
used in this process. Thereafter, Twitter API returned the tweets containing specific
keywords. A total of 1262 tweets are collected in all the four phases of processing.
The daily status of tweet collection is listed in Table 1.

The collected tweets were pre-processed to organise the requisite data into an organised manner in order to get the standardized, uniform and clean data for opinion mining. The normalized tweets were then classified by using various machine learning techniques. The preprocessing procedure comprises of various steps. Tweets duplication may possible while collecting data. Therefore, removal of duplicate tweets is the primary step in data preprocessing followed by (a) removal of numbers and special characters such as #, @, etc. (b) removal of URL links and stop words like is, are, the, etc. Thereafter, stemming has been performed to reduce crumpled words in their root form. The resultant data is finally pre-processed for feature selection and categorized into polarity classification, i.e. positive, negative and neutral. Some examples of the attribute selected are: Lighting, Shedding, Connectivity, Families, Homes, Poor, Scheme, Bulb, Crackdown, etc.

3.2 Machine Learning Techniques for Performance Evaluation Measure

The collected and cleansed data is then used to classify the sentiment polarities (positive, negative) of opinions. Various approaches are used for the purpose of sentiment classification where machine learning and lexicon based approaches are more popular amongst others. Machine learning techniques helps in making data driven predictions or decisions by the use of various computational methods. In this paper, we empirically analyse five standard machine learning algorithms (namely NB, SVM, MLP, kNN and DT) and comparatively analyse them over three parameters of evaluation measures (Precision, Recall and Accuracy) used to assess the performance of overall sentiment classification.

A *Naive Bayesian* model [16, 17] belongs to the class of simple probabilistic classifiers by applying Bayes theorem with strong or naive independence assumptions. NB classifier assumes that a particular feature value is independent of any other feature value. A small amount of training data is required to estimate the parameters necessary for classification. A *Support Vector Machine* [16, 18] is a discriminative classifier based on supervised learning which identify the classification pattern and categorizes the data with an optimal hyperplane. A hyperplane is a line dividing the group of instances having different classes in an infinite dimension space, mainly used for classification and regression analysis. A *Multilayer Perceptron* [16, 19] is a network of neurons called perceptron belongs to the class of feed forward artificial neural network. MLP networks comprises of input layer (source nodes), hidden layers (computation nodes) and output layer (destination nodes) where signal propagates in one direction (forward) layer-by-layer. They are mainly used for pattern classification, recognition, prediction and approximation. A *k-Nearest Neighbours* [16, 20] is one of the most simplest and essential algorithm in machine learning with a non-parametric feature (i.e. does not make any underlying assumptions about data distribution). It uses training data to classify objects into classes identified by an attribute. The object is classified on the basis of voting done by its neighbours and the object being assigned to the class most common among its k-nearest neighbours. A *Decision Tree* [16, 21] is decision support tool which uses tree like model of decisions and their outcomes. It is drawn upside down with its root/single node at the top, branches into possible outcomes ending with leaf nodes. Classification rules can be represented by the path from root node to leaf node.

Precision, Recall and Accuracy [16] are the standard evaluation parameters to assess the performance of sentiment classification. *Precision* describes the correctness or quality of the classifier; how precisely the classifier measures with the ratio of true positives to predicted positives. *Recall* signifies the sensitivity of the classifier; how well the classifier can measure the completeness or quantity of the ratio between true positives and all the actual positives. *Accuracy* describes the closeness of a measurement to the true value. It is the proportion of true results among the total number of cases examined.

3.3 Opinion Mining of Tweets

The process of opinion mining has been implemented for the recognition and classification of opinions or sentiments of each tweet [22]. The mechanism of opinion mining aims to understand the emotions or feeling of an individual writing text about any specific interest area [23]. In this paper, opinion mining is applied on collected tweets of Saubhagya scheme and distributes them in three categories: Positive (P), Negative (N) and Neutral (Nu). Table 2 represents the mining results of all the four phases of data collection on daily basis.

Figure 3 represents the phase wise distribution of positive, negative and neutral tweets. From the graphical representation of the stats it is evident that over a count of 281 tweets collected during first phase, approximate 10% of the tweets were neutral and remaining 90% of the tweets were positive resulting in favourable reaction of people about the scheme. No single negative tweet has been recorded in phase one.

Second phase reflects a mixed reaction of citizen, but still the percentage of positive tweets (63%) are much higher than that of negative (9.6%) and neutral (26.6%) ones. Similar scenario has been reported for phase three with almost 78% of the positive tweets on record along with approximate 6% of negative and 16% of neutral tweets. Phase four signifies the hike in percentage of neutral tweets with 71% in contrast to positive and negative tweets with 17% and 12% respectively. The reason behind this apparent variation is that most of the people express their views in first three phases of evaluation, whereas the last phase reported a lot of informational tweets about the scheme with the latest update of actions carried out for its implementation posted by media or civic.

Table 2 Opinion mining results of four phases

Date	N	Nu	P	Total	Date	N	Nu	P	Total
Phase I					*Phase II*				
9/10/2017	0	4	8	12	16/10/2017	0	3	40	43
10/10/2017	0	5	14	19	17/10/2017	4	25	16	45
11/10/2017	0	3	1	4	18/10/2017	12	6	9	27
12/10/2017	0	3	1	4	19/10/2017	4	1	7	12
13/10/2017	0	4	28	32	20/10/2017	1	3	13	17
14/10/2017	0	0	135	135	21/10/2017	0	4	23	27
15/10/2017	0	8	67	75	22/10/2017	0	16	31	47
Total	0	27	254	281	*Total*	21	58	139	218
Phase III					*Phase IV*				
23/10/2017	8	25	19	52	30/10/2017	3	45	4	52
24/10/2017	5	29	243	277	31/10/2017	26	80	5	111
25/10/2017	9	16	48	73	1/11/2017	3	5		8
26/10/2017	3	1	32	36	2/11/2017	2	4	2	8
27/10/2017	1	0	2	3	3/11/2017	0	1	1	2
28/10/2017	0	1	6	7	4/11/2017	0	0	0	0
29/10/2017	1	2	12	15	5/11/2017	1	0	26	27
Total	27	74	362	463	6/11/2017	1	31	6	38
					7/11/2017	1	44	5	50
					8/11/2017	0	3	1	4
					Total	37	213	50	300

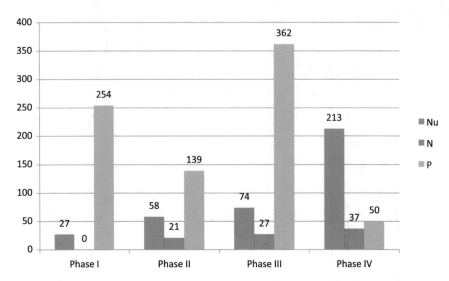

Fig. 3 Phase wise distribution of positive, negative and neutral tweets of Saubhagya scheme

Table 3 Count of standard measure of evaluation for five machine learning techniques

ML technique(s)	Precision	Recall	Accuracy
Naive bayes	84.1	77.7	77.73
SVM	91.5	91.8	91.77
Multilayer perceptron	91	91.5	91.47
K-nearest neighbour	91.6	91.5	91.47
Decision tree	90.5	90.7	90.67

4 Results and Findings

The data set consists of 1262 tweets collected from Twitter and opinion mining is used to find their sentiment polarity about Saubhagya scheme. Following machine learning approaches are used for tweets classification of this scheme: Naive Bayes, SVM, Multilayer Perceptron, K-nearest neighbour and Decision Trees. In this research work, calculation of the standard measure of evaluation i.e. Precision, Recall and Accuracy for every approach has been done and the experimental results are listed in Table 3.

A comparative analysis is performed for all the approaches in order to seek the best approach. NB performed with least accuracy of 77.73% whereas SVM classifier performed the best among all with 91.77% accuracy. MLP depicts promising results by reflecting 91.47% accuracy, almost equivalent to it is kNN having an accuracy of 91.47% followed by DT with an accuracy of 90.67%. Analysing the resulting figures, we found that support vector machine was better than all other approaches in terms of precision, recall and accuracy. Figure 4 represents the graphical illustration

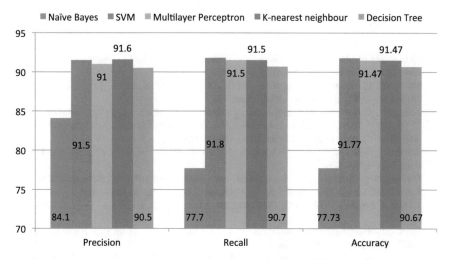

Fig. 4 Standard measure of evaluation for five machine learning techniques

of the comparative performances of all the five ML algorithms used for opinion classification of Saubhagya yojna.

5 Conclusion

Pradhan Mantri Sahaj Bijli Har Ghar Yojana or Saubhagya has been launched by government of India in order to elevate the social and economical curve of the nation by means of electrifying the rural and urban areas across country. Every governmental scheme or policy when put into effect has some beneficial and adverse impact over society. Therefore, the objective of this paper was to evaluate this scheme from common man's perspective by analysing public reaction using opinion mining techniques with the help of tweets. The study reflects that a large number of indian citizen are delighted with the scheme and their opinion are in a favourable direction. Almost, 63% of the tweets are identified to be in favour of this scheme, whereas the ratios of negative and neutral tweets are nearly, 7% and 30%, respectively.

Few opinion mining techniques namely, Naive Bayes, SVM, Multilayer Perceptron, K-nearest neighbour and Decision Trees have been compared on account of precision, recall and accuracy. Based on the calculated results, we can conclude that SVM obtained the best accuracy followed by MLP, kNN, DT and NB which reflects lowest accuracy. Based on all the three parameters of standard measure of evaluation, Support Vector Machine shown the best performance among all the techniques.

As a potential direction of future work, more policies can be evaluated by utilizing the concept of opinion mining. The process of policy evaluation may become better by optimising the task of sentiment classification. More computing techniques and hybrid classifiers may be analysed and validated in order to enhance the evaluation measures.

References

1. Kumar A, Sharma A (2016) Paradigm shifts from e-governance to s-governance. The human element of big data: issues, analytics, and performance. p 213
2. National e-Governance Plan. http://www.mit.gov.in/content/national-e-governance-plan
3. Digital India. http://www.cmai.asia/digitalindia/
4. National e-Governance Plan. http://meity.gov.in/divisions/national-e-governance-plan
5. Bhatia MPS, Kumar A (2010) Paradigm shifts: from pre-web information systems to recent web-based contextual information retrieval. Webology 7(1)
6. Bhatia MPS, Khalid AK (2008) A primer on the web information retrieval paradigm. J Theor Appl Inf Technol 4(7)
7. Saubhagya scheme. All you need to know. http://www.livemint.com/Industry/6agWD5oCYBi6jokfKc3QjP/Saubhagya-scheme-All-you-need-to-know.html
8. Kumar A, Sebastian TM (2012) Sentiment analysis on twitter. IJCSI Int J Comput Sci Issues 9(4):372
9. Press Information Bureau, Government of India, Ministry of Power. http://pib.nic.in/newsite/PrintRelease.aspx?relid=171148
10. Dayal D, Gram U, Yojana J. https://en.wikipedia.org/wiki/Deen_Dayal_Upadhyaya_Gram_Jyoti_Yojana
11. Integrated Power Development Scheme. A priority programme of Govt of India. http://www.ipds.gov.in/Form_IPDS/About_IPDS.aspx
12. Press Information Bureau, Government of India, Ministry of Power. http://pib.nic.in/newsite/PrintRelease.aspx?relid=123595
13. Press Information Bureau, Government of India, Cabinet. http://pib.nic.in/newsite/PrintRelease.aspx?relid=111621
14. What is Saubhagya Yojana. http://indianexpress.com/article/what-is/what-is-saubhagya-yojana-pradhan-mantri-sahaj-bijli-har-ghar-yojana-4860947/
15. Saubhagya Scheme: Features, Benefits and Challenges. http://www.jagranjosh.com/current-affairs/saubhagya-scheme-features-benefits-and-challenges-1507620376-1
16. Kumar A, Jaiswal A (2017) Empirical study of twitter and tumblr for sentiment analysis using soft computing techniques. In: Proceedings of the world congress on engineering and computer science vol 1
17. Naive Bayes classifier. https://en.wikipedia.org/wiki/Naive_Bayes_classifier
18. Support vector machine. https://en.wikipedia.org/wiki/Support_vector_machine
19. Multilayer perceptron. https://en.wikipedia.org/wiki/Multilayer_perceptron
20. K-nearest neighbour algorithms. https://en.wikipedia.org/wiki/K-nearest_neighbors_algorithm
21. What is a Decision Tree Diagram. https://www.lucidchart.com/pages/decision-tree
22. Amolik A, Jivane N, Bhandari M, Venkatesan M (2016) Twitter sentiment analysis of movie reviews using machine learning techniques. Int J Eng Technol 7(6):1–7
23. Kumar A, Teeja MS (2012) Sentiment analysis: a perspective on its past, present and future. Int J Intell Syst Appl 4(10):1

A Brief Survey on Mass-Based Dissimilarity Measures

Suman Garg and S. K. Jain

Abstract Mass-based dissimilarity measures, an alternative to distance-based measures are effective in several data mining tasks. The performance of distance-based dissimilarity measures changes significantly as the data distribution changes. This is due to dependence of distance-based measures only on geometric positions of the given instances. But mass-based dissimilarity measures calculate dissimilarity by considering the data distribution which makes them effective in high dimensional data sets where generally used distance-based measures like l_p norm are not so effective. This paper discusses some widely used distance-based measures and some existing data-dependent dissimilarity measures. It discusses different mass-based dissimilarity measures in detail and compares them based upon some characteristics.

Keywords Data-dependent dissimilarity measure · Mass-based · Nearest neighbor

1   Introduction

Dissimilarity measures are the methods to measure how like or alike two given objects are. Data mining algorithms use dissimilarity measures to find out the closest matches of given instances in a data set. Distance measures have been a core process for many of the data mining applications to compute the dissimilarity between the given instances because they involve simple calculation. But many researches have shown that the performance of distance measures rely upon the data distribution. Particularly in a high dimensional space, the distribution of data becomes sparse and concept of distance turn out to be meaningless.

S. Garg (✉) · S. K. Jain
National Institute of Technology Kurukshetra, Kurukshetra, India
e-mail: grgsuman04@gmail.com

S. K. Jain
e-mail: skj_nith@yahoo.com

© Springer Nature Singapore Pte Ltd. 2019
S. Bhattacharyya et al. (eds.), *International Conference on Innovative Computing and Communications*, Lecture Notes in Networks and Systems 56,
https://doi.org/10.1007/978-981-13-2354-6_41

Many distance-based algorithms like Minkowski distance (l_p norm), Euclidean distance (l_2 norm), etc., are commonly used for dissimilarity computation in data mining applications. From low to moderate dimensional data sets, these algorithms work well. But finding the reliable closest matches turn out to be a stimulating task as the number of dimension increases. But data-dependent dissimilarity measures produce reliable results in high dimensional data sets also because they overcome the key weaknesses of geometric- based dissimilarity measures, i.e., they are not entirely based upon geometric positions.

Mass-based dissimilarity measures are the data-dependent measures based upon human judgment to calculate dissimilarity, i.e., two instances at same interpoint distance are less similar in a dense region and more similar in a sparse region. Since these measures consider the data distribution to compute the dissimilarity, they over perform the distance-based measures, especially in high dimensional data sets.

In Sect. 2, we discuss some generally used geometric-based dissimilarity measures. In Sect. 3, we discuss some existing data- dependent dissimilarity measures and the mass-based dissimilarity measures. In Sect. 4, we compare the measures based upon some characteristics.

2 Geometric-Based Dissimilarity Measures

There is a large number of geometric (distance)-based measures in the literature discussed in [1]. Here, we discuss two commonly used measures which are Minkowski distance and cosine distance.

2.1 Minkowski Distance (l_p Norm)

The dissimilarity of two given d dimensional instances x and y can be calculated as [1]

$$l_p(x, y) = \left(\sum_{i=1}^{d} (| x_i - y_i |)^p \right)^{\frac{1}{p}} \tag{1}$$

where p>0 , || . || is the p order norm of a vector, x_i and y_i are ith component of vector x and vector y, respectively. Manhattan distance (l_1), Euclidean distance (l_2), and Chebysev distance (l_∞) are special cases of l_p norm and widely used measures.

2.2 Cosine Distance

In text mining applications, cosine distance is the popular choice to calculate dissimilarity between query and documents due to sparseness in high dimensional data sets. Because only a few terms of the dictionary are present in a document making most of the entries 0 that results in sparse distribution. And, Minkowski distance is not a good option in a high dimensional sparse distribution. So, cosine distance is used which can be calculated as [1]

$$d_{cos}(x, y) = 1 - \frac{x \cdot y}{|x| \times |y|} = 1 - \frac{\sum_{i=1}^{d} x_i \times y_i}{\sqrt{\sum_{i=1}^{d} x_i^2} \times \sqrt{\sum_{i=1}^{d} y_i^2}} \tag{2}$$

3 Data-Dependent Measures

Researches have shown that the performance of distance-based measures varies with data distribution and tasks in hand. The reason can be the dependence of distance-based measures on geometric positions of two instances. Many psychologists [2, 3] have argued that the judged similarity between two data instances is influenced by the measurement perspective and other instances in the vicinity. Data- dependent dissimilarity measures based upon mass estimation considers the data distribution to estimate the dissimilarity between two instances. Thus, it can overcome the weaknesses of distance-based measures and can be taken as an alternative to widely used distance- based measures. First, we discuss some existing data-dependent dissimilarity measures. Second, we discuss the mass estimation techniques so that reader can be familiar with mass estimation of a single instance. Third, we discuss the dissimilarity measures based upon mass which extends the concept of mass estimation of a single instance to dissimilarity calculation between two given instances.

3.1 Mahalanobis Distance

Mahalanobis distance of two d dimensional points x and y in the d dimensional space R_d is given as [4]

$$MD(x, y) = \sqrt{(x - y)^t S^{-1}(x - y)} \tag{3}$$

where S is covariance matrix of R_d and t is transpose operator. It considers the correlation between the different attributes of a data set and computes the covariance matrix across different dimensions to capture the variance across dimensions. Although it captures variation across dimensions but it does not capture the local distribution, i.e., distribution in one dimension. Also, it includes the calculation of inverse of covariance matrix which is expensive to compute, especially in high dimensional data sets.

3.2   Rank Transformation

In rank transformation, ranks are assigned to the instances in a dimension in increasing order starting from 1, 2, and so on. If n instances have the same value in a dimension and available rank is r, then the same rank calculated as average of rank to be assigned to n instances is given to them. The next instance having a different value assigned the rank as r+n. Finally, the dissimilarity of the two given data instances are calculated as a power mean of dissimilarity across all dimensions as in l_p norm given as in [5]

$$d_{rank}(x, y) = \left(\frac{1}{d} \sum_{i=1}^{d} (| x_i^r - y_i^r |)^p \right)^{\frac{1}{p}} \tag{4}$$

where x_i^r is the rank assigned to instance in ith dimension. It is a data-based measure because in the dense region, $| x_i^r - y_i^r |$ will be high and in sparse region, it will be small while the geometric distance can be same in both cases.

3.3   Lin's Probabilistic Measure

Dissimilarity of two data points x and y by using Lin's probabilistic measure can be calculated as in [6]

$$d_{lin}(x, y) = 1 - \frac{1}{d} \sum_{i=1}^{d} \frac{2 \times \log \sum_{z_i=min(x_i, y_i)}^{max(x_i, y_i)} P(z_i)}{\log P(x_i) + \log P(y_i)} \tag{5}$$

where $P(x_i)$ is the probability of x_i which can be computed from D as $P(x_i) = (f(x_i)+1)/N+u_i$. Here, $f(x_i)$ is the occurrence frequency of x_i in D and u_i is total number of distinct values in dimension i.

3.4   Mass Estimation

In 2010, the concept of mass estimation to be used in data mining applications was introduced [7]. It is based upon the idea of converting the original data space into mass space by calculating the mass around each instance and then using that mass space in various data mining algorithms. In [7], three algorithms have been proposed which are:

Mass Estimation   This algorithm calculates the mass of each data point of a small subsample sampled from a large original dataset. In this way, a large number of mass distributions are created by using different subsamples.

Mass Mapping   This algorithm takes the output of Mass Estimation algorithm and then converts the original data into mass space. The mass space has high dimensions than original data space.

Task Specific   In the third algorithm, the task-specific algorithm can use this mass space. Anomaly detection can avoid the Mass Mapping algorithm and directly use the Mass Estimation approach to rank the instances.

In this paper, Ting and Wells focus on one-dimensional mass distribution only. In 2013, the generalized version [8] for multidimensional mass estimation was proposed by using multiple random regions covering a point. Although one-dimensional mapping can be used to solve multidimensional problems but multidimensional mapping reduces its time complexity by using the half-space trees instead of look-up tables. In this paper, it has been shown that multidimensional mass estimation is a more effective approach than one- dimensional mass estimation.

Although mass estimation is a useful concept for many data mining applications, it is a unary function as it calculates the mass distribution around a single instance. It does not calculate the dissimilarity between two points. So, new algorithms need to be developed for the implementation of mass estimation in data mining applications. But dissimilarity measures can directly replace the existing measures while the algorithms can be left as it is.

3.5 m_p Dissimilarity

In [9], a new data-dependent dissimilarity measure named as m_p dissimilarity has been proposed. It is a binary function which computes the dissimilarity between two given points based upon the mass (number of data instances) around them. In each dimension, it is calculated as the probability mass of a region enclosing the two given instances x and y. Instead of distance calculation, it calculates the relative position of two points from the data distribution around them. If there are many other instances around the two given instances, then they are more dissimilar in that particular dimension and vice versa. The final dissimilarity is calculated by power mean of dissimilarities in each dimension. This is same as l_p norm. The difference is only that m_p dissimilarity is based upon probability mass while l_p norm is based upon geometric difference. m_p dissimilarity is given as [9]

$$m_p(x, y) = \left(\sum_{i=1}^{d} \left(\frac{|R_i(x, y)|}{N} \right)^p \right)^{\frac{1}{p}} \tag{6}$$

where $|R_i(x, y)|$ is number of instances in region $R_i(x,y)$, $R_i(x,y) = [\min(x_i,y_i)\text{-}\delta, \max(x_i,y_i)+\delta]$, $\delta \geq 0$, and n is total number of instances in data set.

m_p dissimilarity is based upon the concept that two data instances in a dense area seems to be less similar as compared to the sparse area. This paper [9] has shown that m_p dissimilarity produces more correct nearest neighbors than l_p norm and cosine distances in high dimensional data sets in information retrieval and classification applications. But the calculation of number of data instances in between the two given data instances is not so straightforward. It involves a range search in each dimension. So, it has high time and space complexity than l_p norm. It has O(d log n) time complexity while l_p norm has O(d) time complexity.

3.6 Mass-Based Dissimilarity

In [10], a generic data-dependent dissimilarity measure called as mass-based dissimilarity has been proposed. It extends the concept of half-space mass [11]. It is calculated as the probability mass of the region enclosing the two given data instances x and y where iForest [12] has been used as a scheme to define the regions in iForest, each tree is built from a small subsample of data $(D_i \subset D, | D_i |= \psi >> N)$ with random non-empty partitioning of the space until instances in D_i are isolated or the tree height reaches the maximum of $\log_2 \psi$. Once a tree is built, the data mass in each node is calculated from the entire data D. Using a collection of t trees, the dissimilarity of x and y is estimated as in [10]

$$m_e(x, y) = \frac{1}{t} \sum_{i=1}^{t} \left(\frac{| R_i(x, y) |}{N} \right) \tag{7}$$

where R_i(x, y) is the deepest node where x and y appears together in ith tree. The authors in [10] have shown that mass- based dissimilarity overcomes the key weaknesses of three data mining algorithms. First considering clustering, the generally used algorithm DBSCAN suffer from the problem of not identifying the clusters of varying densities. But the proposed mass-based algorithm MBSCAN is able to find the clusters of varying densities. Second, the KNN anomaly detectors are unable to detect local anomalies but the proposed MkNN algorithm address this problem. Third, the M-MlkNN algorithm improves the multi-label classification result of MlkNN (distance based).

Mass-based dissimilarity with iForest implementation is a tree-based implementation, so a large number of trees need to be constructed to approximate the consistent results. This can increase the running time of the algorithm.

3.7 m_p Dissimilarity with Histogram

An earlier version of m_p dissimilarity [9] outperforms the distance measures in information retrieval and classification applications but at the cost of high time complexity. Range search is used to find the instances in between two given instances and increases the computation time. Other implementations [13] using histogram avoids this range search. In m_p dissimilarity using histogram, the whole data set is divided into multiple bins. Binning can be performed in two ways: equal width and equal frequency. Each equal width bin is of the same size and each equal frequency bin has the same number of instances. Number of data instances in each bin can be estimated in preprocessing step. So, the calculation of the number of data instances in between the two given data instances involves the detection of bins in between them and bins containing them. Number of instances in between the two instances is just the sum of number of instances in each detected bin. Thus, time complexity of the dissimilarity calculation reduces to 0(d) which is the same as l_p norm.

4 Characteristics

All the measures discussed above can be compared based upon following characteristics in Table 1. Here, self-dissimilarity is a data-dependent feature which should be

Table 1 Comparison of dissimilarity measures based upon some characteristics

Name of measure	Type of measurement	Dependency on distribution of data set?	Constant self dissimilarity?	Time complexity	
				Preprocessing	Execution
Minkowski distance	Distance	No	Yes	O(Nd)	O(d)
Mahalanobis distance	Data covariance	Yes	Yes	$O(Nd^2)$	O(d)
Rank transformation	Rank difference	Yes	Yes	$O(Nd\eta + d\eta^2)$ [14]	$O(d \log_2 \eta)$
Lin's probabilistic measure	Data mass	Yes	Yes	$O(Nd\eta + d\eta^2)$ [14]	$O(d \log_2 \eta)$
m_p dissimilarity	Data mass	Yes	No	O(Nd)	$O(d \log_2 N)$
Mass-based dissimilarity	Data mass	Yes	No	$O(tN \log_2 \psi + t\psi^2)$	$O(t \log_2 \psi)$
m_p dissimilarity with histogram	Data mass	Yes	No	$O(Nd\eta + d\eta^2)$ [14]	$O(d \log_2 \eta)$

variable based upon the data distribution. Time complexity of rank transformation, Lin's measure, and m_p dissimilarity are based upon the one-dimensional implementation as in [14]. N is the total number of instances in a data set. d is number of dimensions in data set. η is average number of intervals overall dimensions in the equal frequency division of bins. ψ is subsample size to build trees and t is number of trees in iForest in mass-based dissimilarity. d(x,y) is dissimilarity of two given instances x and y.

5 Conclusion

This paper studies existing data-dependent dissimilarity measures and different mass-based data-dependent dissimilarity measures and concludes that mass-based computation is an efficient approach to gather the reliable nearest neighbors in data mining tasks. Mere replacement of distance-based measures with mass-based dissimilarity measures in many data mining algorithms like clustering, classification, anomaly detection, and content-based information retrieval have shown significant improvement over the existing measures, especially in high dimensional data sets. In the future, it is also important to detect and work upon the cases where mass-based measures do not provide correct results. It can be very useful to find out other applications where mass-based dissimilarity measures can improve the performance of existing algorithms.

References

1. Deza M, Deza E (2008) Encyclopedia of distances, Springer
2. Krumhansl CL (1978) Concerning the applicability of geometric models to similarity data: the interrelationship between similarity and spatial density
3. Tversky A (1977) Features of similarity. Psychol Rev 84(4):327
4. Mahalanobis PC (1936) On the generalized distance in statistics. National Institute of Science of India
5. Conover WJ, Iman RL (1981) Rank transformations as a bridge between parametric and nonparametric statistics. Am Stat 35(3):124–129
6. Lin D (1998) An information-theoretic definition of similarity. InIcml 98(1998): 296–304
7. Ting KM, Zhou G-T, Liu F, Tan JSC (2013) Mass estimation. Mach Learn 90(1):127–160
8. Ting KM, Zhou, G-T, Liu F, Tan JSC (2010) Mass estimation and its applications. In: Proceedings of KDD10: The 16th ACM SIGKDD international conference on Knowledge discovery and data mining, pp 989–998
9. Aryal S, Ting KM, Haffari G, Washio T (2014) mp-dissimilarity: a data dependent dissimilarity measure. In: 2014 IEEE International conference on data mining (ICDM), pp 707–712. IEEE
10. Ting KM, Zhu Y, Carman M, Zhu Y, Zhou ZH (2016) Overcoming key weaknesses of distance-based neighbourhood methods using a data dependent dissimilarity measure. In: Proceedings of the 22nd ACM SIGKDD international conference on knowledge discovery and data mining, pp 1205–1214. ACM
11. Chen B, Ting KM, Washio T, Haffari G (2015) Half-space mass: a maximally robust and efficient data depth method. Mach Learn 100(2–3):677–699

12. Liu FT, Ting KM, Zhou ZH (2008) Isolation forest. In: ICDM'08. Eighth IEEE international conference on data mining, 2008, pp 413–422. IEEE
13. Aryal S, Ting KM, Washio T, Haffari G (2017) Data-dependent dissimilarity measure: an effective alternative to geometric distance measures. Knowl Inf Syst 53(2):479–506
14. Aryal S (2017) A data-dependent dissimilarity measure: An effective alternative to distance measures. https://doi.org/10.4225/03/5a2f0721de429

Improving Recognition of Speech System Using Multimodal Approach

N. Radha, A. Shahina and A. Nayeemulla Khan

Abstract Building an ASR system in adverse conditions is a challenging task. The performance of the ASR system is high in clean environments. However, the variabilities such as speaker effect, transmission effect, and the environmental conditions degrade the recognition performance of the system. One way to enhance the robustness of ASR system is to use multiple sources of information about speech. In this work, two sources of additional information on speech are used to build a multimodal ASR system. A throat microphone speech and visual lip reading which is less susceptible to noise acts as alternate sources of information. Mel-frequency cepstral features are extracted from the throat signal and modeled by HMM. Pixel-based transformation methods (DCT and DWT) are used to extract the features from the viseme of the video data and modeled by HMM. Throat and visual features are combined at the feature level. The proposed system has improved recognition accuracy compared to unimodals. The digit database for the English language is used for the study. The experiments are carried out for both unimodal systems and the combined systems. The combined feature of normal and throat microphone gives 86.5% recognition accuracy. Visual speech features with the normal microphone combination produce 84% accuracy. The proposed work (combines normal, throat, and visual features) shows 94% recognition accuracy which is better compared to unimodal and bimodoal ASR systems.

Keywords Automatic speech recognition · Hidden Markov model
Multimodal system · Visual lip reading · Discrete cosine transform
Discrete wavelet transform

N. Radha (✉) · A. Shahina
Department of Information Technology, SSN College of Engineering, Chennai, India
e-mail: radhan@ssn.edu.in

A. Shahina
e-mail: shahina@ssn.edu.in

A. Nayeemulla Khan
School of Computer Science and Engineering, Vellore Institute of Technology,
Chennai, India
e-mail: nayeemulla.khan@vit.ac.in

© Springer Nature Singapore Pte Ltd. 2019   397
S. Bhattacharyya et al. (eds.), *International Conference on Innovative Computing
and Communications*, Lecture Notes in Networks and Systems 56,
https://doi.org/10.1007/978-981-13-2354-6_42

1 Introduction

ASR systems are sensitive to changes in the environmental conditions. Standard (close-talk) microphones (referred to as normal microphone or NM) are susceptible to all types of noises and the amount of interfering noises impedes the performance ASR systems. One approach to enhancing the performance of an ASR system is the use of multiple sources of information. Speech captured from an alternative speech sensor and visual lip reading are the two sources of information that are complementary to the NM speech and could aid in building a multimodal ASR system. Alternate speech sensor (throat microphone) is a transducer that worn on the neck and absorbs the vibrations directly from the speaker's throat. It picks up intelligible speech sounds from the pharynx through the throat tissue. Throat Microphone (TM) has voicing and is significantly immune to environmental noise, due to the close contact with the throat skin. TM provides a high Signal-to-Noise Ratio (SNR) over large part (from 0 to 3500 Hz) of the audio frequency range. The throat microphone signals were used for speech, speaker recognition [1], and language identification [2] studies. A robust ASR system built by combining normal and throat microphone signals at feature level was proposed in [3, 4]. In [5], the probabilistic optimum filter was used to estimate the speech features of the normal microphone to build a robust ASR system. The experiments were conducted in both clean and noisy environmental conditions. An improved robust speech recognition system was proposed using the combination of close-talk and throat microphone features under nonstationary background noise [6, 7]. The existing works show that the recognition error rate yet needs to further improve for ASR systems.

The phonemes from the audio data are easily affected by the noise with respect to the place of articulation. However, these phonemes seem to be distinguishable in the visual speech. The visual information comes from the speaker's mouth movements and is called as Visual Lip Reading (VLR). This lip reading technique promises the enhanced to enhance the performance of ASR systems, if used in tandem with audio signals [8, 9]. This is because the visual information acts as an another source of information and is used to build a speech system. VLR-based ASR systems became important for audio-visual speech recognition and human–computer interaction domain, since it is a simplest way to interact with the computer. The application of VLP includes speaker recognition, video surveillance, HCI, etc. Lip tracking is the main part of the VLR system followed by feature extraction, which plays a vital role in the VLR-based speech system. Lip tracking is also useful for the hearing-impaired people.

Feature extraction methods are classified into pixel-based, geometric, and motion-based in [8]. In this work, the pixel-based method is used for feature extraction. To extract features (pixel-based) for visual speech, various image transform methods such as DCT, Principal Component Analysis (PCA), DWT, and Linear Discriminant Analysis (LDA) have been employed [10]. Among these, DCT and DWT have been shown perform well than other methods and the same used in this work. Visual features extraction using the pixel-based method provides the high dimensional feature vectors. These high dimensional features are reduced by PCA. The visual feature extraction is performed by applying the DCT/DWT transformation techniques, followed by PCA for dimensionality reduction.

The following systems are built: The unimodal systems NM-based ASR system (λ^a), TM-based ASR system (λ^t), and visual speech recognition (λ^v), the unimodal combinations NM-TM (λ^{at}), NM-VLP (λ^{av}), TM-VLP (λ^{tv}), and the fixed combination of three modes, NM-TM-VLP (λ^{atv}). All the seven systems have been evaluated. From the unimodals, likelihood scores (acoustic/visual) are computed based on the given feature observations of normal, throat, and visual speech mode, respectively. Acoustic features are extracted using MFCC. Visual features are extracted using the DWT and DCT transformation. Tied-state HMM models are built and used to train the datasets. In testing, all the unimodals are evaluated. A multimodal system for the video was proposed in [11, 12]. In this multimodal recognition, the features of $\lambda^a(f_a)$, $\lambda^t(f_t)$, and $\lambda^v(f_v)$ systems are concatenated using a simple feature concatenation. The combined feature vector of each frame for a multimodal system is given as

$$f = [f_a, f_t, f_v]$$

For the given sequence of observed features f, models are trained using HMM. This paper is organized as follows. In Sect. 2, analysis of the acoustic and visual signal is discussed. Feature extraction, modeling, experimental results, and the performance evaluation are discussed in Sect. 3. Summarization and conclusion of the work are presented in Sect. 4.

2 Acoustic and Visual Analysis

This section provides some unique acoustic characteristics which include spectrogram and resonant frequency of the Throat Microphone speech (TM) as compared with the Normal Microphone speech (NM) as well as that of the accompanying visual lip movements are discussed. From the acoustic characteristics, the major differences in speech between the TM and NM are observed.

In the case of back vowels /u/ (as in "two") and /o/ (as in "zero"), they are articulated near the end of the vocal cavity as shown in Fig. 1. In the normal microphone speech where the F2 is lowered because of lip rounding, whereas, this lowering effect is not present in Throat Microphone speech. In the case of front vowels as /i/ as in "five" and also /e/ as in "seven", are articulated near the front of the oral cavity (NM) as shown in Figs. 2 and 3. This indicates the F2 formants have high energy distribution and low F1 formants articulation. In the TM speech, the energy distribution happens within 3000 Hz whereas the formants F1 and F2 does not give much energy difference with respect to the location of the formants as compared to NM speech.

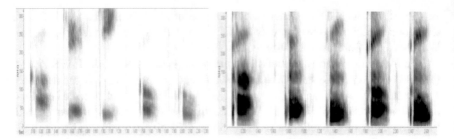

Fig. 1 The spectrogram for the vowels /a/, /i/ (as in "nine"), /e/ (as in "ten"), /o/ (as in "zero") and /u/ (as in "two")

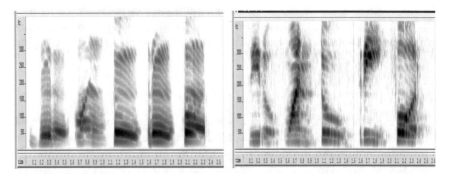

Fig. 2 The spectrogram for the digits (0–5) recorded simultaneously from TM (left) and NM (right)

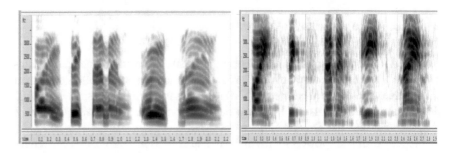

Fig. 3 The spectrogram for the digits (5–9) recorded simultaneously from TM (left) and NM (right)

In the case of central also mentioned as mid vowel, /a/ in both the NM speech and TM speech, the formants F1 and F2 vary within a particular range of 2000 Hz due to vowels articulation on the measurements with the jaw positions relatively low. It is observed that, the F2 is close to F1 in the NM speech but the closeness is not observed in the TM speech.

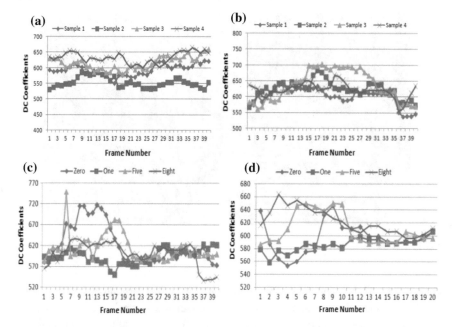

Fig. 4 Visual speech analysis for visemes of same speaker (**a**), (**b**) and different speaker (**c**), (**d**)

Figure 4 shows the variation among the different DC components (arithmetic mean of the gray-scale image) for every sequence frames corresponding to the utterances of each digit. This shows that each viseme looks different in visual domain for different uttered visemes of digits and looks similar for the same utterances. Among all the speaker's, this component shows the high variation. The next section discusses the various process involved visual lip reading and the experimental results.

3 Performance Evaluation

3.1 Database for the Study

Multimodal audio–video recording of digits using three modes are collected from 10 (2 male and 8 female) speakers and is used as a database for this study. Recordings are simultaneously obtained using a standard microphone, a throat microphone, and a camera. The recordings are obtained under laboratory conditions. The camera used is a 50fps SONY Handycam HDR-PJ660/B Camcorder. 50 utterances of each digit are collected from every speaker, out of which 35 utterances are used for training purpose, and the remaining 15 utterances are used for data testing purpose [13]. The size of the training and testing dataset are the same for the normal, throat, and visual

Fig. 5 Recording setup

speech data. The speech data's are recorded under normal laboratory environmental conditions and with the same lighting throughout the recordings. The camera is placed at a distance of about 30 cm and from the speaker at height of 65 cm as shown in Fig. 5.

3.2 Audio-Visual Feature Extraction and Modeling

Figure 6 shows a block diagram of the proposed multimodal speech recognition system using three modes of acoustic (NM, TM) information and visual lip reading for building a multimodal ASR system. The multimodal speech system involves the following steps starting from data collection, feature extraction from acoustic and visual signals, modeling, recognition of individual modes (unimodal), and combined features for multimodal recognition. From the video data, the face detection is followed by extraction, if the lip region is done using Viola–Jones algorithm [14]. This algorithm shows robustness to variation in pose, orientation, and illumination conditions [15]. The sequences of images are then classified into speaking and nonspeaking frames. Region of Interest (ROI) is normalized into 64×40 frame size. From the normalized ROI, the DCT and DWT coefficients are extracted.

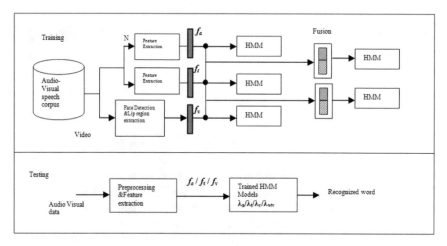

Fig. 6 Block diagram of the proposed multimodal speech system

DCT decomposes the input sequence into a weighted sum of basic cosine sequences. Block-based DCT is used in this work. It divides the ROI image into 8 non-overlapping blocks and then converts the block into DCT coefficients that are defined as

$$f_v = [\{f_1 \cdots f_{64}\}]^v$$

DCT computation for a lip frame is given as

$$F_{xy} = \alpha_x \alpha_y \sum_{i=0}^{N-1} \sum_{j=0}^{N-1} I_{i,j} \cos \frac{(2i+1)y\pi}{2N} \cos \frac{(2j+1)x\pi}{2N} \quad 0 \leq k \leq N-1 \quad (4)$$

where

$$\alpha(0) = \sqrt{\frac{1}{N}}, \, \alpha(k) = \sqrt{\frac{2}{N}} \quad 1 \leq k \leq N-1$$

The dynamic lip variations are well captured by DCT coefficients and it is varied to lighting and illumination conditions [16]. DWT has inherent multi-resolution property. DWT decomposes (sub-band—4) the visual speech signal into low frequency as approximation coefficients and high frequency as detailed coefficients in an iterative processing manner. Only approximation coefficients are used as features in this work because these coefficients contain significant information about visual lip as compared to other coefficients. The DWT-based feature coefficients per visual lip frame are obtained as

Fig. 7 Left-to-right single
stream HMM

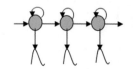

$$f_v = [\{f_1 \cdots f_{256}\}]^v$$
$$f(t) = \sum_k a_k \psi_k(t) \quad a_{jk} = \sum_t f(t)\varphi_{jk}^*(t)$$

where the * is the complex conjugate notation symbol and ψ is some wavelet function. The wavelet transforms visual image into wavelet coefficients. The visual information is highly represented in a compact manner with low pass approximation band [9].

The 39 dimensional mel-frequency cepstral coefficient feature vector (13 energy coefficients, 13 delta coefficients, and double delta coefficients) is extracted from acoustic (f_a) as well as throat signal (f_a) is given as follows.

$$f_a = [\{f_1 \cdots f_{13}\}, \{f_{14} \cdots f_{26}\}, \{f_{27} \cdots f_{39}\}]^a$$
$$f_t = [\{f_1 \cdots f_{13}\}, \{f_{14} \cdots f_{26}\}, \{f_{27} \cdots f_{39}\}]^t$$

The DCT coefficients are wavelet coefficients and mel-frequency cepstral coefficients, are modeled by using HMM [17]. Figure 7 shows the representation of viseme model by a left-to-right Gaussian HMM model (single stream). Each state in HMM is represented a Gaussian Mixture Model (GMM).

A GMM is a parametric model characterized by a mean and variance. The common method for computing viseme likelihood is a GMM probability density function of the observed feature vector, o is given by

$$b_i(o) = \frac{1}{(2\pi)^{\frac{N}{2}} \left| \Sigma_i^{-1} \right|^{\frac{1}{2}}} e^{-\frac{1}{2}(o-\mu_i)^{\mathsf{T}} \Sigma_i^{-1}(o-\mu_i)}$$

where μ_i is the state mean vector, Σ_i is the covariance matrix, and N is the dimension of feature vector o. The probability of the feature vector o being in any of I viseme models denoted by λ, is shown as

$$p(o|\lambda) = \sum_{i=1}^{I} w_i b_i(o)$$

For a large set of training feature vectors, $o = \{o_1, o_2, o_3 \ldots o_n\}$ on a particular viseme model, the GMM denotes the union of Gaussian pdf's, which is estimated by maximum

$$P\left(\{o_1, o_2, o_3, o_4 \ldots o_{m-1}|\lambda_j\right) = \max_{m-1} P\left(o_n|\lambda_j\right) P\left(\lambda_j\right)$$

where max is represents the overall possible paths from $q_i \ldots q_{t-1}$ and selects the best path end time t. The resultant viseme selection is the viseme utterance associated with visemes model with the highest probability. The training and the testing model algorithm for visual speech is given below:

Algorithm 1 Visual lip reading for training	Algorithm 2 Visual lip reading for testing
1:Begin	1:Begin
2:DECLARE Speakers, Videos v, Word w, Feature f, Frames fr.	2:READ Videos v_k of the speaker
3:READ Videos v_k, where $k = 1\ to\ 50$ from the each Speakers$_n$ where $n = 1\ to\ 10$	3:For each v_k
4:For each s_n	$\{$
$\{$ For each v_k	For each w_l1 where $l = 0\ to\ 9$
$\{$ For each w_l1 where $l = 0\ to\ 9$	$\{$
$\{$	3.1. DETECT FACE from the Video v_k
4.1. DETECT FACE from the Video v_k	3.2. fr_i Conversion $i = 1\ to\ 45$
4.2. fr_i Conversion $i = 1\ to\ 45$	3.3. Locating LIP ROI from the Face region
4.3. Locating LIP ROI from the Face region	3.4: Extract f_{dct} and f_{dwt} features.
4.4. Perform scaling and normalization on ROI	3.5: Create HMM model h_l for the word w_l
4.5. Block-based f_{dct} , f_{dwt} extracted	3.6: HMM model h_l probability estimation.
$\}$	$\}$
$\}$	4: Comparing h_l model with trained model.
$\}$	5: Recognized word w_l is
5:Create HMM model h_l for each w_l	5.1: Maximum likelihood value with a
6:Estimation of probability values.	model associated with that word.
7:End	6: End

3.3 Experimental Results

The performance of each λ^v system is measured in terms of the percentage of the number of test utterances identified correctly out of the total test utterances. Comparison of DCT- and DWT-based viseme recognition varying number of states is shown in Fig. 8. DCT- and DWT-based viseme recognition rate are high in the state of $S = 7$, $S = 9$, respectively.

Fig. 8 Viseme recognition rates of λ^v system for varying states

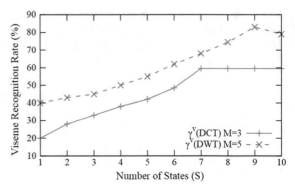

Fig. 9 Recognition rates of the system λ^v for varying Gaussian mixtures

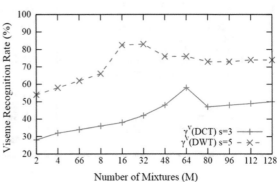

Table 1 N-best performance of the visual speech recognition the DWT and DCT features

N-best	N = 1	N = 2	N = 3	N = 4	N = 5
DWT	82	92	94	96	98
DCT	58	64	68	70	74

Figure 9 gives the comparative performance of the VSR λ^v system using the DCT and DWT coefficients for different Gaussian mixture components with fixed five states. DCT-based viseme recognition rate is high when the Gaussian mixture components become 64, while the recognition rate of DWT-based λ^v system is higher with 16 and 32 mixture components.

The overall DWT-based visual speech recognition gives better performance than DCT-based visual speech system (refer Table 1). Hence, DWT features are used to develop multimodal speech recognition system λ^{atv}. The performance of the λ^v system is evaluated again by using N-score level combination as shown in Table 1. For the different N-gram values, the performance is measured against with image transform approaches (DCT/DWT). N-gram scores at N = 5 give higher recognition accuracy for both image transform approaches.

Figure 10 shows that phoneme and viseme recognition rates for varying number of states for normal (λ^a), throat (λ^a),visual (λ^v), combined (λ^{at}), (λ^{av}), (λ^{tv}), and

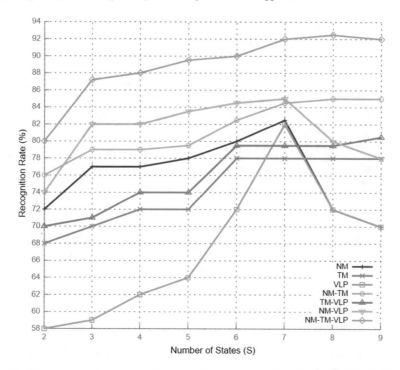

Fig. 10 Phoneme and viseme recognition rates for varying number of states for λ^a, λ^t, λ^v, λ^{at}, λ^{av}, λ^{tv}, and λ^{atv} systems

multimodal systems $\left(\lambda^{atv}\right)$. Among the individual systems, λ^a has the best performance (82%), followed by λ^v (78%) and λ^t (72%) in that order. This shows that the speech through lip radiation has more speech information than the speech through throat skin vibration or lip movements. However, the performance of the combined features shows an improvement over λ^a, when one of the features is fa. The λ^{av} and λ^{av} (84.5%) and λ^{at}(85%) systems performances are better than λ^{tv} (80.5%). This shows that there is more complementary information in the normal speech and lip movements as well as in normal speech and throat speech. However, when all the features are combined [18], then the corresponding system, λ^{atv} (92.5%) outperforms all other systems showing the presence of complementary information in all the three modalities.

Figure 11 shows phoneme and viseme recognition rates for varying number of Gaussian mixtures for the systems λ^a, λ^a, λ^v, λ^{at}, λ^{av}, λ^{tv}, and λ^{atv}.

The best recognition rate for each system is 83%, 82%, and 82.5% for λ^a, λ^a, and λ^v,respectively, and 86%, 85.5%, and 85% for λ^{at}, λ^{av}, and λ^{tv}, respectively. The multimodal system λ^{atv} has the highest recognition accuracy rate of 93.5%.

The acoustic and viseme performance analysis are implemented in HTK. Viseme feature using DCT and DWT are converted into HTK portable format. NM-based

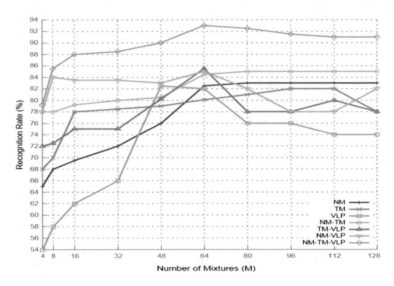

Fig. 11 Phoneme and viseme recognition rates for varying number of Gaussian mixtures for λ^a, λ^t, λ^v, λ^{at}, λ^{av}, λ^{tv}, and λ^{atv} systems

```
================================================================
[root@radha-pc HResult]# HResults -I refs.mlf wlist results.mlf
===================== HTK Results Analysis =====================
  Date: Sat Apr 22 17:58:44 2017
  Ref : refs.mlf
  Rec : results.mlf
---------------------- Overall Results ----------------------
SENT: %Correct=0.00 [H=0, S=2, N=2]
WORD: %Corr=80.00, Acc=80.00 [H=16, D=0, S=4, I=0, N=20]
================================================================
```

Fig. 12 Viseme recognition using HTK

digit recognition results are given in Fig. 12. Figure 13 shows the confusability matrix of datasets of the throat microphone.

4 Summary and Conclusion

An improved ASR system using multimodal approach is proposed in this work, which combines the acoustic and visual speech information. A feature level integration is performed and HMM modeling is used to model isolated digits. MFCC features are extracted from the acoustic signals of the normal and throat microphones while DCT and DWT coefficients are extracted from the visual signal by using DWT and DCT methods. The viseme recognition rate is high for DWT coefficients when compared to DCT coefficients. The combined modes of normal, throat, and visual speech gives

Fig. 13 Confusability
matrix of digits

```
[root@radha-pc HResult]# HResults -p -I refs.mlf wlist results.mlf
==================== HTK Results Analysis ====================
Date: Sat Apr 22 17:58:47 2017
Ref : refs.mlf
Rec : results.mlf
--------------------- Overall Results ---------------------
SENT: %Correct=0.00 [H=0, S=2, N=2]
WORD: %Corr=80.00, Acc=80.00 [H=16, D=0, S=4, I=0, N=20]
--------------------- Confusion Matrix ---------------------
        o   t   f   s   s   e   n   z
        n   w   o   i   e   i   i   e
        e   o   u   x   v   g   n   r
                r       e   h   e   o
                        n   t           Del [ %c / %e]
 one    2   0   0   0   0   0   0   0   0
 two    0   2   0   0   0   0   0   0   0
 thre   0   0   0   0   0   0   2   0   0 [ 0.0/10.0]
 four   0   0   2   0   0   0   0   0   0
 five   0   0   0   2   0   0   0   0   0 [ 0.0/10.0]
 six    0   0   0   2   0   0   0   0   0
 seve   0   0   0   0   2   0   0   0   0
 eigh   0   0   0   0   0   2   0   0   0
 nine   0   0   0   0   0   0   2   0   0
 zero   0   0   0   0   0   0   0   2   0
 Ins    0   0   0   0   0   0   0   0
====================================================================
```

better accuracy results when compared to individual modes and bimodal recognition. From this, the visual speech recognition acts as a complementary source for ASR, and helps the recognition accuracy ASR systems.

Acknowledgements We acknowledge Ms. P. Prabha and Ms. B. T. Preethi, PG graduate, SSN college of Engineering, India, for revealing their human image towards this contributing research work.

References

1. Shahina A (2007) Processing throat microphone speech. Ph.D. Thesis, Indian Institute of Technology, Chennai
2. Shahina A, Yegnanarayana B (2005) Language identification in noisy environments using throat microphone signals. In Proceedings of the international conference on intelligent sensing and information processing, pp 400–403
3. Graciarena M, Franco H, Sommez K Bratt H (2003) Combining standard and throat microphones for robust speech recognition. IEEE Signal Process Lett 10(3):72–74
4. Radha N, Shahina A, Nayeemulla Khan A (2015) A person identification system combining recognition of face and lip-read passwords. In: The proceedings of international conference on computing and network communications, pp 882–885
5. Kittler J, Hatef M, Duin R, Matas J (1989) On combining classifiers. IEEE Trans Pattern Anal Mach Intell 20(3):226–239
6. Dupont S, Ris C (2004) Combined use of close-talk and throat microphones for improved speech recognition under non-stationary background noise. In: Proceedings of workshop (ITRW) robustness issues in conversational interaction
7. Erzin E (2009) Improving throat microphone speech recognition by joint analysis of throat and acoustic microphone recordings. IEEE Trans Audio Speech Lang Process 17(7):1316–1324
8. Neumeyer L, Weintraub M (1994) Probabilistic optimum filtering for robust speech recognition. In: Proceedings of the IEEE international conference on acoustics, speech, signal process (ICASSP'94), pp 417–420

9. Neti C et al (2000) Audio visual speech recognition. No. EPFL-REPORT-82633. IDIAP
10. Hong X, Yao H, Wan Y, Chen R (2016) A PCA based visual DCT feature extraction method for lip-reading. In: International conference on intelligent information hiding and multimedia signal processing, pp 321 – 326
11. Kaklauskas A, Gudauskas R, Kozlovas M, Peciure L, Lepkova N, Cerkauskas J, Banaitis A (2016) An affect-based multimodal video recommendation system, studies in informatics and control 25(1):5–14. ISSN: 1220-1766
12. Radha N, Shahina A, Prabha P, Preethi Sri BT, Nayeemulla Khan A (2017) An analysis of the effect of combining standard and alternate sensor signals on recognition of syllabic units for multimodal speech recognition. https://doi.org/10.1016/j.patrec.2017.10.011
13. Radha N, Shahina A, Vinoth G, Nayeemulla Khan A (2014) Improving recognition of syllabic units of Hindi language using combined features of throat microphone and normal microphone speech. In: The proceedings of international conference on control instrumentation communication and computational technologies, pp 1498–1503
14. Viola Paul, Jones Michael J (2004) Robust real-time face detection. Int J Comput Vis 57(2):137–154
15. Zhao W et al (2003) Face recognition: a literature survey. ACM Comput Surv (CSUR) 35(4):399–458
16. Neumeyer L, Weintraub M (1994) Probabilistic optimum filtering for robust speech recognition. In: Proceedings of the IEEE international conference on acoustics, speech, signal process (ICASSP'94), pp 417–420
17. Radha N, Shahina A, Nayeemulla Khan A (2016) An improved visual speech recognition of isolated words using combined pixel and geometric features. Indian J Sci Technol 9(44):1–6
18. Heckmann M et al (2002) DCT-based video features for audio-visual speech recognition. INTERSPEECH

Halftoning Algorithm Using Pull-Based Error Diffusion Technique

Arvind Bakshi and Anoop Kumar Patel

Abstract Despite several improvements in the field of digital halftoning, there is still a scope of improvement. Halftoning is widely applied for applications like printing, efficient transmission, and storage, etc. Halftoning process reduces 256 levels of a grayscale image to just 2 levels. Three major categories of halftoning are (1) Dithering, (2) Error Diffusion, and (3) Iterative algorithms. A concern always remains regarding the visual perception of an image's halftone output and for obtaining a good visual perception, we have proposed an error diffusion algorithm that applies pulling technique. The perception of output image generated by our algorithm is similar to the input. The SSIM (structure similarity index map), PSNR (peak signal-to-noise ratio), RMSE (root mean squared error), and MSE (mean squared error), of the output image are 0.1426, 7.2672, 110.4531, and 1.2200e+04, respectively.

Keywords Error diffusion · Filter · Forward processing · Grayscale · Halftoning Pull method

1 Introduction

Halftoning is a technique through which a continuous tone image is mimicked using pattern of binary pixels [1]. It is required for displaying of continuous tone images onto binary devices that have a limited color palette for generation of images. For example, devices like printers, digital typesetters, scientific and medical instruments, etc. use halftoning. Halftone image data has lesser file size thus enables faster transmission over the network and consumes less space for storage. Halftoning assumes eye to be a low-pass filter. The aim of halftoning is to display an original multi-level

A. Bakshi (✉) · A. K. Patel
National Institute of Technology Kurukshetra, Kurukshetra, India
e-mail: arvind_31603127@nitkkr.ac.in

A. K. Patel
e-mail: akp@nitkkr.ac.in

© Springer Nature Singapore Pte Ltd. 2019
S. Bhattacharyya et al. (eds.), *International Conference on Innovative Computing and Communications*, Lecture Notes in Networks and Systems 56,
https://doi.org/10.1007/978-981-13-2354-6_43

image on a two-level medium, keeping its perception as good as the original image. Halftoning approaches can be largely grouped below as three classes [2]:

1. **Dithering**: The process of dithering involves attachment of some noise to image's pixel value before its quantization [3]. The pixel's spatial coordinates decide the amount of noise added. Dithering techniques have some advantages like being simpler, faster execution speeds, and cheaper in implementation but also have some drawbacks like generating poor quality output images in comparison to those generated by error diffusion methods.

2. **Error Diffusion**: Error diffusion quantizes a pixel value that generates some error, which is then propagated to neighbouring pixels [3]. In quantization, the pixel value can be quantized to either (0/1) pair or (0/255) pair and error generated is the variation between the original pixel value and the quantized pixel value. Positive error indicates quantization of a pixel to value 0, therefore the value of its neighboring pixels should be increased so that when they are processed their quantization will result in value 1 or 255 (as mentioned previously depends on pair selected (0/1) or (0/255). Similarly, negative error value will result in quantization to 1 or 255 and its neighboring pixels values will then be increased to achieve quantization of 0 on their processing turn. Now, the selection of neighboring pixels to whom error is diffused is decided using weight matrix. For example, Floyd and Steinberg, Jarvis, Judice, and Ninke etc., are some of the well-known weight matrices used in halftoning [4]. Figure 1 depicts the basic principle of error diffusion.

In Fig. 1,

$$B(r, c) = \begin{cases} 255, if\ O(r, c) \geq T \\ 0, otherwise \end{cases} \tag{1}$$

$$E(r, c) = O(r, c) - B(r, c) \tag{2}$$

$$O(r, c) = A(r, c) + \sum_{k,l=\epsilon S} H(k, l)E(r - k, c - l) \tag{3}$$

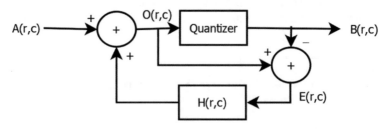

Fig. 1 Filter view of error diffusion

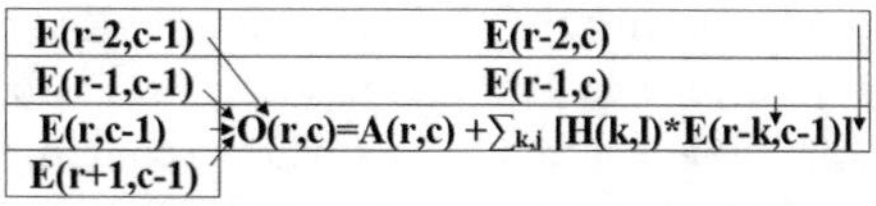

Fig. 2 Pulling error forward

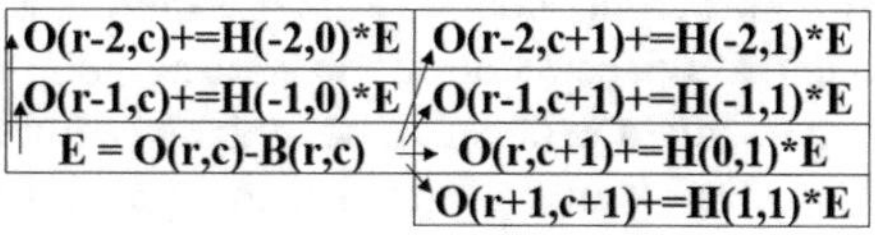

E(r-2,c-1)	E(r-2,c)
E(r-1,c-1)	E(r-1,c)
E(r,c-1)	$O(r,c)=A(r,c) +\sum_{k,l} [H(k,l)*E(r-k,c-1)]$
E(r+1,c-1)	

Fig. 3 Pushing error forward

$O(r-2,c)+=H(-2,0)*E$	$O(r-2,c+1)+=H(-2,1)*E$
$O(r-1,c)+=H(-1,0)*E$	$O(r-1,c+1)+=H(-1,1)*E$
$E = O(r,c)-B(r,c)$	$O(r,c+1)+=H(0,1)*E$
	$O(r+1,c+1)+=H(1,1)*E$

Here, T = threshold and in general case, its value is 127. The values of H(k,l) are positive, obtained from weight matrix like Floyd Steinberg and sum to 1.
The error diffusion can be applied in two ways as shown in Figs. 2 and 3 [4].

   a. Pulling error forward.
   b. Pushing error forward.

3. **Multi-pass or Iterative algorithms**: In this technique, haltoning of an input image is done in multiple iterations [2]. It attempts to find the optimal configuration of binary values in output image, so as to achieve minimum error between output and input image. For example, global search, binary search algorithms.

Among the above three algorithms discussed, dithering approach is the fastest but has output image quality that is not usable practically in bi-level devices such as printers [2]. Multi-pass algorithms are the best in terms of output image quality but are quite complex thus require some processing time making them impractical for real-time and commercial usage. Error diffusion algorithms are the mediocre one that balances both the processing time and quality. Although output image quality of error diffusion is poor than multi-pass ones but better than dithering ones.

Our aim here in this article is to propose an improved version of the error diffusion algorithm that can transform a multi-level grayscale image to a bi-level image (0/255) without loss in its general perception. To perform the above task, we have defined an error diffusion algorithm that uses pulling technique. This article is categorized into five sections. Section one talks about the fundamentals of halftoning as introduction. Section two quickly reviews some existing halftoning techniques. Section three gives our proposed approach and its explanation with details. In section four, the performance of our approach is analyzed and results are shown. Conclusion and future scope are listed in section five.

2 A Brief Review of Some Existing Techniques

Pappas et al. proposed a halftoning technique that tries to produce optimal halftone image by reducing squared error among visual model of binary image, output of a series of printer, and of visual model to the original image [1]. The method produces only the local optimum and is known as LSMB (least squares model based) approach. The method considers both printer and visual perception model. The 1D (one-dimensional) least squares problem where an image's rows and columns are halftones separately is extended to 2D (two-dimensional) using iteration techniques. A major component of the algorithm is a precise printer model. The output image quality is dependent on the starting point and optimization strategy. The technique can be tuned in accordance with an individual device like printer. It achieves higher resolution than traditional techniques by exploiting distortions in devices such as printer and generates images having better textures. Shen et al. proposed a halftoning technique that does halftoning progressively, instead of doing it in single run [5]. The algorithm uses anisotropic edge-adaptive diffusion model of Persona and Malik (PM) as base, uses a stochastic strategy for binary flipping and results in adaptive image de-noising and enhancement. Since it uses PMs model, error diffusion is adaptive and exchange of errors does not happen between different objects in the image. The pattern of error flow, distribution of weights, and tiling all are done dynamically. It uses diffused error for flipping operation and is free from use of thresholds used in traditional error diffusion. Zeng et al. proposes a classification for the output images generated from application of error diffusion [6]. To perform this, image patches are extracted based on their statistical characteristics, used for designing class feature matrices, which forms the first step. In the next step, feature dimension reduction is done by applying spectral regression kernel discriminant analysis. Finally, a technique like nearest centroids classifier is applied for classifying halftone images. The approach has advantages like high accuracy rate in classification, robustness against noise and high execution speed. Kovacs et al. analyze some of the scanning orders used for images when they are treated as arrays to get maximum scientific returns on output [7]. It also compares scanning orders like Lissajous, billiard, spiral, DREAM, etc. The metrics used for analysis includes noise resistance measured using phase-space moments, large-scale sensitivity measured using short-timescale scanning ranges. Fung et al. proposed a multi-scale error diffusion technique for generating halftone image [8]. The approach exploits green noise characteristics of output image. It allows adjustable cluster size and has a linear relationship between input gray level and cluster size. It uses close to the isotropic ring-shaped filter having adjustable outer and inner radii for diffusion that eliminates pattern and directional artifacts while preserving details in the original image. Pnueli et al. gives a new approach of gridless halftoning that considers the problem of halftoning on continuous 2D plane rather than conventional discrete pixel grid [9]. The gray-level images are defined as functions from $R^2 \rightarrow [0, 1]$, grid images as functions from $Z^2 \rightarrow [0, 1]$, and gridless images as functions from $Z^2 \rightarrow \{0, 1\}$. But the above definition of gridless is impractical to implement, so the authors impose a restriction called covering condition to make it feasible. The basic idea is to mimic techniques used by graphic artists for generating halftone image. A Digi Durer system is proposed for this which produces quality halftone based on curve evolution solution to a suitably postulated eikonal equation which incorporates some extra information about the rendered scene. The advantages of this approach includes improved quality, supports a variety of output devices, robust against printing errors, better content understanding, freedom of style, and user adjustable. Liu et al. proposes an error diffusion-based halftoning technique that preserves structure of objects in image and tones using blue-noise property [10]. It uses entropy for measuring the impact of intensity to adaptively constraint threshold modulation strength. To perform this, an entropy-constrained threshold modulation function is constructed. It also uses latest variable coefficient methods having excellent tone reproduction resulting in overall improved tone quality. It is free from error diffusion coefficients and processing pattern. Damera-Venkata et al. proposed another algorithm for error diffusion that uses adaptive threshold modulation that adjusts parameters of error diffusion for improvement in output image quality [11]. All existing quantizers can be used with it. Two existing halftoning algorithms are also improved using it: (1) Edge enhancement halftoning. (2) Green-noise halftoning. An optimization of hysteresis coefficient of green noise halftoning is done by the author's technique. For minimization of linear frequency distortion or sharpening, scalar gain (L) is adjusted by it.

3 Proposed Approach

A novel error diffusion algorithm for halftoning of grayscale images is proposed. The algorithm takes as input a grayscale image and outputs binary image (halftone). To perform halftoning operation, the algorithm proposes a filter that is shown in Fig. 6. The algorithm is explained using block diagrams shown in Figs. 4 and 5.

Fig. 4 Proposed halftoning

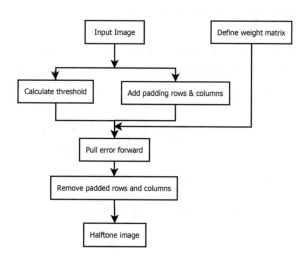

Input image
Weight matrix
Declare error
Execute weight matrix clockwise
Output image

Fig. 5 Proposed error diffusion

P5
e(i-2,j)
D4
e(i-1,j-1)
P1
e(i-1,j)
D1
e(i-1,j+1)
P7
e(i,j-2)
P3
e(i,j-1)
X
img=img+∑(e*w)
P4
e(i,j+1)
P8
e(i,j+2)
D3
e(i+1,j-1)
P2
e(i+1,j)
D2
e(i+1,j+1)
P6
e(i+2,j)

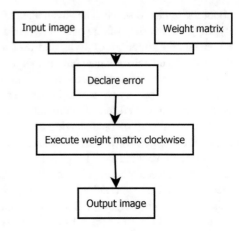

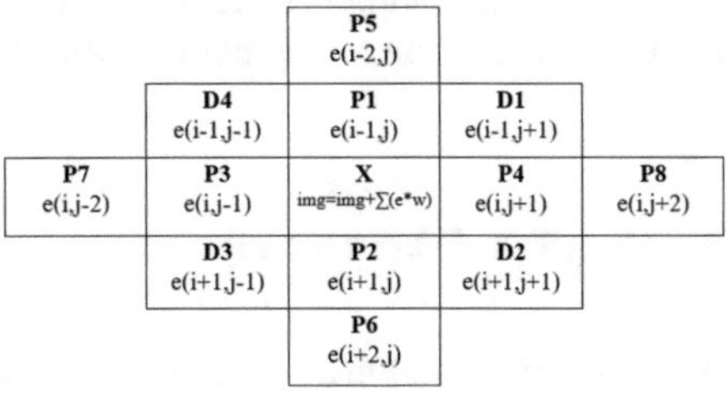

Fig. 6 Proposed filter

4 Analysis and Results

4.1 Performance Analysis

The approach has been implemented in MATLAB R2017a. The performance of the approach has been analyzed using below metrics [12]:

1. **MSE (Mean square error)**:

$$MSE = \frac{1}{u \times v} \sum_{i=1}^{u} \sum_{j=1}^{v} (O(r, c) - A(r, c))^2 \tag{4}$$

   Here A = reference image, O = image to be compared.

2. **Root Mean Square Error (RMSE)**:

$$RMSE = \sqrt{MSE} \tag{5}$$

3. **Peak signal-to-noise ratio (PSNR)**:

$$PSNR = 20 \log_{10} \left(\frac{255}{RMSE} \right) \tag{6}$$

4. **Structure similarity index map (SSIM)**: It compares luminance, constrast and structure of two different images. It defines similarity between two images.

$$SSIM(u, v) = \frac{(2\mu_u\mu_v + C_1) \times (2\sigma_{uv} + C_2)}{(\mu_u^2 + \mu_v^2 + C_1) \times (\sigma_u^2 + \sigma_v^2 + C_2)} \tag{7}$$

Here, μ_t = mean intensity, σ_t = standard deviation, t = u or v, and C_t = constant to avoid instability when $\mu_u^2 + \mu_v^2$ approaches zero which equals $C_t = (k_t L)^2$, where $k_t \ll 1$ and L = dynamic range of pixel values, e.g., L = 255 for 8-bit grayscale image.

Table 1 shows the result of application of these metrics onto the proposed approach using Figs. 7 and 8.

4.2 Comparison with Existing Techniques

Table 2 depicts the proposed approach's performance in comparison to some existing techniques. The comparisons are made with respect to the images in the papers [9], [11], and [5], respectively.

Table 1 Performance analysis

S. No	Metric	Result
1.	MSE	1.2200e + 04
2.	RMSE	110.4531
3.	PSNR	7.2672
4.	SSIM	0.1426

Fig. 7 Original image

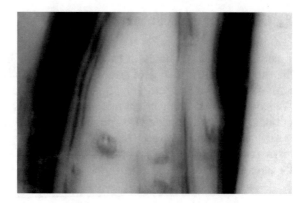

Fig. 8 Halftone image

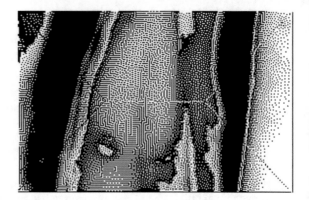

Table 2 Comparison with some existing techniques

S. No	Technique	MSE	RMSE	PSNR	SSIM
1.	Fung's Green noise halftoning	2.1630e + 04	147.0719	4.7802	−0.1168
2.	Proposed	1.6753e + 04	129.4316	5.8900	0.0687
3.	Liu's structure aware using Ostromoukhov's technique	1.8018e + 03	42.4471	15.5738	0.3419
4.	Liu's structure aware using Zhou and Fang's technique	1.9390e + 03	44.0339	15.2551	0.3312
5.	Proposed	1.4940e + 04	122.2286	6.3873	0.1114
6.	PMSF halftoning	823.7951	28.7018	18.9726	0.2919
7.	Proposed	1.6839e + 04	129.7635	5.8678	0.0433

4.3   Result

The original image is represented in Fig. 7 and its halftone in Fig. 8. The halftone obtained in Fig. 8 has clear visual perception as the original image.

5   Conclusion and Future Scope

An error diffusion algorithm for halftoning grayscale images has been proposed in this article. Mostly, pushing of error-based implementation is used while applying error diffusion, but the perception it generates has the possibility of improvement. We have proposed an algorithm that propagates error by pulling them rather than pushing, in a single direction thus producing better results. A circular-shaped filter is designed that processes each pixel by spiral scanning order rather than traditional raster scanning. The algorithm makes use of dynamic threshold calculated from the input image, unlike traditional approach that uses fixed threshold. The halftone images produced by the algorithm has good visual perception as the original. The quality of halftone image obtained by application of the proposed algorithm may be enhanced further by computation of threshold value and filter weights using learning-based approaches.

References

1. Pappas TN, Neuhoff DL (1999) Least-squares model-based halftoning. IEEE Trans Image Process 8(8):1102–1116
2. Chang J, Alain B, Ostromoukhov V (2009) Structure-aware error diffusion. ACM Trans Graph 28(5): 162:1–162:8
3. Katsavounidis I, Jay Kuo CC (1997) A multiscale error diffusion technique for digital halftoning. IEEE Trans Image Process 6(3):483–490
4. Bouman CA (2011) Image Halftoning, Digital Image Processing Laboratory, Purdue University
5. Shen J (2006) Progressive halftoning by Perona-Malik error diffusion and stochastic flipping, SPIE. Image Process Algorithms Syst Neural Netw Mach Learn 6064(3)
6. Zeng Z, Wen Z, Yi S, Zeng S, Zhu Y, Liu Q, Tong Q (2016) Classification of error-diffused halftone images based on spectral regression kernel discriminant analysis. Adv Multimed Hindawi 2016(4985313)
7. Kovacs A (2008) Scanning strategies for imaging arrays. SPIE Proc Soc Photo Opt Instrum Eng 7020(5)
8. Yik-Hing F, Yuk-Hee C (2010) Green noise digital halftoning with multiscale error diffusion. IEEE Trans Image Process 19(7):1808–1823
9. Yachin P, Bruckstein Alfred M (1996) Gridless halftoning: a reincarnation of the old method. Graphical Models Image Process 58(1):38–64
10. Lingyue L, Wei C, Wenting Z, Weidong G (2014) Structure-aware error diffusion approach using entropy-constrained threshold modulation. Vis Comput Springer 30(10):1145–1156
11. Niranjan D-V, Evans Brian L (2001) Adaptive threshold modulation for error diffusion halftoning. IEEE Trans Image Process 10(1):104–116
12. Srivastava R, Performance measurement of image processing algorithms. Professor, Department of Computer Science & Engineering, IIT BHU, India

Regression Analysis for Liver Disease Using R: A Case Study

Nagaraj M. Lutimath, D. R. Arun Kumar and C. Chetan

Abstract Decision Trees are a significant approach for decision-making in data mining. They are easy and efficient classification method. The use of decision tree as a classification method as an important paradigm is its capability to classify the vital attributes of a given problem. The liver disorder is difficult to detect in the early stages. With important symptoms and parameters, the liver disease can be found. The aim of the paper is to study the analysis of liver disorder utilizing regression tree technique using R programming language.

Keywords Decision tree · Classification · Medical data mining · R studio

1 Introduction

The largest organ of the human body is the liver. It carries a weight around 3 lb (1.36 kg) and is reddish brown in color. It is segregated into four lobes of uneven size and shape. It faces to the right side of the abdomen and below the diaphragm. Blood flows to the liver by means of two large vessels. They are the hepatic artery and the portal vein. Oxygen-rich blood flows from the aorta using the hepatic artery. Blood consisting of digested food flows from the small intestine. It is carried by the portal vein [1, 2].

N. M. Lutimath (✉) · D. R. Arun Kumar · C. Chetan
Department of Computer Science and Engineering, Sri Venkateshwara College of Engineering, Kempegowda International New Airport Road, Vidyanagar, Bengaluru 562157, Karnataka, India
e-mail: nagarajlutimath@gmail.com

D. R. Arun Kumar
e-mail: arunkumardr1987@gmail.com

C. Chetan
e-mail: chetanc123@gmail.com

S. Bhattacharyya et al. (eds.), *International Conference on Innovative Computing and Communications*, Lecture Notes in Networks and Systems 56,
https://doi.org/10.1007/978-981-13-2354-6_44

Liver tissue contains several lobules, and each lobule has many hepatic cells which are the typical metabolic cells of the liver. Too much consumption of alcohol can cause an acute or chronic infection to the liver and can even damage other organs in the body. Alcohol-provoked liver disorder resides a major problem for the body.

To detect the liver disease, patient's blood test is taken to determine the content of alkaline phosphatase, alamine aminotransferase, and aspartate Decision Trees are vital procedures for classification and are utilized for decision-making for test dataset. Unlike a decision tree, it searches the data attributes using the best split threshold to separate the dataset into classes, and random forests are a collection of Decision Trees. Other approaches are the boosted decision tree, ID3 and C4.5 tree that learn from their accuracy using the dataset. Cross-validation with leave out can be applied for prediction accuracy. The best outcome was achieved utilizing SVM with a radial basis kernel. 73.20% prediction accuracy was attained [3]. C4.5 and Chi-Square Automatic Interaction Detector (CHAID) procedures are also used to predict accuracy and framing rules for liver disorder [4]. Fuzzy Min-Max neural network technique, Classification and Regression Tree (CART) method with Random Forests were proposed, and the model's efficiency in decision-making are done by examining the medical dataset [5]. Back-propagation network (BPN) classification method was used to classify both the regular and irregular liver disorders with good prediction accuracy [6].

In Decision Trees, the best splits of the dataset occurs recursively until data has been split into homogeneous group of data. Any data mining technique executes in two stages, the training stage and the testing stage. At the training stage, the decision tree is constructed using the current examples of the dataset and at the testing stage, the resent examples are classified using the model developed using the training dataset.

In this paper, a dataset of the number of patients suffering from liver disease is taken and regression analysis is made.

2 Classification

Data mining is the process of extraction of hidden knowledge from the datasets. There are many techniques for data mining. They are the classification, clustering, association set data mining, and neural networks.

Classification is an important approach for data mining. There are three approaches for classification. They are the supervised classification, unsupervised, and semi-supervised classification. Supervised classification classifies the dataset using the labeled classes, while the unsupervised classification uses the method by first converting the classes into labeled classes and then classifies the dataset.

Two types of attributes of the class are taken for classes for classification. They are the categorical values and numerical values. Categorical values are used for the construction of general decision tree. Numerical values are used for the regression analysis and utilizing them designing regression decision trees. Semi-supervised classification uses both labeled and unlabelled examples for classification.

Classification technique of data mining can be thought of as a supervised classification method where each example relates to class attribute called the target attribute. The target attribute takes categorical values, each of them corresponding to a class. Every instance of the dataset consists of two types of attributes, namely the set of predictor feature values and a target feature value. Predictor feature is used to predict the value of the target feature. The predictor feature must be related to predict a class of a given instance.

In classification technique, the dataset of the instance from which the knowledge being derived is divided into two disjoint datasets called the training dataset and the test dataset. The classification activity is, respectively, separated into two stages, the training stage where the classification model is constructed using the training set, and testing stage where the model is tested on the test dataset.

During the training stage, the classification procedure accesses both the predictor features and the target features of training dataset, to construct the classification model. This model is used effectively to predict the accuracy of the class.

This is done by constructing the relationship between predictor feature values and target classes that allow the prediction of the class of that instance knowing the predictor attribute values. In the testing stage, the class values of the test dataset are predicted using the classification model.

An important aim of classification procedure is to increase the predictive accuracy while classifying instances in the data test that are uncovered at the training stage utilizing the class model. The knowledge thus obtained by a classification approach can be represented in numerous ways like association rules, Decision Trees, neural networks, etc.

3 Classification Methods

3.1 Decision Tree

Decision Trees are an important classification technique in data mining. They have two unique nodes, the root node with internal nodes and the leaf nodes. The root node and the internal node connected by the attributes of the dataset. A leaf node is related to the class attribute.

Typically, each and every internal node has an outgoing branch with feasible value attribute. To find the class for a new example of the dataset utilizing the decision tree, we start from the root node and then, subsequently visit the internal nodes until we reach the leaf. The test condition is applied starting from the root node and at every internal node that comes between the root node and leaf node. The result of the test finds the branch traveled between the root node and leaf node visiting every next node between them. The class for the example is the class of the leaf node visited.

3.2 Random Forests

The Random Forests are important entity for classification. Random Forests consists of a group of Decision Trees. Bagging method is used as selection parameter in these forests. The approach is used in a cycle with random forests. Random feature selection with replacement is used taking the training dataset. Thus, the tree grows with the new training dataset. The concept of building forests is to construct many Decision Trees classified by an instance of the class. Giving a suitable vote to each class of the tree, the root of the random forest is selected. Generally, the decision tree class that gets the highest vote is selected as the root [5, 6]. Random forests are normally unpurned. They are used to predict the prediction accuracy.

4   Data Definition, Feature Engineering, and Result Set

In order to study the regression process, dataset from UCI machine learning repository sets for liver disorder for "Diagnosis data for patients suffering from liver disorder" is taken. The dataset is segregated into two sets the training data and the test data. The procedure and the associated attribute engineering are worked on the training dataset, and model so acquired is used on the test data to predict the outcome. The problem statement for the following the problem is stated as follows:

Problem Statement: *"To find the predicted value for the patients suffering from liver disease"*.

In other words, it is stated to combine historical usage patterns with liver disease predictable data in order to forecast number of patients suffering from liver disease. Dataset consists of the diagnosis data of the patients suffering from liver disease. The attributes are defined as follows.
    Data Attributes are:

a_mcv: attribute for mean corpuscular volume
a_alkphos: attribute **for** alkaline phosphatase
a_sgpt: attribute for alanine aminotransferase
a_sgot: attribute for aspartate aminotransferase
a_gammagt: attribute for gamma-glutamyl transpeptidase
a_drink: attribute for number of half-pint equivalents of alcoholic beverages drunk per day
a_selector: attribute created for predicting.

    The liver dataset containing 345 tuples is segregated into 242 tuples for training set and remaining 103 tuples into test dataset. The sample for training is executed in R and is taken using the formula

$$\text{training} \ < \ -\text{sample} \ (1 : 345, \ 242, \ \text{replace} = \text{FALSE}) \tag{1}$$

Data Analysis

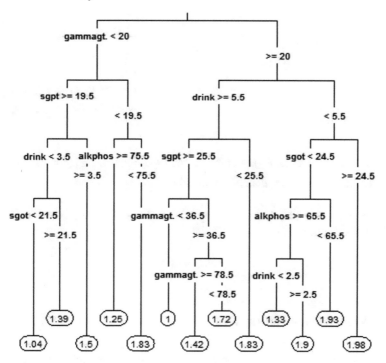

Fig. 1 Regression tree for liver disease

The training set and test set are then calculated. They are represented by training set and test data, respectively.

The regression tree is then constructed using the formula

fit < −rpart (a_selector ∼ a_mcv + a_alkphos + a_sgpt + a_sgot + a_gammagt + a_drink, data = trainingset, method = "anova").   (2)

The fit variable is used to plot the regression tree as follows:

rpart.plot (fit, type = 3, digits = 3, fallen.leaves = TRUE)   (3)

The regression tree for the liver dataset is shown in Fig. 1 below with the attributes a_selector, a_mcv, a_alkphos, a_sgpt, a_sgot, a_gammagt, and a_drink represented as selector, mcv, alkphos, sgpt, sgot, gammagt, and drink, respectively.

We use the random forest for further analysis and a random forest is constructed using the formula in R as

Fig. 2 Error versus trees for
liver disease dataset

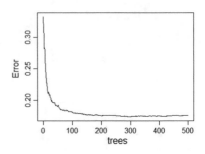

$$xyz.rf = randomForest \, (a\_selector \sim a\_mcv + a\_alkphos + a\_sgpt + a\_sgot$$
$$+a\_gammagt + a\_drink, data = traindata) \quad (4)$$

In the above formula, data is training dataset. Then, the graph below in Fig. 2 is plotted using the formula

$$plot \, (xyz.rf) \quad (5)$$

4.1 Performance Parameters

There are three performance parameters which are utilized for analysis. They are the Mean Absolute Error (MAE), Sum of Squared Error (SSE), and Mean Squared Error (MSE). MAE is defined as the mean of the absolute difference between the actual and predicted values of the instances in the dataset. SSE is defined as the sum of the squares of the actual and predicted values of the instances in the dataset. MSE is defined as the mean of the squares of the actual and predicted values in the dataset.

4.2 Data Analysis

Figure 1 is the regression tree obtained using the training set to predict the test dataset. Figure 2 is the graph obtained using the random trees decision-making technique. We observer from the figure that as the number of trees increases the error decreases. The value of error decreases from 0.35 and it comes to 0.08 for 100 trees. It further decreases to 0.08 for 200 trees. Thus, from the observations, we see that that as the number of trees increases the error decreases moderately. The MAE, SSE, and MSE of overall dataset considering 70% of the dataset as the training set and 30% of the dataset as the test dataset are shown in Table 1. The mean of a_mcv is 90.16. MAE and SSE considering the mean of a_mcv are calculated and listed in Table 2. Similarly, the mean of the attributes of a_alkphos, a_sgpt, a_sgot, and a_gammagt are calculated. They are 69.87, 30.4, 24.64, and 38.24, respectively. Their corresponding

Table 1 MAE, SSE, and MSE for the overall dataset

Type of error	Value of error
MAE	0.41
SSE	28.38
MSE	0.28

Table 2 MAE, SSE, and MSE using the mean of a_mcv

Type of error	Value of a_mcv ≤ 90.16	Value of a_mcv > 90.16
MAE	0.27	0.32
SSE	27.78	30.44
MSE	0.16	0.18

Table 3 MAE, SSE, and MSE using the mean of a_alkphos

Type of error	Value of a_alkphos ≤ 69.87	Value of a_alkphos > 69.87
MAE	0.28	0.31
SSE	31.55	26.67
MSE	0.16	0.18

Table 4 MAE, SSE, and MSE using the mean of a_sgpt

Type of error	Value of a_sgpt ≤ 30.4	Value of a_sgpt > 30.4
MAE	0.28	0.32
SSE	34.93	23.28
MSE	0.15	0.20

MAE, SSE, and MSE are calculated and listed in the Tables 3, 4, 5, and 6. The MAE, SSE, and MSE for the patients who do not drink and who drink are calculated and are listed in Table 7. The value of MAE for who do not drink is more than those who drink. We see that the model predicts better in case of patients who do not drink considering the value of MAE. Now looking at Tables 3 and 4, we see that that for value MAE and MSE are lesser in the case for value a_alkphos and a_sgpt lesser than or equal to 69.87 and 30.4, respectively. Thus, the model predicts better for these instances. The value of SSE varies with the instances in the dataset. In the Table 5, we observe that the MAE is less but MSE is more in the case for value a_sgot lesser than or equal to 24.64. Considering MSE as the better measure, we see than the model predicts when the values of _sgot lesser than or equal to 24.64. In the Table 6, we see that both MAE and MSE are less, for the value of a_gammagt lesser than or equal to 38.28. Thus, the model predicts better when the instances with the values of a_gammagt are comparatively lesser than 38.28. Thus, we see that the model predicts for the cases considered effectively.

Table 5 MAE, SSE, and MSE using the mean of a_sgot

Type of error	Value of a_sgot ≤ 24.64	Value of a_sgot > 24.64
MAE	0.30	0.29
SSE	24.54	33.67
MSE	0.16	0.18

Table 6 MAE, SSE, and MSE using the mean of a_gammagt

Type of error	Value of a_gammagt ≤ 38.28	Value of a_gammagt > 38.28
MAE	0.29	0.34
SSE	44.95	13.26
MSE	0.16	0.21

Table 7 MAE, SSE, and MSE using the mean of a_drink

Type of error	Value of a_drink == 0	Value of a_drink! = 0
MAE	0.33	0.29
SSE	1.88	56.34
MSE	0.34	0.17

5 Conclusion

In this paper, the regression analysis for liver disorder dataset using UCI machine learning data repository dataset is done. The mean absolute error, sum of the squared error, and mean squared error for each of the instances are calculated. The analysis is made by observing these values. Instances of the datasets are predicted using these. Random forests are also used together with the regression decision tree for predicting the accuracy of the instances in the dataset. In the future, prediction accuracy will be improved using data mining techniques such as neural network, deep learning, and association rule analysis.

References

1. Kaur P, Khamparia A (2015) Classification of liver based diseases using random tree. Int J Adv Eng Technol 8(3):306–313. ISSN: 22311963
2. Bahramirad S, Mustapha A, Eshraghi M (2013) Classification of liver disease diagnosis: a comparative study. in: IEEE second international conference on informatics & applications (ICIA), September 23–25, Lodz, Poland, pp 42–46
3. Ribeiro R, Marinho R, Velosa J (2011) Chronic liver disease staging classification based on ultrasound, clinical and laboratorial data. In: IEEE international symposium on biomedical imaging: from nano to macro, 30 March–2 April 2011, Chicago, IL, USA, pp 707–710
4. Abdar M, Zomorodi-Moghadam M, Das R, Ting, I-H (2017) Performance analysis of classification algorithms on early detection of Liver disease. Expert Syst Appl 67:239–251. Elsevier

5. Seera M, Lim CP (2014) A hybrid intelligent system for medical data classification. Expert Syst Appl 41(5):2239–2249. Elsevier
6. Saba L, Dey N, Ashour AS, Samanta S, Nath SS, Chakraborty S, Sanches J, Kumar D, Marinho R, Suri JS (2016) Automated stratification of liver disease in ultrasound: an online accurate feature classification paradigm. Comput Methods Progr Biomed 130:118–134. Elsevier

Web Page Segmentation Towards Information Extraction for Web Semantics

Pooja Malhotra and Sanjay Kumar Malik

Abstract Today, web is a large source of information which may be structured or unstructured. The need is efficient information extraction from various unstructured sources on the web. Therefore, information extraction is playing a prominent role in the current scenario. It focuses on automatically extracting structured information from unstructured distributed resources on the web and is based on several approaches. Web page segmentation is one of the most significant techniques where a web page is broken down into semantically related parts. There are various approaches to Web page segmentation. In this paper, the first information extraction has been explored, discussed and reviewed. Second, a revisit has been done on web page segmentation and its various approaches where a comparative analysis has been made. Third, various phases of vision-based web page segmentation have been presented and reviewed along with a flowchart. Finally, the results and conclusions have been presented along with the future work.

Keywords Web page segmentation · Information extraction
Vision-based web page segmentation

1 Introduction

Web has become one of the major sources of knowledge or information in modern times. Various search engines are used for collecting the information from different websites of the distributed web. But extracting relevant information from the unstructured content of web pages has become a challenge. Many times, irrelevant information such as advertisement etc., is also retrieved along with the desired results on the web. The vast text on the web often contains a lot of information in an

P. Malhotra (✉) · S. K. Malik
USIC&T, GGSIPU, Dwarka, India
e-mail: poojajind@gmail.com

S. K. Malik
e-mail: sdmalik@hotmail.com

© Springer Nature Singapore Pte Ltd. 2019
S. Bhattacharyya et al. (eds.), *International Conference on Innovative Computing and Communications*, Lecture Notes in Networks and Systems 56,
https://doi.org/10.1007/978-981-13-2354-6_45

431

unstructured way. The requirement is information extraction in a structured way either from texts or from pre existing knowledge. Information extraction (IE) is the process of automatically extracting information in a structured form from the available unstructured or semi-structured documents [1]. The basic task of IE involves identifying a set of concepts that are predefined in a specific domain, which has a well-defined information need [2].

Various applications of Information extraction like de-duplication, content extraction and keyword-based web search use web page segmentation as an important step [3]. Web page Segmentation is a process to partition a web page into coherent segments or sections which comprises of semantically related items [4]. Many applications like mobile devices [5], archiving [6], web accessibility [7] etc., use web page segmentation for detecting different blocks in a web page. For every user, only a part of the information is useful while the rest of it can be considered as noise. Filtering out this noise from web pages is a tedious task which can cause difficulty in segmentation. Web applications should understand the user's need and remove irrelevant and redundant information; otherwise, this can seriously harm the process of extracting information from the web.

Web page segmentation has its application in Mobile Web and when we view information on mobile phones, many websites are not displayed properly. It requires the user to zoom in for the information which he or she is searching for and web page segmentation may be helpful in it. Another application where web page segmentation is used is web crawling [8] in which the web page structure is analysed automatically and different parts of web pages are estimated [9], saving the important parts during crawling and ignoring the other parts.

Besides various applications, the web page segmentation has various approaches like DOM-Based, Text-Based and Vision-based. Each approach has its own advantages and disadvantages. The paper is structured as follows: In Sect. 2, information extraction and web page segmentation are explored, reviewed and discussed. In Sect. 3, the approaches of web page segmentation, are revisited along with their advantages and disadvantages are presented. Section 4 reviews and describes about the existing algorithms, and their comparative analysis has been made.

2   Information Extraction and Web Page Segmentation

2.1   Information Extraction

Information extraction extracts data from unstructured text written in the natural language to give structured documents to the user [1]. Input text is analysed by the information extraction module for entities and relations [10]. The process of extracting information from an unstructured text involves pipelines of activities shown below in Fig. 1 [1].

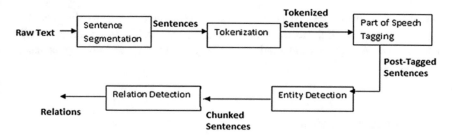

Fig. 1 Process of information extraction [1]

The process of information extraction as presented in Fig. 1 is described as follows: First, raw text is segmented into sentences with the help of sentence segmenter. Next, each sentence is further again divided into tokens using tokenizer. To identify named entities, tokens are tagged with POS (part of speech) tags. Chunking is used for named entity recognition. It is the process of segmenting and labelling multi-token sequence into an entity. Chunking is of two types—low-level chunking and high-level chunking. The low-level chunking takes the token and POS information as input and high-level chunking uses low-level chunks like POS and other information. The chunk level parsing also extracts the named entities and the relation existing with the entities. The IE system may also include a dictionary lookup, to identify the domain-specific entities such as place, person and organization names. This process could be done using Stanford's CoreNLP parser [1], an efficient statistical NLP tool, to identify the sentence dependency and coreference resolution.

2.2 Web Page Segmentation

Web page segmentation is a prominent research area in the domain of information extraction. It focuses on splitting a web page into small fragments based on various efficient criteria so as to uncover the information present in the web page. Just as a web designer divides a web page into lines, and each line is further divided into lines or vertical division of page, this splitting is done in a recursive manner for web page segmentation.

Information in a web page is structured into different parts, which represent specific elements in HTML. When a URL is opened by a user, the HTML file and other related resources of the web page gets displayed in the web browser, from which readers can easily recognize these semantic blocks of information. This is still a challenge though as the web page source code is not encoded in a way such that the semantic blocks can be differentiated.

Precision and generality are one of the main factors in web page segmentation. A segmentation is said to be precise, if its granularity is equal (or very close) to the granularity of an ideal segmentation, where granularity refers to the extent to which a segmentation divides a web page into blocks. A segmentation is said to be generic, if it performs well on all the different types of web pages such as forms, blog, website, archives etc. Web page segmentation may be performed by various approaches as given below.

3 Literature Review

Gupta et al. (2003) proposed a DOM-based approach, a DOM tree is looked for cues on how to divide a page. The crux behind DOM is that semantics of a web page is depicted by HTML structure but this is not always the case. So, the quality depends on the underlying HTML [9]. Kohlschütter and Nejdl (2008) proposed a text-based approach. In text-based approach, the tree structure of HTML is not considered and it only takes into account the textual content of the web page and analyses it for features such as link density or text density of parts of the web page [3]. Sanoja and Stéphane Gançarski (2015) presented a framework for measuring the performance of the segmentation algorithm. This framework describes a model that contains different metrics which evaluates the quality of segmentation obtained from a given algorithm. This model is applied to Block fusion, Block-O-matic, VIPS and jVIPS. All the four algorithms worked nice with respect to text extraction but it does not work well for geometry extraction [11]. Cormier et al. (2017) describes an edge-based segmentation algorithm which is designed for web pages. In this approach, each page is considered as an image. Segmentation and region classification are done at initial level [12]. Cormier et al. (2017) proposed a system which analyse the structure of web page based on visual properties and not on implementation details. As implementation details are hidden and it also reduces the complexity. It produces a segmentation which is hierarchal [13].

4 Approaches for Web Page Segmentation

Th approaches for Web Page Segmentation are broadly divided into four categories [3, 9, 14, 15]:

- DOM-based approaches [9]
- Text-based approaches [3]
- Vision-based approaches [14]
- Hybrid approaches [15].

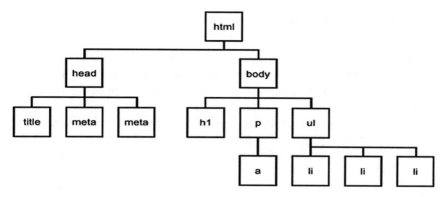

Fig. 2 DOM tree of the web page

4.1 DOM-Based Approach

In DOM-based approach [9], a DOM tree is looked for cues on how to divide a page, as shown in Fig. 2. DOM is the application program interface for XML and HTML document. The crux behind DOM is that the semantics of a web page is depicted by HTML structure but this is not always the case. So, the quality depends on the underlying HTML.

DOM tree consists of various tags like head, body, title, meta, h1, p etc., as shown above, and it enables in segmenting the web page which has various advantages and disadvantages.

Advantages

- Since only HTML code is needed to parse and rendering is not required, this approach is easy to implement.
- It is suitable for segmenting large segments and is easy to implement because no browser engine is involved.
- It takes only structured information into account.

Disadvantages

- It assumes that the HTML document reflects the semantics of the content, which is not necessarily true.
- There are multiple ways of building the HTML document structure while the semantics remain the same.
- It disregards layout and styling information by design.
- The pages which are built using JavaScript are not segmented by this approach.

4.2 Text-Based Approach

In text-based approach, the tree structure of HTML is not considered and it only takes into account the textual content of the web page and analyses it for features such as density of link or density of text in parts of the web page [3]. Techniques based on this approach are similar in results as obtained from quantitative linguistics. This statistically indicates that the blocks of text that have similar features have high chances of belonging together and therefore can be fused together. The threshold is predetermined for calculating this similarity and it depends on the required granularity level.

Advantages

- It is fast, since DOM does not need to be built.
- It is easier to implement since no DOM access is necessary.

Disadvantages

- It does not work with pages built via JavaScript.
- It does not consider structural and visual cues.
- Arbitrary changes to the text-density threshold are required during recursive application on sub-blocks.

4.3 Vision-Based Approach

Visual-based approaches work on the rendered web page [1]. The way the human segments a web page, visual approach works in a similar way, as shown in Fig. 3 below. They are computationally expensive as they have most of the information available. Vision-based approach often divides the web page into separators, like whitespaces, lines, images and content, and using this information construct a content structure. Visual features like location, font size, background colour and type on the web page are also taken into consideration.

It has various advantages and disadvantages.

Advantages

- Takes styling and layout information into account.
- Works similar to how a human performs the task.
- Can take implicit separators like vertical/horizontal lines and whitespaces into account.

Disadvantages

- More complex because they require a browser engine to render the page first.
- Computationally expensive because the page needs to be rendered.
- External resources like .css files and images are required for correct working.

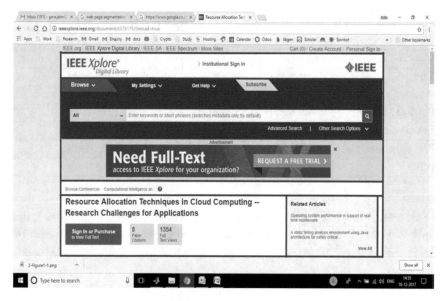

Fig. 3 Vision-based segmentation of web page

4.4 Hybrid Approach

A mix of techniques based on the previously mentioned approaches can be used for efficient web page segmentation purpose and referred to as a hybrid approach.

5 Segmentation Algorithms

There are various algorithms for web page segmentation. Some algorithms are DOM-based while some are text-based and vision-based or a hybrid combination of these. In this section, various segmentation algorithms are discussed, followed by a comparison as shown in Table 1.

5.1 Block Fusion Algorithm

Authors of Block fusion algorithm proposed a new method for segmenting HTML pages which are based on techniques of computer vision [3]. The main idea behind this algorithm is that the number of tokens or token density is the main parameter for taking segmenting decisions. For identifying segments based on text density, block fusion algorithm is presented. Text density is defined as the ratio between count of

Table 1 Comparison of various algorithms of web page segmentation

Algorithm	Approach	Year	Strategy
Block fusion	Text-based	2008	Bottom-up
VIPS	Vision-based	2003	Top-down
Modified VIPS	Vision-based	2013	Top-down
Block-O-Matic	Vision-based	2013	Top-down
Box clustering	Vision-based	2015	Top-down
Page segmenter	DOM-based	2005	Top-down
jVIPS	Vision-based	2012	Top-down

words present in a block and the height of printed blocks in inches. In block fusion algorithm two adjacent blocks are fused by comparing their text densities instead of intensity of pixel. Slope delta between two adjacent blocks r and s, $\Delta\rho(r, s)$ is given as [3]

$$\Delta\rho(r, s) = \frac{\rho(r) - \rho(s)}{\max(\rho(r), \rho(bs))} \tag{1}$$

where $\rho(r)$ and $\rho(s)$ is the block density of block r and block s, respectively. Block density [3] of a block is defined as [3]

$$\rho(r) = \frac{\text{count of tokens in block r}}{\text{count of lines in block s}} \tag{2}$$

Two blocks are fused, if the $\Delta\rho(r, s)$ is less than some threshold value v_{max}. This process keeps on iterating until there exists no pair of adjacent block exists which satisfies the threshold constraint. Also, if the text densities of the previous block and next block are same or greater than its own densities, then the previous block, next block and block itself will be fused.

5.2 Vision-Based Page Segmentation Algorithm (or VIPS)

This algorithm is a vision-based algorithm that retrieves the semantic structure of a web page which is hierarchical based [14]. Each node of the semantic structure maps to block. The coherency of a block is checked by a value that is degree of coherency and it depends on visual perception. Segmentation of a page in this algorithm consists of the following three phases [14]

1. Extraction of visual block.
2. Detection of separator.
3. Construction of content structure.

S. no.	Phase	Description
1.	Extraction of visual block	In this phase, all visual blocks contained in current subpage are found. The degree of coherency value is set to each of the extracted nodes which represent a visual block. This process keeps on iterating until all appropriate nodes are found
2.	Detection of separator	The next step is detection of separator. Visual separators are horizontal or vertical lines that do not cross with blocks visually. They are used to differentiate blocks with varying semantics. In this step, separator weights are also assigned to separator set
3.	Construction of content structure	The last step is constructing the content structure. In this step separator with lowest weight are selected and the new blocks are formed by combining the blocks beside the separators with lowest weight. The process keeps on repeating until the separator having maximum weight is found. Based on the highest weight of separator in the block region, the degree of coherence for each new block is set. The granularity requirement of each leaf node is checked. If a node fails, then extraction of visual block is redone. The iterative process is stopped if every node meet the granularity requirement

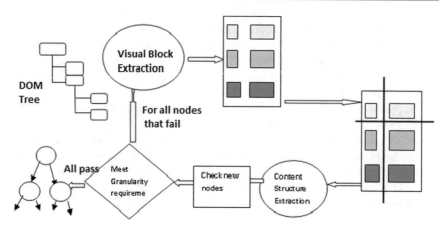

Fig. 4 Vision-based page segmentation flowchart [14]

The above phases may be represented in the form of a flowchart as shown in Fig. 4.

5.3 Vision-Based Page Segmentation Algorithm: Extended and Perceived Success [16]

This algorithm overcomes the shortcomings of VIPS algorithm. One of the drawbacks of VIPS is that the threshold and the rule set for extracting visual block is not properly defined [16]. Authors have implemented a modified version of the VIPS algorithm on Java platform. Another problem with VIPS is that HTML 5 tags are not mentioned in the tag set. First, the set of tags is extended by adding HTML5 tags and they are organized under tag set as mentioned in VIPS algorithm. Then, line break nodes are classified with respect to the space produced by these nodes. For detecting distinguished contents in a web page, visual properties like background colour, font colour and font size are also added.

5.4 Block-O-Matic [11]

Authors of Block-O-Matic have suggested this hybrid approach, a framework for segmenting web page which is triggered by visual-based content web segmentation techniques and automated document processing methods [11]. This framework combines visual, logical and structures features of the web page for analysing and understanding the contents of web page. It considers the flow order while segmenting the web page. This framework contains three phases analysing, understanding and reconstructing. For analysing the web page DOM tree is involved, for understanding the web page content structure is involved and for reconstructing the web page logical structure is involved. The final result is a semantic tree which consists of semantic structure. The model proposed by this framework is shown below in Fig. 5.

Using rendering of the web browser, DOM tree is retrieved. Output of the DOM tree is analyzed by D2C algorithm and content structure $W_{content}$ is obtained.D2C algorithm takes DOM tree as input and its output is $W_{content}$. Content structure represents the placement and classification of the objects on the web page. Document understanding is the mapping of $W_{content}$ to $W_{logical}$. and for understanding the web page c21 algorithm is used. Finally, the *Rec* function transforms logical structure into semantic tree W'.

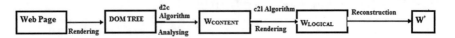

Fig. 5 Web page segmentation model [11]

5.5　*Box Clustering Segmentation [17]*

A different method for vision-based web page preprocessing [17] was proposed, which uses a rendering engine to get an image of the web page. It takes the smallest rendered elements of that image, performs clustering using a custom algorithm and produces a flat set of segments of a given granularity. For the clustering metrics, purely visual properties—the distance of elements and their visual similarity are used. It has been shown that its performance is superior to the VIPS performance and has two major advantages over the existing algorithms. First is the strict usage of visual information only, which makes this method transferrable to other document types, and also makes it resilient to changes in HTML, DOM and other technologies used on the web. The second advantage is the output structure that is more comprehensive and convenient for further processing. A comparison has been made of the following algorithms as shown below in Table 1.

6　Results and Conclusions from Web Page Segmentation

- Approaches of web page segmentation have been analysed and concerned algorithm based on these approaches have been compared.
- The above may be useful in drawing various concerned inferences.
- It may be used for retrieving sentiments from a web page, i.e. moving towards sentiment analysis which may be milestones in web semantic research.
- It may be used for separating/filtering noise from the concerned useful content.
- The study may be a useful reference for the information extraction technique derivation.

7　Future Work and Conclusion

In this paper, the role of Web page Segmentation towards Information Extraction for Web Semantics has been revisited. Various algorithms and approaches for web page segmentation are described along with their advantages and disadvantages, followed by a tabular comparison. Web page segmentation finds its usage in many applications like retrieval of information, classification of web page, archiving, and adaptation etc. It thus offers huge scope in the information extraction domain. In the future, various techniques of web page segmentation may be explored and applied towards the goal of semantic web. Also, the techniques of web page segmentation combined with information extraction may be used to retrieve sentiments from web page, i.e. Sentiment Analysis. The vision-based page segmentation combined with the classification of semi-structured data may be used for information extraction from web documents.

References

1. Bird S, Klein E, Loper E (2009) Natural language processing with Python. O'Reilly Media, Inc.
2. Piskorski J, Yangarber R (2013) Information extraction: past, present and future. In: Multi-source, multilingual information extraction and summarization. Springer, Berlin, Heidelberg, pp 23–49
3. Kohlschütter C, Nejdl W (2008) A densitometric approach to web page segmentation. In: Proceedings of the 17th ACM conference on information and knowledge management, pp 1173–1182
4. Feng H, Zhang W, Wu H, Wang CJ (2016) Web page segmentation and its application for web information crawling. In: Proceedings of the ICTAI-2016, IEEE Computer Society
5. Xiao Y, Tao Y, Li Q (2008) Web page adaptation for mobile device. In: Proceedings of the wireless communications, networking and mobile computing, IEEE Computer Society
6. Saad MB, Gançarski S (2010) Using visual pages analysis for optimizing web Archiving. In: Proceedings of the 2010 EDBT/ICDT workshops
7. Mahmud J, Borodin Y, Ramakrishnan IV (2007) Csurf: a context-driven non-visual web-Browser. In: Proceedings of the 16th international conference on World Wide Web, WWW'07, New York, NY, USA, pp 31–40. ACM
8. Barrio P, Gravano L (2016) Sampling strategies for information extraction over the deep web. Inf Process Manag 53(2):309–331
9. Gupta S, Kaiser G, Neistadt D, Grimm P (2003) DOM-based content extraction of HTML documents. In: Proceedings of the 12th international conference on World Wide Web, May 20–24, Budapest, Hungary, pp 1173–1182
10. Sanoja A, Gançarski S (2015) Web page segmentation evaluation. In: Proceeding of the of the 30th annual ACM symposium on applied computing, pp 753–760
11. Sanoja A, Gançarski S (2014) Block-o-matic: a web page segmentation framework. In: International conference on multimedia computing and systems (ICMCS), pp 595–600
12. Cormier M, Mann R, Moffatt K, Cohen R (2017) Towards an improved vision- based web page segmentation algorithm. In: 2017 14th conference on computer and robot vision computer and robot vision (CRV), pp 345–352
13. Cormier M, Moffatt K, Cohen R, Mann R (2016) Purely vision-based segmentation of web pages for assistive technology. Comput Vis Image Underst 148(3):46–66
14. Cai D, Yu S, Wen J-R, Ma W-Y (2003) Vips: a vision-based page segmentation algorithm. Microsoft technical report, MSR-TR-2003-79
15. Kuppusamy KS, Aghila G (2012) Multidimensional web page evaluation model using segmentation and annotations. Int J Cybern Inf 1(4):1–12
16. Elgin Akpınar M, Yesilada Y (2013) Page segmentation algorithm: extended and perceived success. Curr Trends Web Eng. ICWE 2013; Lect Notes Comput Sci 8295:238–252
17. Zeleny J, Burget R, Zendulka J (2017) Box clustering segmentation: a new method for vision-based web page preprocessing. Inf Process Manag 53(2):735–750

Towards an Evolved Information Food Chain of World Wide Web and Taxonomy of Semantic Web Mining

Priyanka Bhutani and Anju Saha

Abstract The addition of semantic knowledge has completely revamped the World Wide Web (WWW). The taxonomy of Web Mining traditionally classifies it broadly into the three subtypes of Web Content Mining, Web Structure Mining and Web Usage Mining. But with the emerging concepts of semantic web, the boundaries of this classification got more and more blurred. Hence, there is a need to classify Web mining taking into consideration the changing semantics of the web. This paper presents the evolved taxonomy of Semantic Web Mining which has noticeably developed over the years. Along with this, it also presents the Expanded Information Food chain of the WWW in the new era web which has been unfolding over the last two decades. Further, an approach for differentiating between the segments of 'Utilizers' and 'Contributors' to the Semantic Web data is also put forward. This increased awareness would facilitate in increasing 'Contributors' for completing the loop of transforming the Current Web to the Intelligent Web; as envisioned by the founder of World Wide Web.

Keywords Semantic web mining · Taxonomy · Information food chain
World Wide Web · WWW · Intelligent web

1 Introduction

The way, we access information that has rapidly changed since the last few decades with the extensive usage of the World Wide Web; which is undoubtedly the most prevalent, diverse and dynamically growing medium of searching or authoring information today. Although the data mining techniques started being used over the WWW

P. Bhutani (✉) · A. Saha
USICT, GGS Indraprastha University, Government of NCT of Delhi, Sec 16-C, Dwarka,
New Delhi, India
e-mail: priyanka.b@ipu.ac.in

A. Saha
e-mail: anju_kochhar@yahoo.com

© Springer Nature Singapore Pte Ltd. 2019
S. Bhattacharyya et al. (eds.), *International Conference on Innovative Computing
and Communications*, Lecture Notes in Networks and Systems 56,
https://doi.org/10.1007/978-981-13-2354-6_46

443

data, but the users of the web still faced many problems while using the web as the 'First authority' for their day-to-day information needs [1]. These include (though not limited to) the finding of meaningful and relevant information which matches their specific needs; creating useful knowledge out of the information extracted; having personalized information suiting diverse preferences of each user; learning about customers' to retain them or learning about casual visitors to turn them into customers. Towards this direction, the foundation of an *intelligent* web which incorporates the meaningful metadata along with the web data was laid down by the founder of the web, Tim Berners-Lee termed it as 'The Semantic Web' [2].

This paper first reviews these aforesaid changes in a comprehendible and structured manner. Section 2 starts with the background of web mining. Section 3 compiles the timeline of major milestones that have led to the development of semantic web mining.

But, with these changes, the boundaries of the traditional classification of web mining have blurred, leading to the need to classify Web mining taking into consideration the semantic additions to the web. Also, after these developments, the whole changed scenario of information usage and exchange needs to be re-modelled.

This paper addresses the aforesaid needs. Section 4 presents the current evolved taxonomy of semantic web mining. Section 5 presents the expanded Information Food Chain of the World Wide Web, giving some recent research work examples of each part of the chain. The conclusion and observations are given in Sect. 6.

2 Background on Web Mining

It was a widespread belief in the sceptics that the information on the web is too unstructured for data mining techniques to be applied on it for deriving knowledge. It was in the year 1996 that Etzioni [3] first gave the term 'Web Mining' and also explored whether Web mining is effectively possible. He defined Web Mining as usage of the various 'data mining techniques' for automatic discovery of services and extraction of information from data on the World Wide Web.

2.1 Evolution of the Taxonomy of Web Mining

Depending on the data that is used for web mining, the taxonomy of web mining has also evolved over the years as discussed in this section. In the earlier attempts, only two types of web mining were proposed. In one of the first attempts of developing a formal taxonomy of web mining, Cooley et al. [4] gave the taxonomy of Web mining of being composed of Web Content Mining and Web Usage Mining. *Web Content Mining* is the process of using the Web page contents including text (i.e. unstructured or semi-structured), audio, images and video, for the purpose of discovering useful information or knowledge. The basic search engine was the most primitive form of web content mining. Nowadays, it actually goes beyond using basic keywords in the search engine for finding subject directories, developing intelligent agents, forming clusters of similar resources, etc. One of the prototypes for web text mining was given by Jicheng et al. [5]. *Web Usage Mining* uses the usage patterns of the web pages from the usage logs stored in the cookies, browsers, proxy servers or web server; to comprehend user behaviour while interacting with a particular website or the web in general. An architecture for Web Usage Mining was proposed by Cooley et al. [4].

Jicheng et al. [5] proposed the taxonomy of Web mining having the two types of Web Content Mining and Web Structure Mining. *Web Structure Mining* uses the hyperlinks among web documents for modelling web structure—to improve navigability, to discover new resources, to rank web pages or to test broken links. The comprehensive taxonomy of Web Mining was given by Kosala and Blockeel [1] which consisted of Web mining having all subtypes of Web Content Mining, Web Usage Mining and Web Structure Mining.

3 Timeline of Major Milestones Leading to Semantic Web Mining

Historically, the concept of Ontology came from the branch of metaphysics. Gruber [6] introduced the concept of 'Ontologies' in the 'Information Sciences' field for the first time as formalized specification of shareable conceptual knowledge. He also specified principles for Ontology design. The widely cited paper by Fayyad [7] is credited for proposing the step-by-step process for the discovery of knowledge in databases (called KDD) and introducing the term 'Data Mining' in this process. Etzioni [3] coined the term 'Web mining' and the further development of web mining taxonomies progressed as given in Sect. 2.1. Decker et al. [8] modelled the whole scenario of information usage and exchange for the WWW in which agents use metadata of 'ontologies' for solving the problems of the end user. Tim Berners-Lee [2], who is credited for laying the foundation of the World Wide Web, put forward his vision for the future web by introducing the Semantic Web for adding meaning to computers and the web. Berendt et al. [9] rightly observed that with the advent of semantics, the boundaries of Content and Structure mining started vanishing. With the applications like personalization of Web Content, the differentiation among Content, Structure and Usage mining blurred too. Actually, semantics became the bridge between the 'standard' types of Web Mining for the purpose of solving many research problems around the Web. To link various open datasets semantically, the Linked Open data project [10] was started as a community project. Timeline of these events is shown in Fig. 1.

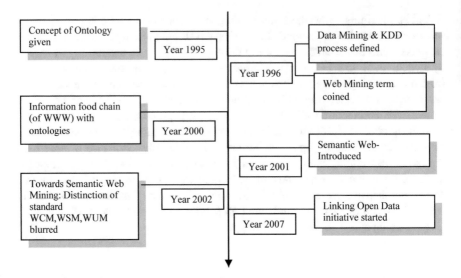

Fig. 1 Timeline of major milestones to semantic web mining

4 Evolved Taxonomy of Semantic Web Mining

The era of semantic web mining has emerged due to the need of making the web both syntactically and semantically smart. The usage of newly added semantic annotations on the web has started to improve the efficiency of web mining [11]; further, web mining is also being used for building the future intelligent web, i.e. Semantic Web using concepts and applications of metadata, ontologies, RDF and other semantic technologies. The work done towards this end began to enhance the domain knowledge of various domains on the web like medical sciences, education, social sciences, etc. The captured semantic metadata began to be shared and reused using initiatives like that of Linked Open Data [10].

Thus, the formalized taxonomy of semantic web mining which has evolved over the years can be seen as being divided into a few broad categories as shown in Fig. 2.

(i) '*Semantic-Driven*' Web Mining: Many new-age mining systems have been developed in which either the semantic structures in the Web or the semantic datasets drive the mining process of web mining. This can be termed as Semantic-driven Web mining. This is essentially using data mining techniques over the current web 2.0 itself but using semantic data too. When these systems only benefit from or utilize the semantic data or structure over the web, these form 'Semantic-*Utilizing*' *Web* Mining systems. But when they contribute the semantic data explored or improved by them to the semantic web domain, these can be termed as 'Semantic (Web)-*Contributing*' *Web* Mining systems.

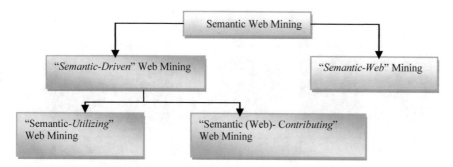

Fig. 2 Evolved taxonomy of semantic web mining

(ii) '*Semantic-Web*' Mining: The development and usage of new improved data mining techniques for mining the semantic web, within the architectural formalizations specified for it, can be categorized as *Semantic Web* Mining.

5 Evolved Information Food Chain of WWW with Advent of Semantic Web Mining

Decker et al. [8] modelled the scenario of information usage and exchange over the World Wide Web in the form of an 'Information Food Chain'. But, the way web information is used, shared and authored has changed rapidly over the last two decades. This section models the expanded information food chain of the WWW which has unfolded in the new-era web (as shown in Fig. 3). Depending on their varying information needs ranging from simple to complex ones, different types of web users like *Generic Web User*, *Sophisticated Web User* and an *Analyst End User* are accessing the WWW today. The *web data* is the data on the web which could be related to the *contents* of the web page, the hyperlinked *structure* of a website in particular or the web in general or *usage* related data stored in the weblogs.

The *Semantic Web Data* includes the type of data which incorporates the machine-understandable concept, context or relationship related data to make the web smart, like that of metadata tags in web pages JSON objects storing RDF triplets, ontologies, etc. This may be *Domain-Specific data* like the data that belongs to medical domain or *Domain-Independent data* like the data that is relevant to the process of data mining itself. The *Linked Open Data* (LOD) [10] consists of machine-interpretable, open, interlinked collection of datasets, which has semantic data spanning over multiple domains medical domain to the government data. *Semantic Annotation systems* are adding meaningful metadata information to given document or web source for the machines to be able to decipher the concepts and context of data as well [12]. These contribute a great extent to the semantic web data. The *Semantic Integration* systems

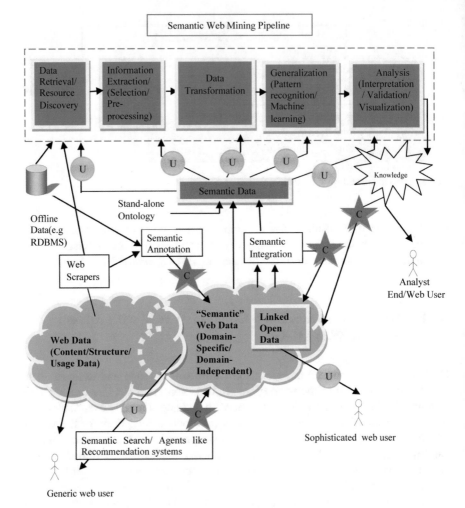

Fig. 3 Evolved information food chain of WWW with the advent of semantic web mining here, 'Utilizers' are marked as 'U' and 'Contributors' as 'C'

are combining the semantically related data from multiple sources into single valuable source by using the semantic matching technology [13].

Semantic Web Mining Pipeline comprises of the various subtasks as proposed in the process of web mining [14]; but with the addition of semantic data. The semantic data is being used at *any* level of this process along with *either* the web data extracted through *web scrapers* or the *offline data* like that of an RDBMS data store; or both. In *Data Retrieval/Resource Discovery* step, semantic data is being used to identify the most relevant data or resource or service for user's need by better matching and removing of ambiguities as in case of Web service discovery [15]. When semantic data is used at the *Information Extraction (Selection/Preprocessing)* stage, the purpose is to do away or reduce the need for a domain expert. The semantic knowledge of the concerned domain is used for interpreting the meaning of different types of data at hand like the structured relational data from databases [16] or semi-structured data from a web table. Various semantic data APIs like DBPedia Spotlight [17] are used for identifying named entities from unstructured web data. Semantic-based approaches are used for data cleaning and data validation like detecting of outliers or identifying and filling missing values [18].The *Data Transformation* step is improved by augmenting a dataset with those features which are found in semantic data or selecting better features from a feature space. Some of the ways for supporting the *Generalization (Pattern recognition/Machine learning)* step include usage of semantic information of a dataset for learning better classification rules or using Data Mining 'Ontologies' for better workflow planning [19]. In the *Analysis (Interpretation/Validation/Visualization)* step, semantic web data is used to explain and interpret results of mining like for understanding the logic behind the data belonging to the same cluster, its LOD graph is explored to find the common paths. Another example of this is building of a concept lattice with LOD data [20] so as to understand how the original data fits in the derived concepts. The semantics of data also helps in visualizing the data mining results like geographic entities' mining results have been visualized with semantics [21].

Semantic search has taken over the search domain by storm by increasing the search accuracy much beyond keyword matching. Semantics generates more relevant results due to better linking to the context of the search [22]. Similar to this is *Semantic Question-answering systems* which aim to understand the searcher's intent too and provide better answers to the questions asked by them [23]. *Recommendation systems* are rapidly benefitting from semantic data to extract useful features that improve the recommendation results [24]. As pointed out by Hans-Jörg Happel [25], despite significant systems supporting either the use of semantic metadata or creation of the same, a notion of recommending the contribution of semantic metadata in the public domain is required. In the view of the same, the above diagram differentiates between the segments of only '*Utilizers*' of semantic data (being marked with a 'U' in a circle) and '*Contributors*' to the Semantic Web data (being marked with a 'C' in a star).

6 Conclusion

By analysing the contributions being made by various communities of researchers in the development of semantic web mining, it is seen that there is a vast amount of work done for emerging needs of the intelligent web. This paper addresses the need to classify Web mining taking into consideration the changing semantics of the web by presenting an evolved taxonomy of semantic web mining. It also models the expanded information food chain of the WWW which covers the latest changes in the way web information is published and used on the web. Furthermore, it proposes an approach for differentiating between the segments of 'Utilizers' and 'Contributors' to the Semantic Web data, so as to support the notion of recommending a contribution of semantic metadata in the public domain.

It is observed that the need of the hour is to encourage the paradigm shift towards more and more contribution to the *shareable* semantic data and knowledge rather than *individual ontology-based* approaches to problem-solving. This would reduce the replication of research works by reusability of semantic datasets; reduce man-hours of developing such systems and lead to the faster fulfilment of the overall objective of an intelligent world wide web.

References

1. Kosala R, Blockeel H (2000) Web mining research: a survey. ACM SIGKDD Explor Newsl 2:1–15
2. Berners-Lee T, Hendler J, Lassila O (2001) The semantic web. Sci Am 284:34–43
3. Etzioni O (1996) The World Wide Web: quagmire or gold mine? Commun ACM 39(11):65–68
4. Cooley R, Mobasher B, Srivastava J (1997) Web mining: information and pattern discovery on the World Wide Web. In: Proceedings ninth IEEE international conference on tools with artificial intelligence, pp 558–567
5. Jicheng W, Yuan H, Gangshan W, Fuyan Z (1999) Web mining : knowledge discovery on the web. In: IEEE international conference on system man, cybernetics 1999. IEEE SMC'99 conference proceedings, pp 137–141
6. Gruber TR (1995) Toward principles for the design of ontologies used for knowledge sharing. Int J Hum Comput Stud 43:907–928
7. Fayyad U, Piatetsky-Shapiro G, Smyth P (1996) From data mining to knowledge discovery in databases. AI Mag, 37–54
8. Decker S, Jannink J, Melnik S, Mitra P, Staab S, Studer R, Wiederhold G (2000) An information food chain for advanced applications on the WWW. In: Proceedings of the 4th European conference on research and advanced technology for digital libraries, ECDL 2000, Lisbon, Portugal, September 2000, pp 490–493
9. Berendt B, Hotho A, Stumme G (2002) Towards semantic web mining. Semantic Web—ISWC 2002, pp 264–278
10. Bizer C, Heath T, Berners-Lee T (2009) Linked data-the story so far. Int J Semant Web Inf Syst 5:1–22
11. Ristoski P, Paulheim H (2016) Semantic web in data mining and knowledge discovery: a comprehensive survey. Web Semant Sci Serv Agents World Wide Web 36:1–22
12. Albukhitan S, Alnazer A, Helmy T(2016) Semantic annotation of Arabic web resources using semantic web services. Procedia Comput Sci, 504–511
13. Bella G, Giunchiglia F, McNeill F (2017) Language and domain aware lightweight ontology matching. J Web Semant 43:1–17
14. Zhang Q, Segall RS (2008) Web mining: a survey of current research, techniques, and software. Int J Inf Technol Decis Mak 7:683–720
15. Chen F, Lu C, Wu H, Li M (2017) A semantic similarity measure integrating multiple conceptual relationships for web service discovery. Expert Syst Appl 67:19–31
16. Kharlamov E, Hovland D, Skjæveland MG, Bilidas D, Jiménez-Ruiz E, Xiao G, Soylu A, Lanti D, Rezk M, Zheleznyakov D, Giese M, Lie H, Ioannidis Y, Kotidis Y, Koubarakis M, Waaler A (2017) Ontology based data access in Statoil. Web Semant Sci Serv Agents World Wide Web 44:3–36
17. Mendes PN, Jakob M, García-silva A, Bizer C (2011) DBpedia spotlight : shedding light on the web of documents. In: Proceedings of the 7th international conference on semantic system (I-Semantics), 95, pp 1–8

18. Matentzoglu N, Vigo M, Jay C, Stevens R (2018) Inference inspector: improving the verification of ontology authoring actions. J Web Semant 49:1–15
19. Miksa T, Rauber A (2017) Using ontologies for verification and validation of workflow-based experiments. J. Web Semant 43:25–45
20. D'Aquin M, Jay N (2013) Interpreting data mining results with linked data for learning analytics: motivation, case study and Directions. In: Proceedings of the third international conference on learning analytics and knowledge—LAK'13, pp 155–164
21. Ristoski P, Paulheim H (2015) Visual analysis of statistical data on maps using linked open data. In: European conference on semantic web, ESWC 2015, pp 138–143
22. Maynard D, Roberts I, Greenwood MA, Rout D, Bontcheva K (2017) A framework for real-time semantic social media analysis. Web Semant Sci Serv Agents World Wide Web
23. Höffner K, Walter S, Marx E, Usbeck R, Lehmann J, Ngonga Ngomo AC (2017) Survey on challenges of question answering in the semantic web. Semant Web. 8:895–920
24. Bischof S, Harth A, Kämpgen B, Polleres A, Schneider P (2018) Enriching integrated statistical open city data by combining equational knowledge and missing value imputation. J Web Semant 48:1–21
25. Happel HJ (2011) Semantic need: an approach for guiding users contributing metadata to the semantic web. Int J Knowl Eng Data Min 1:350

Usability Feature Optimization Using MWOA

Rishabh Jain, Deepak Gupta and Ashish Khanna

Abstract Usability is by far the most prominent term used to specify the quality of a software. It is defined and understood in reference to a hierarchical model called usability model, which combines the proposed seven basic usability factors, attributes and characteristics. In this work, a modified version of whale optimization algorithm evaluation called MWOA (Modified Whale Optimization Algorithm) has been discussed for software usability feature extraction.

Keywords Whale optimization algorithm · MWOA · Usability model
Feature selection

1 Introduction

Evolutionary algorithms belong to a class of algorithms, which mimic biological evolution such as reproduction, mutation, and a methodology of selection to unravel the optimization problems. Derived from evolutionary computation, they are population-based meta-heuristic optimization algorithms which imitate the above mentioned biological evolution. The popularly known evolutionary algorithms are Genetic algorithms (GA), motivated by natural selection [1], Cuckoo search (CS), motivated by cuckoo's deception of making other birds raise her young ones by putting her eggs into host birds' nests [2], and Bat algorithm (BA), inspired by the echolocation behaviors of microbats [3].

In the past few years, the data has been greater than before in the form of the number of features and instances, which produces noisier data. Noisier datasets

R. Jain (✉) · D. Gupta · A. Khanna
Maharaja Agarsen Institute of Technology, Rohini, Delhi, India
e-mail: rishabh.jain1379@gmail.com

D. Gupta
e-mail: deepakgupta@mait.ac.in

A. Khanna
e-mail: ashishk746@yahoo.com

© Springer Nature Singapore Pte Ltd. 2019
S. Bhattacharyya et al. (eds.), *International Conference on Innovative Computing and Communications*, Lecture Notes in Networks and Systems 56,
https://doi.org/10.1007/978-981-13-2354-6_47

increase the computational cost, complexity, and also trains the model slower. Due to which the need for feature selection arises in the field of Machine Learning. It aims to select an optimal subset of number of features from the given dataset, hence upgrading the performance and efficiency of the training models. Some of the evolutionary algorithms for feature selection are Binary Gray Wolf Optimization [4], Binary Bat Algorithm [5], Chaotic Crow Search Algorithm [6], etc.

A new meta-heuristic optimization-based algorithm, whale optimization algorithm [7] was proposed by Mirjalili in 2016. The study of Hof and Van Der Gutch [8] showed that cells in some areas of a whale's brain showed resemblance to the brain cells of humans also known as spindle cells. The spindle cells are generally responsible for the emotional issues, judgmental behavior, and social behaviors in humans. The WOA is motivated by the exclusive hunting characteristic or behavior of the humpback whales. In the above-discussed methodology, the humpback whales swim around the prey and also generate very distinctive bubbles along a circle or nine-shaped track.

In this paper, we have introduced a newfangled natural inspired optimization algorithm called Modified Whale Optimization Algorithm (*MWOA*) for usability feature reduction. WOA is an evolutionary algorithm for solving optimization problems. MWOA is an advancement on the same, with better accuracy and efficiency at the same cost. The feature selection problem with binary datasets can be tackled and solved with this algorithm, highly efficiently. The fitness function is made specifically to work with a dataset of binary type, which runs along the modified function.

The purpose and motivation behind implementing this Algorithm (MWOA) are to highlight the ease and feasibility with which it can be implemented and be comprehended.

In the presented work, Sect. 2 presents the preliminary algorithms are presented; Sect. 3 presents *MWOA* algorithm for feature selection which is followed by its implementation in Sect. 4. The results of MWOA algorithms have been discussed in Sect. 5; the comparison between *WOA* and *MWOA* are discussed in the Sect. 6; which is followed by the conclusion, future scope, and references.

2 Preliminaries

2.1 Software Usability Model Based on Hierarchical Approach

The presented section discusses about the software usability model in a hierarchical manner. The model combines 7 basic usability factors, their further classification into 23 features and 42 characteristics. The detailed taxonomy for this model has already been discussed in [1]. The purpose of this research is to study all the aspects of software usability and to define new features where required. The model consists of adding new features, removing unnecessary features, and gradually categorizing them under suitable factors. The proposed seven basic factors and their further classifications are as follows:

- **Effectiveness**: accomplishment of tasks, reusability, extensibility, operability, and scalability.
- **Efficiency**: resource, time, economic cost, and user effort.
- **Memorability**: learnability, comprehensibility, memorability of structures, and consistency in structures.
- **Productivity**: it is defined as the useful user task output.
- **Satisfaction**: is also known as likeability, convenience, and esthetics.
- **Security**: safety and error tolerance.
- **Universality**: approachability, utility, faithfulness, and cultural universality.

2.2 Feature Selection

Feature selection is a technique used to reduce the total number of features from a complete set of features. This has been done in such a way that the performance and accuracy of the system are improved or remains same on the reduced set of features. The main aim of this algorithm is to produce an optimal set of features from all the features with improved accuracy. Main motives to use feature selection are:

- Enabling the algorithm to train faster.
- Reducing the complexity of a model.
- Improving the accuracy of a model in case of precise selection of the subset.
- Reduces overfitting of the model.

3 Modified *WOA* for Usability Feature Selection

In WOA, since the updated solution is mostly depended on the current optimal candidate solution, we introduced an inertial weight ω in [0, 1] into WOA to obtain the Modified whale optimization algorithm (MWOA). Modified mathematical equations are represented as follows:

1. In encircling prey mechanism, the positions of search agents are updated as follows:

$$V = \left| C \cdot \omega Z^*(t) - Z(t) \right| \tag{1}$$

$$Z(t+1) = \omega Z^*(t) - A \cdot V \tag{2}$$

2. In spiral updating position, the position of the search agents are updated as follows:

$$Z(t+1) = V' \cdot e^{bl} \cdot cos(2\pi l) + \omega Z^*(t) \tag{3}$$

MWOA also follows the same procedure for choosing either of the two mechanisms for prey hunting but with updated mathematical equations. We introduced a variable p whose value lies in [0,1] which eventually decides which mechanism is to be performed. The equations are as follows:

$$Z(t+1) = \omega Z^*(t) - A \cdot V \quad if \ p < 0.5 \tag{4}$$

$$Z(t+1) = V' \cdot e^{bl} \cdot cos(2\pi l) + \omega Z^*(t) \quad if \ p \geq 0.5 \tag{5}$$

However, the humpback whales search for prey randomly, in addition to the bubble-net method, it is worth noticing that the introduced inertial weight remains constant for all search agents for every iteration.

Now, the Modified whale optimization algorithm (MWOA) for feature selection has been proposed.

ALGORITHM 1: *Modified whale optimization algorithm for usability feature selection:*

1. For each feature x_i ($\forall i \leftarrow 1..., m$), do
2.   For each model j ($\forall j \leftarrow 1..., n$), do
3. $x^j_i \leftarrow$ Assign Training Set
4. $v \leftarrow$ Random (0, 1)
5. $x^j_i \leftarrow v * x^j_i$
6.   Leader_score $\leftarrow 0$
7.   Max_leader_score $\leftarrow 0$
8.   Leader_position $\leftarrow$ zero vector order (1,j)
9. For each iteration ($t \leftarrow 1..., T$), do
10.   For each feature x_i ($\forall i \leftarrow 1..., m$), do
11.   fitness_eachi $\leftarrow$ fitness_function(x_i, n)
12.   if (fitness_each > Leader_score)
13.   Leader_score $\leftarrow$ fitness_each
14.   if (fitness_each > Max_leader_score)
15.   Max_leader_score $\leftarrow$ fitness_each
16.   if (fitness_each >= Leader_score / 2)
17.   feature_index $\leftarrow$ append(i)
18. $a \leftarrow 2 - t * ((2) / T)$
19. $a2 \leftarrow -1 + t * ((-1) / T)$
20. For each feature x_i ($i \leftarrow 1..., m$), do
21. $r1 \leftarrow$ Random [0 , 1]
22. $r2 \leftarrow$ Random [0 , 1]
23. $A \leftarrow 2 * a * r1 - a$
24. $C \leftarrow 2 * r2$
25. $b \leftarrow 1$
26. $l \leftarrow (a2 - 1) * Random (0, 1) + 1$
27. $p \leftarrow$ Random (0, 1)
28.   For each model j ($\forall j \leftarrow 1, ..., n$), do
29.   If (p < 0.5)
30.   if (abs (A) >= 1)
*31.   rand_leader_index $\leftarrow$ largest int of (x_i * Random)*
32.   x_rand $\leftarrow$ xi [rand_leader_index , :]
*33.   V_Z_rand $\leftarrow$ absolute (C * Z_rand [j] - x^j_i)*
*34. $x^j_i \leftarrow$ x_rand[j] - A * V_Z_rand*
35.   if (abs (A) < 1)
*36.   V_Leader $\leftarrow$ absolute (C * Leader_pos[j] - x^j_i)*
*37. $x^j_i \leftarrow$ Leader_pos[j] - A * V_Leader*
38.   elif (p>=0.5)
39.   distance2Leader $\leftarrow$ absolute (Leader_pos[j] - x^j_i)
*40. $x^j_i \leftarrow$ distance2Leader * (e^{b*l}) * cos(1 * 2 * pi) + Leader_pos[j]*
41.   no_selected_features $\leftarrow$ append size.(feature_index)
42.   accuracy $\leftarrow$ accuracy_function(no_selected_features , T)
43.   print no_selected_features
44.   print feature_index
45.   print accuracy

The steps of the proposed algorithm are as follows:

1. The population of whales is initialized in the first loop in lines 1–7. The position of whales is then initialized as per the values in the dataset and is multiplied by an inertia weight. The Leader score and Leader position are initialized.
2. The second loop in lines 9–16 fitness of each whale/feature is checked using a fitness function and corresponding whales are added to an array. Also, "a" is decreased linearly.

3. The third loop in lines 17–24 for all the constants are initialized.
4. The fifth loop in lines 25–37, the position of each whale is changed according to the equations mentioned in the paper. In lines 38–42, all the selected whales are stored and accuracy is calculated for each iteration.

Algorithm 2: accuracy(x_i, T)
1. Initialize accuracy vector
2. accuracy ← []
3. For each iteration   (□ ← 1..., □), do
4.   values ← x_j / total no of features that is 23
5.   accuracy ← append values [t]
6. return accuracy

Algorithm 3: fitness_function(x_i,T)
1. s ← number of non zero terms for each x^j_i
2. return s

4   Implementation of *Modified WOA*

In this section, the *Modified WOA* algorithm has been coded using Python. Python is interactive, modular, dynamic, and portable. This section discusses about the experiment related setup, input-related parameters, and the dataset.

4.1   *Experiment-Based Setup*

To test the proposed algorithm, a computational device with 2.2 GHz Core (TM), 2 CPU and 4 GB RAM under Ubuntu 16.04 is being used. The implementation of the algorithms have been coded in Python version 3.6.3:: Anaconda, Inc.

The proposed algorithm determines the optimal set of features and accuracy for each SDLC dataset. This has been performed by the following two stages:

1. Obtaining an optimal set of features using MWOA.
2. Calculating accuracy for each SDLC dataset.

4.2   *Input Parameters*

The following explained the parameters that are computed at the starting of the algorithm:

ω   Random number in (0,1), ω is considered to be inertia weight.
a   2, where a is linearly decreasing from 2 to 0 over the course/flow of iterations.

r   It is random vector in [0,1].
p   It is random number in [0,1].
l   Random number in [−1,1], parameter in Eqs. 11 and 12.2.
b   1, parameter in Eq. 11 and 12.2.

4.3 The Dataset

The dataset used in this research consists of 23 usability features as rows and 6 SDLC models as columns. The binary numbers 0 and 1 have been used to indicate the occurrence and nonoccurrence of these features in the respective SDLC models. The dataset used in this work is taken from [5, 9].

5 Results and Discussion

The MWOA algorithm is applied to the dataset to get a reduced set of features and the results have been validated over the course of 10 iterations. The results are shown in the following figures.

Figure 1 shows that MWOA generates an optimal set of features containing 19 features with an accuracy of 80% over the course of the final iteration. Thus, we have proposed an efficient tool which takes binary datasets as input and generates an optimal set of features out of all the features. Now, according to the accuracy of each SDLC model, we can find out the best model for software development (Figs. 2, 3).

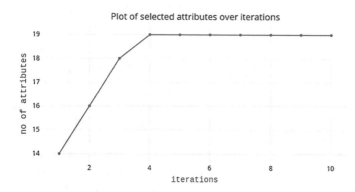

Fig. 1 Plot of selected attributes versus iterations for MWOA

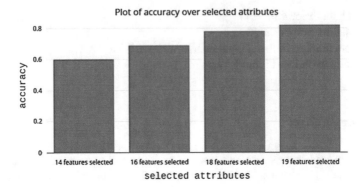

Fig. 2 The plot of accuracy versus the selected attributes for MWOA

Fig. 3 Plot of accuracy versus six SDLC models

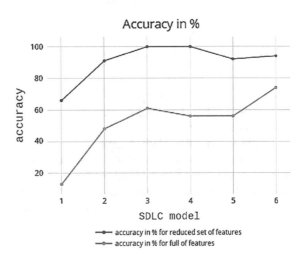

6 *MWOA* Versus Other Optimization Algorithms

The result of Modified WOA has been compared with the results of standard WOA, Modified Binary Bat Algorithm (MBBAT), and Binary Bat Algorithm (BBAT) for the feature selection. The plot of accuracy versus selected attributes over successive iterations of MWOA, WOA, MBBAT, and BBAT is presented in Fig. 4.

As seen from Fig. 4, standard WOA fails to reduce the features and producing irrelevant results over successive iterations and gradually generating 100% accuracy that means it is not reducing the number of features. Meanwhile, other algorithms like MBBAT and BBAT are obtaining a subset of features with less accuracy as compared to MWOA. The number of selected attributes by MWOA is 19 with an accuracy of 80%.

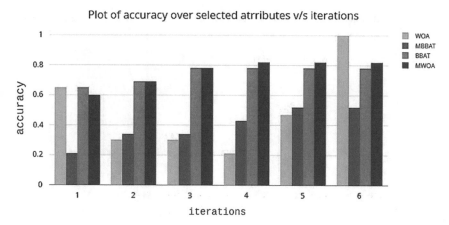

Fig. 4 Plot of iterations versus accuracy MWOA, WOA, MBBAT, and BBAT

7 Conclusion and Future Scope

The term "usability" has been defined by using a hierarchical-based software usability model. In this work, we have introduced a modified version of meta-heuristic optimization algorithm for usability feature selection called Modified whale optimization algorithm (MWOA). The MWOA algorithm is an extension of WOA. The Modified WOA works with the motive of reducing the attributes and retain a subset of relevant features/attributes which best describes the problem and downgrades the performance of the system to a little extent. The Modified WOA has selected minimum number of attributes with an accuracy of 80%.

References

1. Gupta D, Ahlawat A, Sagar K (2014) A critical analysis of a hierarchy based usability model. In: 2014 international conference on IEEE contemporary computing and informatics (IC3I)
2. Yang X-S, Deb S (2009) Cuckoo search via Levy flights. In: Proceedings of world congress on nature & biologically inspired computing (NaBIC), Coimbatore, India
3. Yang X-S (2010) A new metaheuristic bat-inspired algorithm. Department of Engineering, University of Cambridge, Trumping ton Street, Cambridge CB2 1PZ, UK
4. Emary E, Zawbaa H, Hassanien A (2016) Binary gray wolf optimization approaches for feature selection. Neurocomputing 172
5. Gupta D, Ahlawat AK, Sagar K (2017) Usability prediction & ranking of SDLC models using fuzzy hierarchical usability model. Open Eng
6. He YY, Zhou JZ, Li CS (2008) A precise chaotic particle swarm optimization algorithm based on improved tent map, ICNC
7. Mirjalili S, Lewis A (2016) The whale optimization algorithm. Adv Eng Softw
8. Hof PR, Van der Gucht E (2006) The structure of the cerebral cortex of the humpback whale, The Anatomical Record

9. Gupta D, Ahlawat S (2017) Usability feature selection via MBBAT: a novel approach. J Comput Sci 23

A Computational Study on Air Pollution Assessment Modeling

A. Nayana and T. Amudha

Abstract Air pollution is termed as introducing of biological substances, particulate matter, and chemicals to the atmosphere which causes damage to human beings and other living organisms, or cause harm to the natural atmosphere or to built environment. The origin of air pollution is classified into anthropogenic and non-anthropogenic. India is one of the biggest emitters of atmospheric pollutants caused by the road transportation sector. Air pollution modeling describes an arithmetical concept for understanding or predicting how pollutants are affecting the atmosphere. Modeling is also used to evaluate the connection among sources of pollution and their effects and influence on ambient air quality. This paper aims to survey the various techniques used for the assessment of air pollutant emission modeling.

Keywords Air pollution · Geographical information system (GIS)
Land-use regression (LUR) · Air quality modeling

1 Introduction

Environmental pollution is defined as the spread of toxic chemicals into the aquatic and terrestrial habitats which is undesired. Pollutions are classified into air, water, soil pollution, etc. Pollutants are of two types, primary and secondary. Carbon, nitrogen, sulfur, and halogen compounds fall under primary pollutants, whereas secondary pollutants are generated in the atmosphere from the primary pollutants and it is not directly emitted from any of the other sources. For example, NO_2, ozone, etc. Air pollution is the condition in which air contains harmful substances such as chemicals, particulate matter or biological substances, gases, dust, fumes, or odor in harmful

A. Nayana · T. Amudha (✉)
Department of Computer Applications, Bharathiar University, Coimbatore, India
e-mail: amudhaswamynathan@buc.edu.in

A. Nayana
e-mail: nayanavivek2011@gmail.com

© Springer Nature Singapore Pte Ltd. 2019
S. Bhattacharyya et al. (eds.), *International Conference on Innovative Computing and Communications*, Lecture Notes in Networks and Systems 56,
https://doi.org/10.1007/978-981-13-2354-6_48

Table 1 Factors responsible for air pollution [1]

Factors	Sources
Natural factors	**Volcanic eruption**: Volcanoes are responsible for emission of carbon monoxide and sulfur dioxide. On global scale, about 67% of the sulfur dioxide is produced by volcanoes
	Bacterial action: By bacterial action, nitrogen monoxide is emitted in the atmosphere. Methane is released by anaerobic decomposition of organic matters by bacteria that takes place in water sediments and in soil
	Trees and plants: Trees and plants in the atmosphere release huge quantities of hydrocarbons. The methane released from such sources can be remained in atmosphere up to an average of 3–7 years
Anthropogenic factors	**Burning different kind of fuels**: The sources of air pollution are the harmful smoke emitted by stack of power plants, automobiles, and furnaces of industries. Burning fuels emit injurious gases like, nitrogen oxide, carbon monoxide, and Sulfur dioxide
	Use of paints: Causes generation of harmful unwanted fumes
	Chlorofluorocarbons (CFC): Released from refrigerators, air conditioner insect repellent, etc., which will affect the ozone layer
	Waste decomposition in landfills: Emits nontoxic Methane gas, which is inflammable when it is contacted with air
	Military/Defense practices: Military weapons, missiles, rockets, and nuclear weapons produce toxic gases

amounts and causes discomfort to living beings and environment. A factor responsible for air pollution includes various activities or sources those release pollutants into the atmosphere.

2 Factors Responsible for Air Pollution

The factors of air pollution are classified as natural factors and anthropogenic factors. Natural factors are situated naturally in the environment and release pollutants to the atmosphere.

The human activities which contribute pollutants to the atmosphere are called anthropogenic factors/sources. Table 1 shows the factors responsible for air pollution.

3 Air Pollution Trends in India

A recent survey done in 2015 shows that air pollution level is at its peak in India [2]. The survey stated that India is labeled as the world's seventh environmentally unsafe country and the major part of pollutants are emitted from vehicles. Table 2

Table 2 Average pollutant percentage released from various metropolitan cities

Type of pollutant	Percentage value (%)
CO	70
Hydro carbons	50
Oxides	30–40
Suspended particulate matter (SPM)	30

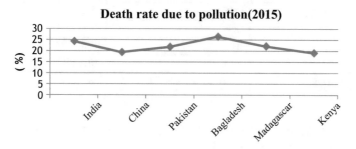

Fig. 1 Proportion of total deaths attributable to pollution in percentage

describes the pollutant percentage level from the study of Central pollution control board (CPCB) held in major cities, i.e., Bengaluru, Chennai, Pune, Delhi, Mumbai, and Kanpur as these cities are found to contribute more than 30% of pollutants in affecting ambient air quality. The tremendous increase in number of vehicles results in increase in air pollution level and greenhouse gas emissions. Vehicular emission plays a lead role in air pollution [3]. In 2015, air pollution is the major contributor, related to 6.5 million deaths in the worldwide. In India, 2.5 million deaths were linked to pollution, and 1.8 million people died only due to air pollution. Figure 1 shows the proportion of total deaths (in percentage) attributable to pollution word wide (Times of India, Oct 21, 2017). The survey done in 2015 shows that India has the second place in death rate due to pollution.

4 Air Quality Modeling

Air quality modeling describes an arithmetical concept for understanding or predicting how pollutants are affecting the atmosphere. Modeling is used to evaluate the connection among sources of pollution and their influence on ambient air quality. Information about pollution sources such as emission rates, stack height, and meteorological data are the inputs for air quality models [4]. The atmosphere is getting polluted from the pollutants that are constantly released from different kind of sources. The pollution sources are stationary (e.g., oil refineries, power plants, factories, and industrial facilities), mobile (e.g., trucks, buses, cars, trains, and planes),

Table 3 Statistical techniques used for calculating emission rate

Method	Location of research	Year and reference
Land-use regression	Hamilton	2006, [8]
	British Columbia, Canada	2007, [9]
	United States	2011, [10]
	India	2015, [3]
Dispersion model	Taiwan	2002, [11]
	India	2011, [12]
	India	2013, [13]
	India	2015, [3]
Regression equation	British Columbia, Canada	2007, [9]

natural (e.g., volcanoes, wildfires, and wind-blown dust) and area (e.g., cities, wood burning fireplaces, and agricultural areas).

5 Statistical Methods Used in Air Quality Monitoring

Statistical methods are used for calculating the emission rates from various sources [5–7]. Table 3 shows some of the statistical methods. Pollutant concentrations in the air at any location, is measured by the environmental characteristics of that area particularly those that reflect emission intensity and dispersion efficiency.

6 Land-Use Regression (LUR) Modeling

At high spatial resolution, ambient air pollutant concentrations are predicted using the Land-use regression technique. It is one of the alternatives for conventional approaches. The relationship between measured pollutants and land cover characteristics are analyzed using statistical techniques [9]. LUR models are used for modeling small-scale special variations of air pollution concentration. It predicts ambient concentration from transportation, population density characteristics, and landcover within the area. Initially, there was a variation in air pollution in the model which is developed as a part of the small area [8]. Multivariate regression technique is used in LUR modeling. In this technique, initially, the independent variable is added which is associated with the dependent variable, and the process continues by selecting the parameters that are regularly associated with the residuals as the next independent variable [10].

Table 4 Factors influencing the pollutants dispersion [3, 4, 14]

Factors	Parameters
Source characteristics	Pollutant emission rate
	Height of stack
	Gas exit velocity
	Gas exit temperature
	Diameter of stack
Meteorological parameters	Velocity of wind
	Direction of wind
	Ambient temperature
	Stability of the atmosphere
	Mixing height of stack and plume rise

7 Dispersion Model

Dispersion modeling is the method to estimate the impact of air pollutant emission that causes future changes, which is an essential part of air quality modeling. It is a mathematical tool that simulates the air pollutant dispersion in the atmosphere [14]. Various factors in the atmosphere that affects the dispersion of the pollutants emitted are shown in Table 4.

Different types of dispersion models are used and they are Gaussian Models, Lagrangian Model, Computational fluid dynamics Models, Box Models, Dense Gas Model, Eulerian Model, and Aerosol Dynamic Models [15].

8 Regression Analysis

In LUR, many regression analysis techniques are used, and Regression equation is used to predict the air pollutants. In this equation, undefined variables are shown in terms of known variables. The term, "U", which is the dependent variable is predicted on the basis of many independent variables. Multiple Linear Regression [16] is represented in Eq. 1.

$$U = a_1 + a_2 J_2 + \ldots + a_k J_k + err \tag{1}$$

where

U	Dependent variable
$J_2, J_3 \ldots, J_k$	Independent variables
$a_1, a_2, \ldots, a_k$	Parameters of linear regression
err	Estimated error term

Table 5 Air quality modeling research with GIS

Research work	Research location	Year and Reference
Analysis of air pollution seasonality in Delhi, India and the role of meteorology	India	2011, [12]
An emissions inventory of spatial resolution using GIS for air quality study in Delhi, India	India	2013, [21]
A review on air pollution due to traffic exposure in urban areas using GIS	India	2015, [3]
Review of air quality monitoring: Case study of India	India	2016, [17]

The term "err" can be calculated from a normal distribution with independent random sampling, constant variance and mean zero. The concept of the regression model is used for calculating the values of $a1$, $a_2 \dots$, a_k, which can be calculated with Minimum Square Error (MSE) method [9].

9 GIS Application

Geographic information system (GIS) is integrated with five interacting parts that include hardware, software, data, procedures, and people. The software consists of applications that help make maps. Data and information are in the form of points, lines, and polygons; we can use these in the map easily [15]. The function of GIS is to capture, store, check, and express the data in relation to the position on Earth's surface. GIS shows various kinds of data on a single map. It compares nonplanar (3D) information to conventional planar (2D) information [17]. The air pollution sources are visualized using GIS and based on the location, editing, and displaying of the road network becomes easier in a particular area [18, 19]. It gives the overall summary of places, where high impacts of air pollution are expected [14]. Humaib Nasir et al. (2016) described about air quality monitoring based on vehicle and automobiles which generates more pollutants to the atmosphere. They discussed about scenario of air quality in various Indian cities [20]. Some of the GIS-related works are indicated in Table 5.

10 Land-Use Regression and GIS

LUR model uses the regression equations. Predictor variables are used in the regression equation. For each site, these values are calculated using site coordinate and digital data sets inside the GIS [10]. GIS is very helpful for the selection of land area in modeling and also used to represent the level of concentration of the pollutants. Through GIS representation, the area-wise pollutant levels can be easily identified.

11 Conclusion

The concentration of air pollutants that are emitted from different sources is estimated by the air quality models. Models are highly helpful in a variety of air quality management decisions such as design of monitoring networks, stacks, and evaluation of potential mitigation strategies, etc. In the present scenario of highly pollutant air which poses a severe threat to the environment, air quality assessment and modeling are used to evaluate the connection among factors of pollution and their effects on ambient air quality. Also, it is inferred from the study that vehicular emission plays a lead role in air pollution. This paper describes a comprehensive review on the modeling techniques and sources of air pollution to assess the air quality.

References

1. Ali LG, Haruna A (2015) Effects of primary air pollutants on human health and control measures-a review paper. Int J Innov Res Dev 4(9):45
2. Sangeetha A, Amudha T (2016) An effective bio-inspired methodology for optimal estimation and forecasting of CO2 emission in India, innovations in bio-inspired computing and applications. Springer, Cham, pp 481–489
3. Kumar A, Mishra RK, Singh SK (2015) GIS application in urban traffic air pollution exposure study: a research review. Suan Sunandha Sci Technol J 2(1)
4. Modi M, Venkata Ramachandra P, Liyakhath Ahmed SK, Hussain Z (2013) A review on theoretical air pollutants dispersion Models. Int J Pharm Chem Biol Sci 3(4):1224–1230
5. Guidelines for manual sampling & analyses national ambient air quality series: NAAQMS, 36, (2012-13)
6. Major Andrew Ross Pfluger PE (2014) U.S. Military Academy: a GIS-based Atmospheric dispersion modeling project for introductory air pollution courses
7. Sangeetha A, Amudha T (2016) A study on estimation of CO2 emission using computational techniques. In: IEEE international conference on advances in computer applications (ICACA), pp 244–249
8. Jerrett M, Beckerman B (2006) A land use regression model for predicting ambient concentrations of nitrogen dioxide in Hamilton, Ontario, Canada. Air Waste Manag Assoc 56:1059–1069
9. Henderson SB, Beckerman B, Jerrett M, Brauer M (2007) Application of land use regression to estimate long-term concentrations of traffic related nitrogen oxides and fine particulate matter. Env Sci Technol 41:2422–2428
10. Novotny EV, Bechle MJ, Millet DB, Marshall JD (2011) National satellite-based land-use regression: NO2 in the United States. Envi Sci Technol 45
11. Diema JE, Comrieb AC (2002) Predictive mapping of air pollution in volving sparse spatial Observations. Env Pollut 119:99–117
12. Guttikunda SK, Gurjar BR (2011) Role of meteorology in seasonality of air pollution in megacity Delhi, India
13. Singha V, Vaishyaa RC, Shuklab AK (2013) Analysis & application of GIS based air quality monitoring- state of art Mamta Pandeya. Int J Eng Res Technol (IJERT) 2(12)
14. Angelevska B, Markoski A (2009) Decision making through integrated system for air quality assessment
15. Law M, Collins A (2016) Getting to know ArcGIS PRO. Esri press
16. Sangeetha A, Amudha T (2018) A novel bio-inspired framework for CO 2 emission forecast in India. Procedia Comput Sci 125:367–375. Elsevier

17. Wang G, van den Bosch FHM, Kuffer M (2008) Modelling urban traffic air pollution dispersion. Int Arch Photogramm Remote Sens Spat Inf Sci XXXVII, Part B8
18. Prasad Raju H, Partheeban P, Rani Hemamalini R (2012) Urban mobile air quality monitoring using GIS, GPS sensors and internet. Int J Env Sci Dev 3(4)
19. Pummakarnchanaa O, Tripathia N, Duttab J (2005) Air pollution monitoring and GIS modeling: a new use of nanotechnology based solid state gas sensors. Sci Technol Adv Mater 6:251–255
20. Nasir H, Goyal K, Prabhakar D (2016) Review of air quality monitoring: case study of India. Indian J Sci Technol 9(44)
21. Guttikunda SK, Calori G (2013) A GIS based emissions inventory at 1 km _ 1 km spatial resolution for air pollution analysis in Delhi, India. Atmos Env 67

Sentence Similarity Using Syntactic and Semantic Features for Multi-document Summarization

M. Anjaneyulu, S. S. V. N. Sarma, P. Vijaya Pal Reddy, K. Prem Chander and S. Nagaprasad

Abstract Multi-Document Summarization (MDS) is a process obtaining precise and concise information from a specific set of documents which are on the same topic. The generated summary makes the user to understand the content in a set of documents. The existing approaches suffer with the lack of establishment of semantic and syntactic relationship among the words within a sentence. In this paper, a novel unsupervised MDS framework is proposed by ranking sentences using semantic and syntactic information embedded in the sentences. Empirical evaluations are carried using lexical, syntactic, and semantic features on DUC2002 dataset. The experimental results on DUC2002 prove that the proposed model is comparable with existing systems using various performance measures.

M. Anjaneyulu (✉) · K. Prem Chander
Department of Computer Science, Dravidian University, Kuppam, India
e-mail: anjan.lingam1@gmail.com

K. Prem Chander
e-mail: kpc.1279@gmail.com

S. S. V. N. Sarma
CSE Department, Vaagdevi Engineering College, Warangal, India
e-mail: ssvn.sarma@gmail.com

P. Vijaya Pal Reddy
CSE Department, Matrusri Engineering College, Hyderabad, India
e-mail: drpvijayapalreddy@gmail.com

S. Nagaprasad
Computer Science Department, S.R.R. Govterment Arts & Science College, Karimnagar, India
e-mail: nagkanna80@gmail.com

© Springer Nature Singapore Pte Ltd. 2019
S. Bhattacharyya et al. (eds.), *International Conference on Innovative Computing and Communications*, Lecture Notes in Networks and Systems 56,
https://doi.org/10.1007/978-981-13-2354-6_49

471

1 Introduction

The flooding of information on numerous topics onto the World Wide Web makes the people all over the world able to access the content with no time. The information overload onto the internet results in difficulty of concise the meaning information on a topic. Automatic text summarization is the process of gist generation to address the specified problem. MDS is a procedure of comprehending the meaningful content form a set of related documents on the same topic. The basic purpose of a summary generation system is to derive most useful and precise important information from the input document. The focus of an MDS system is to produce the precise and concise information from a set of text documents. Identification of key sentences that carry important information and arrange the most meaning sentences in a particular order for not to miss the meaning of content in the document is an important research problem. Multi-Document Summarization problem can be defined as concatenating of all documents pertaining to a topic and notify the set as D. The number of sentences in the set D is m, represented as $S = \{s_1, s_2, ..., s_n\}$ and set of words in all sentences are $W = \{w_1, w_2, ..., w_n\}$, the MDS problem is to identify a set of sentences from S that concisely represents the information from all the documents in the set D.

The approaches for Multi-Document Summarization are broadly classified into extractive and abstractive approaches. Most of the research is towards extractive approaches that directly identifies the most meaningful sentences from whole set of sentences formed with multiple documents and combine these sentences to form a summary. In this paper, an attempt is made to generate extractive summarization from multiple documents. As in [1], the effectiveness of the summarization approach can be measured using four parameters, namely informative, length, relevancy, and diversity. The concise and precise summary should be covering all fundamental themes of the documents as much as possible. The information in the sentences should not be repeated. It should contain the optimal information with suitable number of sentences in the summary that aptly represents the whole set of documents. The abstractive approaches generate the sentences for the summary by learning the meaning of all sentences from the document set. The produced sentences in the summary are not as same as in the original sentences.

The extractive summarization approaches are of two types such as supervised methods and unsupervised approaches. The supervised approaches depend on the training data set. The model can be learned with the suitable dataset. This approach is used to generate the summary for a given test document set. As in [2], this method treats the problem of MDS as a categorization problem. For these approaches, a large set of training data is required to learn the model. The process of learning is time consuming. The unsupervised methods derive summary for a given document set based on the inferences generated from the documents. The unsupervised approaches identify the most meaning sentences-based statistical, syntactic, and semantic relations that exists among the words and sentences within the set of sentences and ranks the sentences based on the information that carry by the sentences.

Semantic Textual Similarity (STS) is one of the core disciplines in NLP. STS assess the level of semantic similarity between two textual segments. The textual segments are phrases, sentences, paragraphs, or documents. The objective of the STS is to measure the similarity in the range [0, 5] between a sentence pair where 0 indicates both the sentences are on irrelevant topics and 5 indicates both the sentences mean the same thing [3]. STS system is trying to emulate the idea of similarity, thus replicating human language understanding.

The approaches for measuring STS has been categorized into alignment-based approaches, vector space models, and machine learning approaches [4]. Alignment approaches compute the similarity between the words or phrases in a sentence pair and align the words or phrases that are most similar [5]. Vector space approach is a traditional NLP feature engineering approach represents the sentence as bag of words, and the similarity is evaluated according to the occurrence of words or co-occurrence of words or other replacement words. Machine learning approaches use supervised machine learning models to combine syntactic, semantic and lexical features of sentence pair [6]. However, estimating the STS is difficult if both the sentences have no words in common. Text summarization as an application of NLP can benefit from effective STS techniques in grouping of semantic similar sentences.

The proposed framework extracts sentences which are more important for summary generation. The proposed work is organized into five sections. Section 2 presents the existing work related to the automatic text summarization and multi-document summarization. Section 3 presents the proposed MDS approach and the summary generation process using syntactic and semantic sentence similarity. The proposed framework is evaluated on the benchmark of DUC2002 data set from Document Understanding Conference (DUC) for the task of MDS. The conclusions observed from the proposed work and the possible extensions that can be made on the proposed model are presented in Sect. 5.

2 Related Work

Most of the researchers proposed models for multi-document summarization using Topic model approaches using Bayesian theorem and graph-based approaches [7, 8]. The graph-based approaches follow bottom-up strategy from sentence selection to summary generation. These approaches identify the sentence similarity based on the content presented in the sentences. Hence, graph-based approaches are more familiar in text categorization and web content mining using searching. Bayesian topic model approaches follow top-to-bottom strategy for summary generation. These approaches depend on the semantic information that is embedded in the sentences to identify informative sentences. The similarity between two sentences can be measured using various distance-based techniques such as cosine similarity, Jaccard similarity, and square distance-based similarity. The advanced methods for finding similarity between two topics such as probabilistic distributions-based and WordNet-based approaches. Topic model-based methods do not establish relationships among the sentences and hierarchies, whereas using graph-based models can perform the links among the sentences and form an hierarchy among them. The graph-based models are also useful in performing hierarchical multi-document summarization. Both the models cannot consider the lexical co-occurrence relationships that exist among the words and within the sentences. As in [9], the lexical, syntactic and semantic co-occurrences among the words are most important for identification association among the words and correlations among the topics. These features are useful in speech recognition, text mining, and syntactic parsing.

The MDS approaches can be classified based on various dimensions such as extractive and abstractive approaches and also as supervised and unsupervised approaches. The training dataset is used to learn the model in supervised models. The various learning models are into existence HMM approach [10] to learn the relationships that exist within the training data set. As in [11], the image processing technique like sparse coding is also useful in document summarization. Basically, any supervised methods depends on human-annotated data to form a set of rules by using which predictions can be made on unknown data sets. Many researchers proposed various models for unsupervised extracted-based summarization and the common principle to identify the important sentences based on their saliency features, but not depends on the training data sets. The sentences are ordered based on their importance in the descending order to generate final summary. The most useful unsupervised partitioning algorithm for text summarization is clustering algorithm. As in [11], the HITS algorithm is used to identify salient sentences clustering techniques based on the salient scores of the sentences required length summary can be generated. As in [12], the sentences are grouped into clusters by using the symmetric nonnegative matrix factorization (SNMF) approach to generate the summary. As in [13], generative model is used to select important sentences with probability distributions. These salient sentences are used to generate summaries from single and a document set.

In [14], there is an investigation on various empirical approaches to build text summarization systems. In [15], a multilingual multi-document summarization system was presented and based on the machine translated documents content, important sentences from English text documents can be identified. A semantic graph-based approach is proposed in [16] for document summarization. A novel approach of document summarization as in [17], is used to generate semantic relations within the document and perform machine learning techniques to extract a semantic structure in the form of a subgraph to generate document summary. As in [18], n-gram model is used to constitute the fine-grained evidence to distinguish coherence from incoherence. As in [19], some research has done using Onto WordNet to summary generation using the kinds of relationships that exist in the WordNet.

The methods for measuring the similarity among texts are classified into vector-, corpus-, hybrid-, and feature-based methods. Vector-based models are used in IRSystems [21]. The corpus-based methods include LSA [20] to explore semantics of text, computing the word similarity and passages by analyzing the large natural language text corpus to generate a semantic space based moving window concept [22]. As in [23], hybrid methods are a combination of knowledge- and corpus-based approaches.

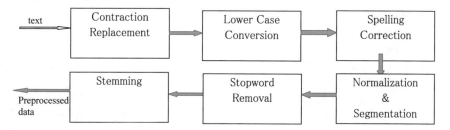

Fig. 1 Text preprocessing steps

Feature-based methods represent a sentence by generating a set of features using syntactic and semantic information embedded in the sentence. The primary and composite features are introduced to build the feature vector of a text [24]. Primary features compare individual items of a text unit. Composite features are formed by combining two or more primary features. The challenging task in this method is finding the effective features that aids in measuring the semantic similarity and a classifier is required to build the model upon these features. Corley [25] has proposed two corpus-based measures and six knowledge-based measures for finding the semantic similarity between word and a method which combines the information extracted from the similarity of component words to compute semantic similarity between two texts. Li and Mclean [26] has proposed an unsupervised method which computes the similarity between two texts by combining both syntactic and semantic information. For obtaining the syntactic information, the measure used is word order and for syntactic information is measured with the aid of knowledge-base and corpus-base. Inkpen [27] proposed a method that measures the similarity between two texts by normalizing three features string similarity, common word order and semantic similarity. The first two features emphasis on syntactic information, whereas the semantic similarity emphasis on semantic information and it is calculated using corpus statistics (Fig. 1).

3 Proposed Approach

The various phases in summary generation are preprocessing, sentence scoring using statistical, semantic, and syntactic features, sentence extraction based on ranking among the sentences and finally, multi-document summary generation. These phases are explained in the following sections and presented in the Fig. 2.

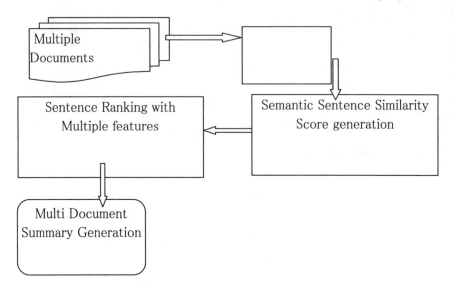

Fig. 2 Proposed framework

3.1 Preprocessing

Data is preprocessed in preprocessing stage using various preprocessing techniques. Preprocessing stage contains six steps such as contractions replacement, lowercase conversion, and spelling corrections. The text contractions are replaced with full text. Lowercase conversion is used for standardizing the text in checking the string equivalence and part of speech tagging. Misspelled and wrongly spelled words are corrected. The remaining three steps are normalization to remove the unnecessary symbols which is not useful to identify the content words and meaning sentences, segmentation which separate all the words on the basis of space, stopword removal to eliminate all the stopwords from the list of words and stemming which replace all the words by its root words. The steps are shown in Fig. 1.

D represents the document set and k is the number of documents, N be the number of all sentences in document collection, m be the number of words in each sentence, d_i be the ith document in D, $S_{i,k}$ be the ith sentence in document d_k, and w be a word.

3.2   Summary Generation with Statistical Features

The four statistical features are as follows:

(a) **Sentence Length Feature**

$$SL(S_{i,k}) = \frac{N \times length(S_{i,k})}{length(d_k)}$$

where SL ($S_{i,k}$) is the sentence length, the word count in a sentence is represented by length($S_{i,k}$), whereas word count in a document is denoted with length(d_k).

(b) **Sentence Centrality Feature**

$$SC(S_{i,k}) = \frac{words(S_{i,k}) \cap words(others)}{words(S_{i,k}) \cup words(others)}$$

where SC ($S_{i,k}$) is the sentence centrality can be measured as the ratio between the number of common words in the sentence i of the document d_k and the words in the remaining words in the document set D with the number of words that exist in the sentence $S_{i,k}$ and the words in the other documents.

(c) **Sentence Frequency Score Feature**

To determine sentence frequency score, the following is used. To calculate, equation is used.

$$SFS(S_{i,k}) = \frac{\sum_{i \in s} SFS(W_i)}{|s|}$$

where SFS(W_i) is the importance of individual word in a sentence as in |s| sentence count in a document.

$$SFS(w) = \frac{(|s| : w \in s)}{|N|}$$

where N issentence count in the document set.

(d) **Document Frequency Score Feature**

$$DFS(S_{i,k}) = \frac{\sum_{i \in d} DFS(W_i)}{|d|}$$

where DFS (W_i) is the importance of individual word in a document as in (4). |d| sentence count in a document. |D| document count in a set.

Fig. 3 Semantic sentence
similarity model

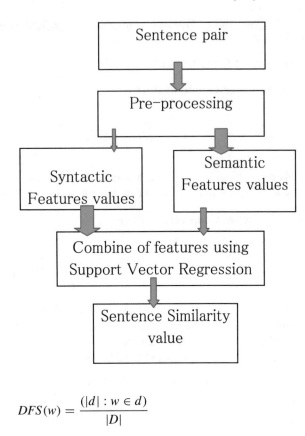

Fig. 3 Semantic sentence
similarity model

$$DFS(w) = \frac{(|d| : w \in d)}{|D|}$$

3.3   The Proposed Model: Semantic Sentence Similarity Model

The function of this model as shown Fig. 3 are to find sentence similarity. A pair of sentences is inputed into the system. Then the sentences are preprocessed using various steps as shown in Fig. 1. The syntactic similarity values and semantic similarity values are combined using support vector regression techniques. Semantic similarity score for a sentence is calculated by combining similarity scores of a sentence with all other sentences.

(1) **Syntactic features**

Syntactic features are used to measure the structure and stylistic similarity between a sentence pair. The various features are:

(a) **Longest common sequence (LCS):**

LCS is the ratio between the length of the longest common word sequence (lcws) and the length (len) of the shorter sentence.

$$Lcs(s_1, s_2) = \frac{len(lcws(s_1, s_2))}{min(len(s_1), len(s_2))}$$

(b) **n-gram features:**
n-grams at character and word level are generated. For 2, 3, and 4—character and word n-grams are generated. The Jaccard similarity between these n-grams is calculated.

$$J(X, Y) = \frac{|X \cap Y|}{|X \cup Y|}$$

n-gram set is represented by X and Y.

(c) **Word order similarity:**
Word Order Similarity (WOS) is measured between two sentence vectors. Unique Word Vector (UWV) is formed from the unique words contained in the sentence pair. Alignment between the UWV and the words in the sentence is made with WordNet. UWV is constructed to make sentence vector1 (SV1) and sentence vector2 (SV2) of equal dimensions. To form SV1, if the unique word (UW_i) of UWV is present in sentence1 at jth position then the ith entry of SV1 is the value j. Otherwise, the similarity between unique word (UW_i) and all the words (W_j) in sentence1 is calculated from WordNet using S (UW_i, W_j). Then, the most similar word position is assigned to the ith entry in SV1.

$$S(UW_i, W_j) = e^{\alpha l} \cdot \frac{e^{\beta h} - e^{-\beta h}}{e^{\beta h} + e^{-\beta h}}$$

where l is the shortest path between the words UW_i and W_j. h is the depth measured in the WordNet and α, β are the constants. The α value and β values are 0.2 and 0.45, respectively, which are found to be best by Li [17] (Fig. 4).

$$WOS(SV1, SV2) = 1 - \frac{|SV1 - SV2|}{|SV1 + SV2|}$$

(2) **Semantic features**
Semantic features deal with the meaning of the words in the sentences.

(a) **Knowledge based feature**
For finding the semantic similarity between a sentence pair, each sentence is mapped to the unique word vector (UWV) to form semantic vectors. If the unique word (UW_i) is present in the sentence, then the ith entry in the semantic vector (SV) is 1 otherwise, ith entry in the semantic vector (SV) is the highest semantic similarity value computed between the UW_i and every word in the sentence using S (UW_i, W_j). For computing S (UW_i, W_j), lexical database is used, i.e., WordNet.

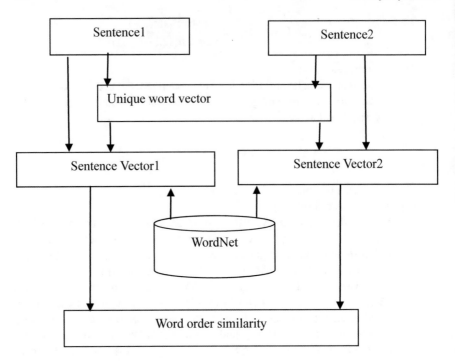

Fig. 4 Word order similarity model

(b) **Sent2Vec features**

Sent2Vec is used to generate feature vectors for a sentence pair with deep structured semantic model (DSSM) [17]. Similarly, CDSSM deep learning model [18] is also used to generate feature vectors of a sentence pair. Cosine similarity measure is used to find the sentence similarity. The syntactic and semantic features are combined using support vector regression technique [20].

Character n-grams, word n-grams, and lemma n-grams are generated using WordNet Lemmatizer and n-gram packages from nltk. The list of characters, words and lemmas are stored as per the order present in the sentences to preserve syntactic structure. Jaccard similarity is computed between the corresponding n-grams generated for a sentence pair. For finding the word order similarity lexical database WordNet is used to align the words in the sentence pair. Sent2Vec tool is used to generate both dssm and cdssm features. The regression model is implemented using R.

The sentence similarity score for a sentence S_i is measured as the sum of sentence S_i similarity scores of a with all other sentence vectors in a document.

$$Score_{f_s}(S_i) = \sum_{j=1, j\neq i}^{n} similarity(S_i, S_j)$$

The step-by-step procedure for summary generation for given text document set is presented below.

Algorithm for multi-document summary generation
Input: The set of text documents, the multi-document summary length "L".
Output: Text summary from the document set

Step1: The set of documents on a topic are combined together to form a single document.
Step 2: Pro-process the sentences in the document.
Step 3: The find the score of each sentence for each statistical feature.
Step 4: Find the sentence similarity score of a sentence using semantic similarity score and syntactic similarity score.
Step 5: Combine the scores obtained from step 3 and step 4 to assign a score to each sentence.
Step 6: Arrange the sentences in the descending according to the score order.
Step 7: Summary = (Topmost scored sentence).
Step 8: Form the summary of the document, until required number of sentences are considered in summary.
Step 9: To maintain coherence with the original documents, rearrange the sentences in the summary according the sentence order in the original documents.

4 Experimental Results and Discussion

4.1 Dataset

The experiments were performed using DUC 2002 [24] for summarization task. DUC 2002 dataset contains 567 text documents and there are a total of 59 clusters and each cluster contains approximately 10 news documents on an average. The task aims to create a short summary with less than 100 words for each cluster. Human-generated summaries are created for evaluation of multi-document summarization.

4.2 Evaluation Metrices

The summary generated by a machine can be evaluated using various performance measures. The evaluation measures are of two types such as intrinsic measures and extrinsic measures. Intrinsic measures are followed in this paper to measure the performance of the proposed system with manually crafted summaries for the actual documents. The performance is also measured by comparing with various existing techniques. The intrinsic measures gages the performance of a target system with reference system based on the percentage of overlap between these two systems. The

Fig. 5 Precision, recall, and F-measures values on DUC2002 data set for MDS using combination of four statistical features

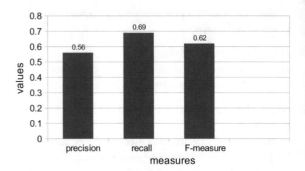

commonly used measures for intrinsic evaluation are Recall (R), Precision (P), and F1 measure. Precision is the ratio of overlapping between summary generated by the target system and the reference system with the number of sentences in the target system. The formula to find the precision is presented below:

$$Precision(P) = \frac{|S \cap T|}{|S|}$$

where S is the summary generated by the target system and T is the summary generated by the reference system.

The recall is the ratio of overlapping between summary generated by the target system and the reference system with the number of sentences in the reference system. The formula to find the recall is presented below:

$$Recall(R) = \frac{|S \cap T|}{|T|}$$

F_1 measure is used to measure the balance between precision and recall. F1 value will reach to maximum value only when both the precision and recall values maintain a balance in the system. F1 value can be calculated using the following formula:

$$F_1 = \frac{2 * P * R}{P + R}$$

4.3 Results and Discussion

The experimental evaluations are carried out on statistical features on the DUC2002 data set. The results are presented in the Fig. 5.

From Fig. 5, it is observed that the average precision on DUC2002. The precision is 0.56, recall is 0.69, and F1 value is 0.62. From the results, it is inferred that the number retrieved documents that are relevant is less compared with the number of

Fig. 6 Precision, recall, and F-measures values for MDS using syntactic and semantic features

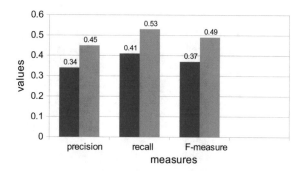

documents that relevant documents that are retrieved. The weighted harmonic mean between precision and recall which balances between precision and recall.

The proposed model is implemented using combination of three syntactic features such as Longest common sequence (LCS), n-gram feature, and word order similarity feature. As shown in Fig. 6, the precision is 0.34, recall is 0.41, and F1 value is 0.37. The model is also evaluated using two semantic features such as knowledge based feature which uses the WordNet to identify the syntactic relationships that exists among the words. The other semantic feature, sen2vec is used to generate feature vectors for a given sentence pair using convolutional deep learning models. The similarity between two feature vectors is measured using cosine similarity.

From the obtained results, it is observed that the precision, recall and F1 values for syntactic features are low compared with semantic features. The probable reason is that the syntactic features does not capture the underlying relationships that exist between two sentences. The two sentences may produce same meaning with different set of words in various forms. But grammatical rules can only captured using semantic features that causes high precision, recall and F1 values.

The comparison among statistical features, semantic, syntactic features and the combination of all three categories syntactic, statistical, and semantic features are evaluated on the proposed model. From Fig. 7, it is observed that the combination of all three categories of features leads to high precision, recall, and F1 values compared with statistical and syntactic and semantic features individually.

The precision is 0.68, recall is 0.76, and F1 value is 0.72 for all feature combination. From the results, it is observed that the identification of similarity between two sentences can be done more effectively by considering all the categories of features.

5 Conclusion and Future Scope

The proposed unsupervised model to generate generic extractive summary from multi-documents. The proposed model ranks the sentences based on the scores obtained using various features. The scores are assigned to the sentences based on

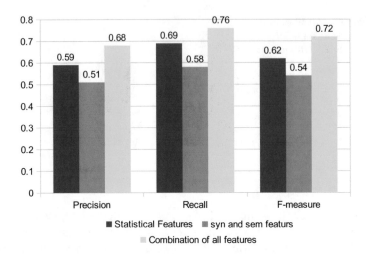

Fig. 7 Comparison of MDS with statistical, syntactic with semantic features and combination of all features using precision, recall, and F-measure

combined score which is resulted from the statistical, syntactic, and semantic features. The proposed method is evaluated on the DUC2002 corpus using precision, recall, and F1 measures. For a multi-document summarization, implementing the proposed model have demonstrated its effectiveness.

In future work, the proposed work will be extended to investigate advanced deep learning techniques such as convolutional neural networks, recurrent neural network to obtain more reliable semantic representations of sentences, summaries, and documents and propose new features based on the new semantic representations.

References

1. Wan X (2010) Minimum distortion with multi document summarization. In: International conference on data mining, pp 354–363
2. Dong L et al (2015) MDS with recursive neural networks. AAAI, pp 2153–2159
3. Agirre E. Spanish semantic textual similarity using interpretability. In: 9th international workshop on semantic evaluation, pp 252–263
4. Hanig C. Semantic textual similarity from word alignments. In: 9th international workshop on semantic evaluation, pp 264–268
5. Arafat Sultan Md. Contextual evidence and word similarity, its exploiting. ACM Trans 2:219–230
6. Daniel (2012) Multiple similarity measures for semantic textual similarity. In: First joint conference on lexical and computational semantics
7. Kononenko I. Multi document summarization with weighted multielement graph. Expert Syst Appl 41(2):535–543
8. Hakkani-Tur. Topically coherent sentences using summarization. In: 49th annual meeting of the HLT-ACM, vol 1, pp 491–499
9. Griffiths, Nonparametric inference for the nested Chinese restaurant process. J ACM 57(7):30

10. Nenkova K (2014) Estimation of word importance for MDS for news. EACL, pp 712–721
11. Yang J (2008) Multi-document summarization using cluster-based link analysis. In: 31st annual international SIGIR, information retrieval, pp 299–306
12. Zhu S (2008) Symmetric matrix factorization and semantic analysis MDS. In: 31st annual international Conference on information retrieval, pp 307–314
13. Zhu S et al (2009) Multi-document summarization using topic modelling. In: Proceeding of the IJCNLP, pp 297–300
14. Martins A (2007) Automatic text summarization, a survey. Literature survey for Language, CMU
15. Klavans J et al (2005) Multilingual MDS for sentence similarity. Report, Columbia University, Apr 2005
16. Milic-Frayling N et al (2004) Semantic graphs for document summarization. In: Workshop on link analysis
17. Grobelnik M et al (2000) Learning semantic graph mapping for text summarization. In: 49th annual meeting of the ACM, USA
18. Kan M (2011) Evaluating text coherence using discourse relations. In: 49th annual meeting of the ACM
19. Navigli R et al (2003) Axiomatization in WordNet, extension and conceptual relations. In: Databases and applications of semantics, Italy, pp 820–838
20. Kintsch W. Measurement of textual coherence using lsa. Disc Proc 25(2–3):285–307
21. Boyce B. Information retrieval systems. Academic Press
22. Livesay K. Explorations in context space using words, sentences. Disc Proc 25:211–257
23. Turney P. Mining the web for synonyms—PMI versus LSA. In: 12th European conference on machine learning
24. Chodorow M. Lexical database using WordNet. MIT Press, word sense identification, 265–283
25. Corley C. Knowledge-based measures for text semantic similarity. American Association for AI
26. Li Y, Mclean D. Corpus statistics for sentence similarity. IEEE Trans. Knowl Data Eng 13:1138–1149
27. Inkpen. String similarity for STS. ACM Trans Knowl Data 2(10)
28. Balasubramanie (2010) MDS using feature specific sentence extraction. IJCSIT 2(4):99–111

Topic Oriented Multi-document Summarization Using LSA, Syntactic and Semantic Features

M. Anjaneyulu, S. S. V. N. Sarma, P. Vijaya Pal Reddy, K. Prem Chander and S. Nagaprasad

Abstract Multi-document Summarization (MDS) is a process obtaining precise and concise information from a set of documents described on the same topic. The generated summary makes the user to understand the important information that is present in the documents. In general, a set of interrelated documents discusses about a subject. The subject contains a set of topics. The description of various topics that belongs to the main subject are presented in the documents. The user wish to consolidate the subject in terms of topics that are covered in the various documents. The existing approaches suffer with identification of the topics within a document and also lack the establishment of semantic and syntactic relationship among the words within a sentence. In this paper, a novel unsupervised model is proposed to generate extractive multi-document summaries by identifying the topics that are present in the documents using Latent Semantic Analysis (LSA) and eliminating the redundant sentences that are describing the same topic that are present in the multiple documents using semantic and syntactic information embedded in the sentences. Empirical evaluations are carried out using LSA, lexical, syntactic and semantic features on DUC2006 dataset. The experimental results on DUC2006 demonstrate

M. Anjaneyulu (✉) · K. Prem Chander
Department of Computer Science, Dravidian University, Kuppam, India
e-mail: anjan.lingam1@gmail.com

K. Prem Chander
e-mail: kpc.1279@gmail.com

S. S. V. N. Sarma
CSE Department, Vaagdevi Engineering College, Warangal, India
e-mail: ssvn.sarma@gmail.com

P. Vijaya Pal Reddy
CSE Department, Matrusri Engineering College, Hyderabad, India
e-mail: drpvijayapalreddy@gmail.com

S. Nagaprasad
Computer Science Department, S.R.R. Government Arts & Science College,
Karimnagar, India
e-mail: nagkanna80@gmail.com

© Springer Nature Singapore Pte Ltd. 2019
S. Bhattacharyya et al. (eds.), *International Conference on Innovative Computing and Communications*, Lecture Notes in Networks and Systems 56,
https://doi.org/10.1007/978-981-13-2354-6_50

487

that the performance of proposed summarization system is comparable with the existing summarization systems in terms of F-measure, recall, and precision values.

Keywords Multi-document Summarization · Topic identification
Latent Semantic Analysis · Semantic features · Syntactic features

1 Introduction

The flooding of information on numerous topics onto the World Wide Web makes the people to access the content with no time. The information overload onto the Internet results in difficulty of accessing concise and meaning information on a topic. Automatic text summarization is the process of summary generation from a document that can address the specified problem. Multiple Document Summary (MDS) generation is a procedure for comprehending the important content from a set of interrelated documents on the same topic. The basic purpose of a summary generation system is to derive most useful and precise information from the input document. Identification of important sentences that carry most useful information and arrange them in a particular order is an interesting and challenging research problem.

The approaches for Multi-document Summarization broadly classified into extractive and abstractive. Most of the existing research is towards extractive an approach that directly identifies the most meaningful sentences that are present in multiple documents and combine these sentences to form a summary. As in [1], the effectiveness of the summarization approach can be measured using four parameters namely informative, length, relevance, and diversity. The concise and precise summary should cover all fundamental themes of the documents as much as possible by eliminating the redundancy. It should contain the optimal information with a suitable number of sentences in the summary that suitably represents the whole set of documents. The abstractive approaches generate the sentences for the summary by learning the meaning of all sentences from the document set. The produced sentences in the summary might not as present in the original documents.

The extractive summarization approaches can be viewed as supervised and unsupervised methods. The supervised approaches depend on the training data set. The model is learned with suitable training data set. The learned model can be used to generate summary for a given test document set. As in [2], these methods treat the problem of MDS as a categorization problem. For these approaches, a large set of training data is required to learn the model results to time consuming. The unsupervised methods derives a summary for a given document set based on the inferences generated from the documents. The unsupervised approaches identifies the most meaning sentences based statistical, syntactic and semantic relations that exist among the words that exist within the sentences and ranks the sentences based on the information that carry by the sentences. Most of the existing successful summarization systems are used in domain of news articles where each document is assumed to have a single topic. It is assumed in these systems that a document has information about a single event.

The literature on document analysis mentions various approaches for machines to learn the associations between words and documents. Among them Latent Semantic Analysis (LSA) is a model derived from mathematics that relates the documents in terms of semantic structure using word co-occurrence relations [3]. This model finds the semantic relations based on input documents only but not depend on knowledge from external sources such as grammar rules, morphological structures, or syntactic parsers. The proposed model is a mathematical model thus it is independent from the underlying language syntactic and semantic structure.

Semantic Textual Similarity (STS) is one of the core disciplines in Natural Language Processing (NLP). STS assess the degree of semantic similarity between two textual segments. The textual segments are phrases, sentences, paragraphs or documents. The objective of the STS is find the similarity between two sentences in the range [0, 5] between a sentence pair where 0 indicates both the sentences are on irrelevant topics and 5 indicates both the sentences mean the same thing [4]. STS system is trying to emulate the idea of similarity, thus replicating human language understanding. The approaches for measuring STS have been categorized into alignment based approaches, vector space models, and machine learning approaches [5]. Alignment approaches compute the similarity between the words or phrases in a sentence pair and align the words or phrases that are most similar [6]. Vector space approach is a traditional NLP feature engineering approach represents the sentence as bag-of-words, and the similarity is evaluated according to the occurrence of words or co-occurrence of words or other replacement words. Machine learning approaches use supervised machine learning models to combine heterogeneous features such as lexical, syntactic and semantic features of sentence pair [7]. NLP techniques are required to use for STS measures as two or more sentences derive similar meaning with no matching of the words. Text summarization as an application of NLP can benefit from effective STS techniques in grouping of semantic similar sentences.

In general, a set of interrelated documents discusses a subject. The subject contains a set of topics. The description of various topics that belongs to the main subject is presented in the documents. The user wishes to consolidate the subject in terms of topics that are covered in the various documents. It is required to identify the topics present in the documents and then forms the summary on the topics by eliminating the redundant sentences from each topic. The proposed approach is a generic extractive summary for multi-topic, multi-document summarization. Domain-independent topics are extracted automatically, and the system ranks sentences by their topic richness. The summary from multiple documents is generated in two phases. In the first phase, the concepts present in the documents are identified using Latent Semantic Analysis (LSA). In the second phase, redundant sentences that describe the topic are removed using statistical, syntactic and semantic features.

This paper is organized into five sections. Section 2 describes the existing research work related to the automatic text summarization and multi-document summarization. The proposed model using LSA, syntactic and semantic features are detailed in Sect. 3. The proposed framework is evaluated on the benchmark of DUC2006 data set from Document Understanding Conference (DUC) for the task of multi-document summarization is presented in Sect. 4. The conclusions observed from the proposed work and the possible extensions that can be made on the proposed model are presented in Sect. 5.

2 Related Work

The MDS approaches can be classified based on various dimensions such as extractive and abstractive approaches and also as supervised and unsupervised approaches. The training dataset is used to learn the model in supervised models. The various learning models are into existence such as Hidden Markov Model (HMM) and conditional random field to learn the relationships that exist within the training data set. As in [8], the image processing technique like sparse coding is also useful in document summarization. Basically, any supervised methods depend on human-annotated data to form a set of rules by using which predictions can be made on unknown data sets. Many researchers proposed various models for unsupervised extracted-based summarization and the common principle to identify the important sentences based on their saliency features, but not depends on the training data sets. The sentences are ordered based on their importance in the descending order to generate final summary. The most useful unsupervised partitioning algorithm for text summarization is clustering algorithm. As in [9], HITS algorithm is used to identify salient sentences by forming clusters as hubs and sentences as authorities and finally based on the salient scores of the sentences required length summary can be generated. As in [10], the sentences are grouped into clusters by using the symmetric nonnegative matrix factorization (SNMF) approach to generate the summary. Generative model is used to find Bayesian probability distributions of selecting salience sentences. These salient sentences are combined with different summarization results to form single document summarization systems.

There is an investigation on various empirical approaches to build text summarization systems. A multilingual multi-document summarization system was presented and based on the machine translated documents content; important sentences from English text documents can be identified. A semantic graph-based approach is proposed in [11] for document summary generation. A novel approach of document summarization as in [12], is used to generate semantic relations within the document and perform machine learning techniques to extract a semantic structure in the form of a subgraph to generate document summary. As in [13], some research has done using OntoWordNet to summary generation using the kinds of relationships that exist in the WordNet.

Latent Semantic Approach was used in many applications. Based on the user interest information is filtered from news articles. In [14], cross-language retrieval has performed using portrayed strategy with LSA. Relationships that exist in human memory has performed using LSA space. A word synonym test has performed using word occurrence. In [15], allocation of reviewers for research articles that are submitted to a conference has been performed using LSA model. In [16], indexing of abstracts by analyzing word-to-word relationships has performed. In [17], automate spell checking and correction has performed using unigrams and bigrams with accurately spelled words. In [18], word sense disambiguation (WSD) has performed. In [19], discriminate analysis has been performed using LSA as initial space. Neural network protein classification has performed. In [20], authorship attribution, authorship profile, and authorship verification have performed. In [21], linear text segmentation has performed using inter-sentence similarity and named entity recognition has performed.

The approaches for similarity measure on texts are categorized into four types such as feature-based methods, vector based, hybrid and corpus-based methods. Vector-based models are used in Information retrieval systems [22]. The corpus-based methods include Latent Semantic Analysis (LSA) [23] is a method for extracting and representing the contextual meaning of text, computing the similarity of words and passages by analyzing the large natural language text corpus and the HAL generates a semantic space based on fixed length moving window concept on co-occurring words [24]. Hybrid methods involve corpus-based [25] and knowledge-based measures.

Feature-based methods represent a sentence by generating a set of features using syntactic and semantic information embedded in the sentence. The primary and composite features are introduced to build the feature vector of a text [26]. Primary features compare individual items of a text unit. Composite features are formed by combining two or more primary features. The challenging task in this method is finding the effective features that aid in measuring the semantic similarity and a classifier is required to build the model upon these features. Li and Mclean [27] have proposed two corpus-based measures and six knowledge-based measures for finding the semantic similarity between word and a method which combines the information extracted from the similarity of component words to compute semantic similarity between two texts. Inkpen [28] has proposed an unsupervised method which computes the similarity between two texts by combining both syntactic and semantic information. For obtaining the syntactic information, the measure used is word order and for syntactic information is measured with the aid of knowledge-based and corpus-based. Balasubramanie [29] proposed a method that measures the similarity between two texts by normalizing three features string similarity, common-word order and semantic similarity. The first two features string similarity and common-word order similarity emphasis on syntactic information whereas the semantic similarity emphasis on semantic information and it is calculated using corpus statistics.

3   The Proposed Model

In this paper, a model is proposed to generate a summary from multiple documents. The documents contain a set of topics that describe a subject. The model is executed in two phases. In the first phase, the sentences representing the topics present in each document are identified. In the second phase, remove the redundant sentences

Fig. 1 Proposed framework

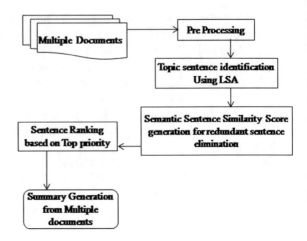

from each topic and produce the summary by combining all the topic representing sentences. These phases are explained in the following sections and presented in the Fig. 1.

3.1 Preprocessing

Data is preprocessed in preprocessing stage using various preprocessing techniques. Preprocessing stage contains six steps such as contractions replacement, lower case conversion, and spelling corrections. The text contractions are replaced with full text. Lower case conversion is used for standardizing the text in checking the string equivalence and part of speech tagging. Misspelled and wrongly spelled words are corrected. The remaining three steps are normalization to remove the unnecessary symbols which as not useful to identify the content words and meaning sentences, segmentation which separate all the words on the basis of space, stop word removal to eliminate all the stop words from the list of words and stemming which replace all the words by its root words. The steps are shown in Fig. 2.

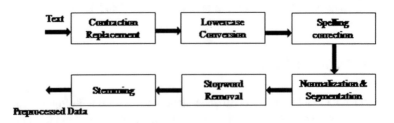

Fig. 2 Sequence of steps in text preprocessing

3.2　Latent Semantic Analysis

Singular Value Decomposition (SVD) divides 'A' matrix as three U, S and V matrices. The theorem is as follows:

$$A_{mn} = U_{mm}S_{mn}V^T nn \tag{1}$$

where S is a diagonal matrix with square roots of Eigen values from U or V which are sorted in descending order and the columns of U and V are orthonormal eigenvectors. A can be used to capture word co-occurrence relations and to find meaningful patterns from it. The A matrix can be approximately constructed from dimensionality reduction matrix Ak_{mn} with U_{mk}, V_{kn}^T, S_{kk} with reduced k dimensions from SVD of A. The reduced model is:

$$AK_{mn} = U_{mk}S_{kk}V_{kn}^T \tag{2}$$

The reduced dimensionality matrix 'AK' represents the semantic relations that exist based on word co-occurrences. The number of dimensions 'k' to be retained is based on the problems to be addressed and dataset size.

3.3　Topic-Oriented Sentence Extraction

To identify unique concepts in document D, an LSA is used to extract k sentences. Concepts are presented in the rows whereas sentences are represented in the columns of V^T. The values presented in the cells shows the strength between the concept and the sentence. The required number of sentences was extracted from the matrix based on the length of required summary length in terms of number of sentences.

3.4　Summary Generation with Statistical Features

The proposed model combines four statistical measures as in [28] with semantic sentence similarity model. The four statistical features are as follows:

(a)　**Sentence Length feature**

$$SL(S_{i,k}) = \frac{N \times length(S_{i,k})}{length(d_k)} \tag{3}$$

where $SL(S_{i,k})$ is the sentence length, the word count in a sentence is represented by length $(S_{i,k})$, whereas word count in a document is denoted with length (d_k).

M. Anjaneyulu et al.

(b) **Sentence Centrality feature**

$$SC(S_{i,k}) = \frac{Words(S_{i,k}) \cap words(others)}{length(S_{i,k}) \cup words(others)} \tag{4}$$

where SC(S$_{i,k}$) is the sentence centrality can be measured as the ratio between the number of common words in the sentence i of the document d$_k$ and the words in the remaining words in the document set D with the number of words that exist in the sentence S$_{i,k}$ and the words in the other documents.

(c) **Sentence Frequency Score feature**

To determine sentence frequency score, the following is used. To calculate, equation is used.

$$SFS(S_{i,k}) = \frac{\sum_{i \in s} SFS(W_i)}{|s|} \tag{5}$$

where SFS(W$_i$) is the importance of individual word in a sentence as in (4). |s| is the number of sentences in the document.

$$SFS(W) = \frac{(|s| : W \in S)}{|N|} \tag{6}$$

where N is the total number of sentences in the document set.

(d) **Document frequency score feature**

$$DFS(S_{i,k}) = \frac{\sum_{i \in d} DFS(W_i)}{|d|} \tag{7}$$

where DFS(W$_i$) is the importance of individual word in a document as in (4). |d| is the number of sentences in the document. |D| is the number of documents in the set.

$$DFS(w) = \frac{(|d| : w \in d)}{|D|} \tag{8}$$

3.5 Semantic Sentence Similarity Model

The function of this model as shown Fig. 3 is to measure the degree of similarity between two sentences. The model takes two sentences as input into the system. Then the sentences are preprocessed using various steps as shown in Fig. 2. The syntactic similarity and semantic similarity between two sentences can be measured using various features as presented below. The syntactic similarity values and semantic

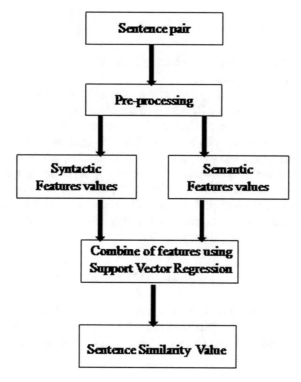

Fig. 3 Semantic sentence similarity model

similarity values are combined using support vector regression technique for STS. The process is repeated for all the sentence pairs in the document set. Finally, semantic similarity score for a sentence is calculated by combining similarity scores of a sentence with all other sentences.

(1) Syntactic Features: Syntactic features are used to measure the structure and stylistic similarity between a sentence pair. The various features are:

(a) **Longest common sequence (LCS):**
LCS is the ratio between the length of the longest common word sequence (lcws) and the length (len) of the shorter sentence.

$$I.Lcs(s_1, s_2) = \frac{len(lcws(s_1, s_2))}{min(len(s_1), len(s_2))}$$

(b) **n-gram features:**
n-grams at character and word level are generated. For 2, 3, 4—character and word n-grams are generated. The Jaccard similarity between these n-grams is calculated.

$$\text{II}.J(X, Y) = \frac{|X \cap Y|}{|X \cup Y|}$$

n-gram sets are represented using X and Y.

(c) **Word Order Similarity:**

Word Order Similarity (WOS) is measured between two sentence vectors. Unique Word Vector (UWV) is formed from the unique words contained in the sentence pair. Alignment between the UWV and the words in the sentence is made with WordNet. UWV is constructed to make sentence vector1 (SV1) and sentence vector2 (SV2) of equal dimensions. To form SV1, if the unique word (UW_i) of UWV is present in sentence1 at jth position then the ith entry of SV1 is the value j. Otherwise, the similarity between unique word (UW_i) and all the words (W_j) in sentence1 is calculated from WordNet using S (UW_i, W_j). Then most similar word position is assigned to the ith entry in SV1.

$$S\left(UW_i, W_j\right) = e^d \cdot \frac{e^{\beta h} - e^{-\beta h}}{e^{\beta h} + e^{-\beta h}}$$

where l is the shortest path between the words UW_i and W_j. h is the depth measure in the wordNet and α, β are the constants. The α value and β values are 0.2 and 0.45 respectively which are found to be best by Li [17] (Fig. 4).

Fig. 4 Word order similarity model

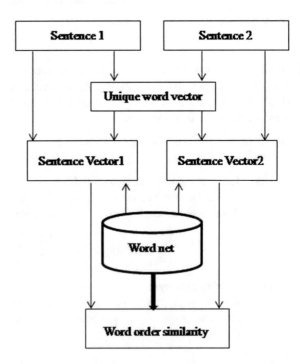

(2) **Semantic features**
Semantic features deal with the meaning of the words in the sentences.

(a) **Knowledge-based feature**
For finding the semantic similarity between a sentence pair, each sentence is mapped to the unique word vector (UWV) to form semantic vectors. If the unique word (UW_i) is present in the sentence then the ith entry in the semantic vector (SV) is 1 otherwise ith entry in the semantic vector (SV) is the highest semantic similarity value computed between the UW_i and every word in the sentence using S (UW_i, W_j). For computing S (UW_i, W_j) lexical database is used, i.e., WordNet.

(b) **Sent2Vec features**
Sent2Vec is used to generate feature vectors for a sentence pair with deep structured semantic model (DSSM) [17]. CDSSM deep learning technique [18] is also used to generate feature vectors of a sentence pair. The similarity between the sentence pair is computed by finding the cosine similarity between the feature vectors. The syntactic and semantic features are combined using support vector regression technique [20].

Character n-grams, word n-grams, and lemma n-grams are generated using WordNet Lemmatizer and n-gram packages from nltk. The list of characters, words, and lemmas is stored as per the order present in the sentences to preserve syntactic structure. Jaccard similarity is computed between the corresponding n-grams generated for a sentence pair. For finding the word order similarity lexical database WordNet is used to align the words in the sentence pair. Sent2Vec tool is used to generate both dssm and cdssm features. The regression model is implemented using R. The sentence similarity score for a sentence S_i is measured as the sum of sentence S_i similarity scores of a with all other sentence vectors in a document.

The score for each sentence is measured using possible feature combinations. The set of scored sentences are picked based on the score value and its presence in the document. The step-by-step procedure for summary generation for given text document set is presented below:

Algorithm for multi-document summary generation
Input: Document Dataset. The required number of topics in the summary 'T'.
  Output: Topic based multi-document summary.

Step 1: Preprocess the sentences in the document.
Step 2: Identify the sentences that represent the topics in a document.
Step 3: Repeat Step 2 for every document in the given document dataset.
Step 4: Combine all the sentences based on the topics identified in Step 2.
Step 5: Calculate the sentence score that is present in each topic using statistical measures.
Step 6: Find the sentence similarity score of each sentence from every topic using semantic similarity score and syntactic similarity score.

Step 7: Combine the scores obtained from step 5 and step 6 to assign a score to each sentence.
Step 8: Delete the sentences that are redundant in each topic.
Step 9: Arrange the sentences in the descending according to the topic priority.
Step 10: Form the summary of the given document dataset based on the number of sentences that are to be considered in summary.

4 Experimental Results and Discussion

4.1 Dataset

The dataset is collected from DUC 2006 with 1250 documents and their corresponding summaries which covers 25 various topics.

4.2 Evaluation Measures

The summary generated by a machine can be evaluated using various performance measures. The evaluation measures can be divided into two categories such as intrinsic measures and extrinsic measures. Intrinsic measures are followed in this paper to measure the performance of the proposed system with manually crafted summaries for the actual documents. The performance is also measured by comparing with various existing techniques. The intrinsic measures gauge the performance of a target system with a reference system based on the percentage of overlap between these two systems. The commonly used measures for intrinsic evaluation are precision (P), recall (R), and F1 measure. The precision is defined as the ratio of overlapping between summary generated by the target system and the reference system with the number of sentences in the target system. The formula to find the precision is presented below:

$$Precision(P) = \frac{|S \cap T|}{|S|}$$

where S and T represent the target system summary and reference system summary.
    The recall is defined as the ratio of overlapping between the summary generated by the target system and the reference system with the number of sentences in the reference system. The formula to find the recall is presented below:

$$Recall(R) = \frac{|S \cap T|}{|T|}$$

F_1 measure is used to measure the balance between precision and recall. F1 value will reach maximum value only when both the precision and recall values maintain a balance in the system. F1 value can be calculated using the following formula:

$$F_1 = \frac{2*P*R}{P+R}$$

4.3   Results and Discussion

The experimental evaluations are carried out on statistical features on the DUC2006 data set. The performance of the model using statistical, syntactic and semantic measures without generating topics using Latent Semantic Analysis is measured using recall, F1-measures, and precision. The results are presented in Fig. 5.

The comparison among statistical features, the combination of syntactic and semantic features and the combination of all three categories of features such as statistical, syntactic and semantic features are evaluated on the model presented in sub Sect. 3.5. From the Fig. 5, it is observed that the combination of all three categories of features leads to better precision, recall and F1 values compared with statistical and syntactic and semantic features individually. The precision is 0.68, recall is 0.76 and F1 value is 0.72 for all feature combination. From the results, it is observed that the identification of similarity between two sentences can be done more effectively by considering all categories of features.

The performance of the model using statistical, syntactic, and semantic measures on the sentences generated based on the topics using Latent Semantic Analysis for

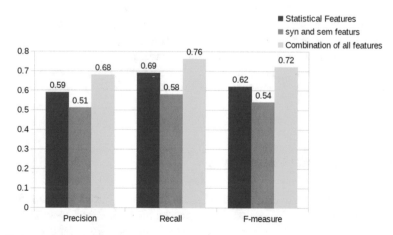

Fig. 5 Comparison of MDS with statistical, syntactic with semantic features and the combination of all features using precision, recall and F-measure

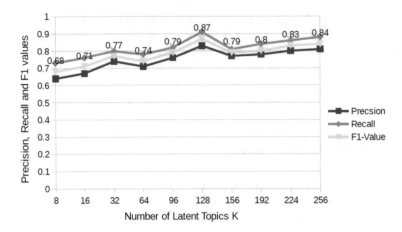

Fig. 6 Recall, F-measure and Precision values for the proposed model for varies the number of latent

various dimensions is measured using precision, recall, and F1 measures. The results are presented in Fig. 6.

From Fig. 6, it is observed that the precision, recall, and F1 values are increased from the latent topics 8 to 32. The F1 value is increased from 0.68 to 0.77 as shown in the figure. But for 64 topics, F1 value is decreased from 0.77 to 0.74. The F1 values are increasing from 0.74 to 0.87 when the numbers of latent topics are increasing from 64 to 128. The F1 score is achieved at the number of latent topics 128. When the number of latent topics is 156 the F1 score is decreased to 0.79. As the number of topics increases from 156 to 256, the F1 score is increased from 0.79 to 0.84.

Figure 7 gives the comparison between the proposed model and three existing models. From the figure, it is observed that the model using syntactic and semantic features is least performed compared with other three systems. The combination of statistical measures with syntactic and semantic measures is performed better when compared with two individual systems. The proposed model with the combination of LSA and the remaining three measures performed well compared with the remaining three systems.

5 Conclusion and Future Scope

This paper proposed a novel approach which is oriented towards topic identification for multi-document summarization. The proposed model is an unsupervised model and generates a generic topic oriented extractive summary from multi-documents. The topics are identified from the documents using Latent Semantic Analysis model. Each topic is represented by a more informative sentence from the document. The topics are identified from all the documents and sometimes the same topic might

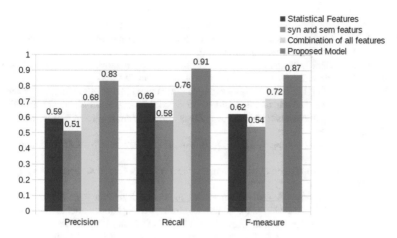

Fig. 7 Comparison among the existing models and the proposed model using precision, recall and F1-measure

be presented in different documents. While generating a summary for the multiple documents, the duplicated topics need to be removed. So it is required to identify the similarity between the sentences that represent topics. The sentence similarity is identified using the combination of various features such as statistical, lexical, syntactic and semantic features. The summary is generated based on the number of topics to be considered in the summary. The performance of the proposed approach is evaluated on DUC2006 dataset with specified three measures as mentioned in the Fig. 7. For a multi-topic multi-document summarization, the proposed model has demonstrated the effectiveness and it is comparable to the state-of-the-art summarization systems on multi-document summarization.

In future work, the proposed work will be extended to investigate the identification topics using various other techniques such as Probabilistic Latent Semantic Analysis (PLSA) and Latent Dirichlet allocation (LDA). The influence of weight measure between words and the sentences that present in the document can be evaluated. The word correlations can be better identified using various word embedding techniques like Word2Vec, Sen2Vec, and Para2Vec which need to be studied.

References

1. Wan X (2010) Minimum distortion with MDS. In: International conference on data mining, pp 354–363
2. Cao Z et al (2015) Multi document summarization with recursive NN. AAAI, pp 2153–2159
3. Deerwester S (1990) Indexing by LSA. American Society for Information Science, pp 391–407
4. Agirre E. Interpretability on Spanish semantic textual similarity. In: 9th international workshop on semantic evaluation, pp 252–263
5. Hanig C. Semantic textual similarity using extensive feature extraction. In: 9th international workshop on semantic evaluation, pp 264–268

6. Arafat Sultan Md. Contextual evidence and its exploiting. Trans ACM 2:219–230
7. Daniel (2012) Similarity measures for computing semantic textual similarity. In: 1st conference on lexical and computational semantics
8. Wan X et al (2008) Link analysis for multi-document summarization. In: 31st annual international ACM conference on information retrieval. ACM, pp 299–306
9. Wang D et al (2008) Semantic analysis and symmetric matrix factorization for multi-document summarization. In: 31st annual international Conference on information retrieval, pp 307–314
10. Zhu S (2009) Topic models for multi-document summarization. In: Proceeding of the IJCNLP 2009, pp 297–300
11. Grobelnik M, Milic-Frayling N, (2000) Document summarization with learning semantic graph mapping
12. Ng H, Kan M, (2011) Automatically evaluating text coherence. In: 49th annual meeting of the ACM, USA
13. Kintsch W. Latent semantic analysis for textual coherence. Disc Proc 25(2–3):285–307
14. Littman LM (1990) Cross-language document retrieval using latent semantic indexing. University of Waterloo Centre for the New Oxford English Dictionary, pp 31–38
15. Nielsen J (1992) Assignment of manuscripts to reviewers. ACM in information retrieval, pp 233–244
16. Phillips VL (1994) Handwritten notes retrieving. In: Conference on information technology
17. Kukich K (1990) A comparison of traditional lexical distance metrics for spell correction. In: IEEE international neural society, pp 309–313
18. Schutze H (1992) Dimension of meaning. In: Supercomputing, pp 787–796
19. Hull D (1994) Text retrieval for the using latent semantic analysis. In: SIGIR conference on information retrieval, pp 82–291
20. Nicholas CK, Ebert DS (1997) Document authorship using n-grams. In: Workshop on information visualization and manipulation, pp 43–48
21. Hastings P (2001) Latent semantic analysis for text segmentation. In: Proceedings in natural language processing, pp 109–117
22. Boyce B. Information retrieval systems, Second edn, Academic Press
23. Livesay K. Words, sentences, discourse explorations in context space. Disc Proc 25(2–3):211–257
24. Turney P. LSA on TOEFL-mining the web for synonyms. In: 12th European conference on machine learning
25. Chodorow M. Electronic lexical database using WordNet. MIT Press, word sense identification, pp 265–283
26. Corley C. Text semantic similarity with knowledge measures. American Association for Artificial Intelligence
27. Li Y, Mclean D. Corpus statistics for sentence similarity. IEEE Trans Knowl Data Eng 18(8):1138–1149
28. Inkpen. Corpus word similarity and string similarity for semantic text similarity. ACM Trans Knowl Discov Data 2(2), Article 10
29. Balasubramanie (2010) Multiple document summarization using feature specific sentence extraction. Comput Sci Inf Technol 2(4):99–111

Text Summarization for Big Data: A Comprehensive Survey

Vanyaa Gupta, Neha Bansal and Arun Sharma

Abstract Availability of large volume of online data has necessitated the need for automatic text summarization. In this paper, a survey of various text summarization accomplishments that have been implemented in the recent past is discussed. An intentional weightage is given to the issue of legal text summarization, as summarization as a tool has a great scope in the legal domain. The paper starts with a brief introduction of automatic text summarization, then succinctly mentions the novel advances in extractive and abstractive text summarization techniques, and then moves to literature survey, and it finally winds up with some future work directions.

1 Introduction

Automatic text summarization has become an important tool in the era of digitalization specially when there is abundance of information available everywhere. Internet today is a great source of information for various stakeholders but the information available online is in such a huge amount that it distresses the user, first for searching relevant information from such an overwhelming number of documents and then absorbing that information.

Automatic text summarization aims at resolving the above problem. When the source document is too long to go through and user is in hurry to get a quick grasp of the information, then the summary is what all is needed. The goal of summarizer is to condense the source document and present a shorter version of it yet preserving the main essence of the source text.

V. Gupta (✉) · N. Bansal · A. Sharma
Department of Information Technology, Indira Gandhi Delhi
Technical University for Women, New Delhi, Delhi, India
e-mail: 2016vanyaa@gmail.com

N. Bansal
e-mail: nehabansal33@gmail.com

A. Sharma
e-mail: arunsharma@igdtuw.ac.in

S. Bhattacharyya et al. (eds.), *International Conference on Innovative Computing and Communications*, Lecture Notes in Networks and Systems 56,
https://doi.org/10.1007/978-981-13-2354-6_51

503

2   Automatic Summarization

This section addresses the following agendas:

- Definition of automatic summarization and a brief about it.
- Process of generating summary.
- Its evaluation techniques.

2.1   Automatic Text Summarization

Automatic text summarization is described as the process of encapsulating the entire document in a form of short text by software, yet capturing the main idea of the original document.

Summary can be categorized into two forms based on what is being focused on in the summary:

2.1.1   Generic Summary

Given a source document, relevant sentences are extracted based on sentence score and are included in the summary without including any extra information.

2.1.2   Query-Oriented Summary

In query-oriented summarization, generation of summary is based on user-given query. The sentences are scored on the basis of relevance with the user given query then the highest scored sentences are selected and included in the summary.

2.2   Process of Text Summarization

The process of Automatic Text Summarization involves three steps [1]. Extraction of relevant text, gathering of information from several sources and finally representation of summary.

2.3 There are Two Types of System to Evaluate Summaries

2.3.1 Intrinsic

Intrinsic systems evaluate only the quality of summary generated.

2.3.2 Extrinsic

Extrinsic system takes values of the user assistance in task performance while evaluating summaries.

3 Summarization Methods

There are two methods of text summarization which are described as follows:

3.1 Extractive Summarization Techniques

Extractive techniques aim at those summarization approaches that deal with sentence-level extraction as a relevant text. For indicating the importance of a particular sentence, several techniques have been implemented in past. In the subsequent section, we would be describing some of the recent extractive summarization approaches. Similar approaches are also described in several review papers [2].

3.1.1 Term Frequency–Inverse Document Frequency Method

It is the most commonly used method for sentence scoring. It is a measure of importance of particular word or sentence in the entire corpus of document.

3.1.2 Cluster-Based Method

Each document is organized in different sections addressing different-different topics. The summarizer takes input as various clusters and each cluster is considered as a particular subject. According to that subject the higher scoring words are included in respective cluster.

3.1.3 Graph-Theoretic Approach

Each sentence is represented as a node. Two nodes are connected with an edge, if they share some common words. Following this approach, one can identify unconnected portions in a graph and various connected portions. For a query-centric summary, the sentences may be chosen from a relevant portion of graph, whereas for generic summary, sentences may be chosen from every subgraph.

3.1.4 Fuzzy Logic Approach

This method takes similarity, sentence length and keywords as input to fuzzifier then certain rules required for summarization based on characteristics of sentences are fed into the knowledge base. The output would result in a vector score between zero and one which in turn indicates the importance of a particular sentence. Important sentences are then included in the summary.

3.2 Abstractive Text Summarization Techniques

Abstractive summary aims at producing a generic summary, displaying the information in a precise way and yet capturing the main idea of the source document text. The following section describes various Abstractive Summarization Techniques. Same techniques were also discussed in various other review papers [3].

3.2.1 Tree-Based Method

This method works on the similarity of sentences. Sentences are represented in subject-predicate form then overlapping algorithms are used to determine the sentences that share common information. These sentences are then accumulated to form a summary.

3.2.2 Template-Based Method

Templates are used to represent a document in this method. Based on linguistic patterns and extraction rules, text snippets are determined which are then projected into the template. These text snippets are indicators of important text that could be included in the summary.

3.2.3 Semantic Graph-Based Method

This method involves a three-step process. A document is represented in form of semantic graph. First, nouns and verbs are represented as nodes and relationship between them is represented through edges. The second step deals with further reduction of original graph based on some heuristic functions and lastly, abstractive summary is generated using reduced rich semantic graph.

3.2.4 Rule-Based Method

In this method, the source document is classified on various aspects. Then, the best sentences are selected among the ones those are shortlisted on the basis of certain criterion listed in extraction rules. These sentences are then accumulated in form of summary.

4 Literature Survey

This section aims at exploring various techniques implemented in the recent past for the aim of text summarization. Different tools and algorithms used in various researches with their implementation are discussed in the respective description of each paper.

Sinha et al. [4] proposed a fully data-bound approach using neural networks for summarization featuring single document. The designed model is flexible as it can produce summary of any size document. It splits source document into definite segments and then recursively feed them into network.

Khan et al. [5] presented a semantic graph-based method for abstractive text summarization. The semantic graph was designed such that nodes of the graph represent PAS (predicate argument structure) from the source document. Then, the nodes are ranked and top nodes are picked to include in the summary.

Abstract argumentation-based approach was presented by Ferilli and Pazienza [6] to select crucial sentences from a text and present them in form of summary.

In 2018, Sehgal et al. [7] published his work on text summarization dealing with graph-based algorithm—Text Rank with a slight modification in similarity function used for sentence scoring. The paper also covers significance of title in scoring of a sentence.

Aggarwal and Gupta [8] used three main techniques k-means clustering, SSHLDA and vector space model to propose a new technique for text summarization.

Alguliyev et al. [9] proposed a two-step sentence selection method for automatic text summarization. The first step deals with discovery of all topic spaces and then sentences are clustered using k-means algorithm. In the second step, relevant sentences are extracted based on relevance with topics and proximity with neighbouring sentences in the cluster. This method generates optimum summary and reduces redundancies in the summary.

Morchid et al. [10] presented a research on automatic text summarization to spur the process of topic modelling in which the authors proposed to build topic spaces from the summarised documents. A comparison was made between complete documents and summarized ones using topic space representation. The motive was to show the effectiveness of the summarised documents over the complete document.

Guran et al. [11] proposed an additive FAHP-based sentence score for text summarization. They used a hierarchical approach to generate sentence scoring function. These functions are used to rank the sentence and the top ranked sentences are then used to generate summary.

A survey on text summarization from the legal document is presented by Kanapala et al. [12]. They have discussed various works that have already been done in recent past focusing on text summarization and compared the efficiencies and performance of various approaches used in them.

Alguliyev et al. [13] proposed a sentence selection model and HLO algorithm for extractive text summarization. This paper aims at scoring the sentences and then selecting them to generate optimal summary. For sentence scoring, an HLO optimization model is used and for sentence selection, the criteria of minimal semantic optimal summary under the assumption that summary is upper bounded by a specific length.

Nallapati et al. [14] presented a research using sequence-to-sequence RNNs on abstractive text summarization. They apply an off-the shelf attentional encoder–decoder RNN to sight that it already outperforms state-of-the-art system on two different English corpus. Motivated by concrete problems in summarization, they proposed a new dataset for task of abstractive summarization of document into multiple sentences and showed a significant improvement from previous works.

Yu et al. [15] proposed a Text Rank-Based Method for automatic text summarization. Text Rank algorithm performs well for large text mining and keywords extraction but it does not consider the basic information about structure and context of text while generating summary. To overcome this shortcoming, the authors of this paper proposed iTextRank which considers sentence similarity and acclimatize weights of nodes in graph by taking into consideration linguistic and statistical features like similarities in title, paragraph structure, etc. When experimented, it performed better than Text Rank.

Text summarization using clustering and topic modelling approach based on semantic similarity between sentences was implemented using MapReduce framework proposed by Nagwani [16].

Kinariwala and Kulkarni [17] performed Text Summarization using fuzzy relational clustering algorithm. The general technique used for text summarization is taken into account the sentence level clustering algorithms. It used fuzzy relational clustering algorithm for it and considers graphical representation of data where graph centrality is used as likelihood with the help of expectation maximization framework which determines cluster membership values. The result of algorithm, when applied on sentence clustering, shows that it is able to identify overlapping clusters. The output cluster membership values can be used for generating extractive summary.

Babar and Thorat [18] put forward the mechanism of summarization using Fuzzy Logic and Latent Semantic Analysis which generates automatic summary as an output for given input. It poses a condition that input should be well structured for this, there is initial preprocessing like removing stop words, tokenization etc., then each sentence is represented by a feature vector namely title feature, term weight, sentence position, etc. Each feature has a value between 0 and 1. After feature extraction, the result is passed to fuzzifier and defuzzifier. After this, each sentence has a score which is arranged in decreasing order. Finally, 20% of sorted sentences produce the resultant summary.

Gupta et al. [19] presented a research which deals with connectivity among sentences based on the semantic relationship (graph-based approach), which in turn is used for extraction of important sentences that are then included in the summary. Ferreira et al. [20] evaluates various sentence scoring methods and thus proposed a thematic summarization system. The study presents the fact that quality of summary generated through sentence scoring methods is based on the context of the text to be summarised. Three different subjects, namely news, blogs and articles were used as datasets. The results obtained show the validity of wok in each domain.

Ferreira et al. [21] researched on experts system with applications. The paper performs quantitative and qualitative analysis of fifteen different sentence scoring methods using dataset of three different contexts, namely news, blogs, and article. Directions were given at the end of paper for further improvement.

Erkan and Radev [22] explored LexRank: Graph-based lexical centrality in text summarization. Cosine similarity was used to figure out similar sentences that are then represented in form of adjacency matrix of graph.

Knight and Marcu [23] performed summarization beyond sentence extraction: The aim of this paper is to focus on sentence compression which basically deals with two goals: compression based on grammar and most relevant part of information should be retained. They used decision tree and noisy channel approach and then evaluated results against manual compression.

Hovy and Lin [24] proposed SUMMARIST system for automatic text summarization. The aim of SUMMARIST proposed in this paper is to produce automatic summary in English and other languages. SUMMARIST is architecture according to the following equation: Summarization = Topic identification + interpretation + generation.

Each paper has been discussed in the subsequent table for comparative analysis of literature survey (Table 1).

Table 1 Various approaches used in research papers present in literature survey

S.no	Author	Technique	Dataset	Tool	Discussion
1.	Sinha [4] 2018	Neural network which contains an input layer, a hidden layer and an output layer	DUC2002, and a separate 35 document dataset	TensorFlow library, Models: ILP, PCFG, NN-SE, TGRAPH, URANK, GENE	They used ROUGE for evaluation and demonstrated that their model outperforms the complex genetic algorithm and also suggested improvement on summary length
2.	Khan [5] 2018	GA, IWGRA, PAS semantic similarity	DUC2002	SENNA SRL	They combined a modified graph-based algorithm with semantic similarity and showcased better results than traditional
3.	Ferelli [6] 2018	Abstract argumentation	English subset of the benchmark Multiling 2015 dataset.	Not applicable	The proposed approach produced great results on the experimented dataset
4.	Sehgal [7] 2018	TextRank, PageRank, Lexemes	English articles	Not applicable	The paper proposes variation of sentence similarity function and lays emphasis on role of title in sentence scoring
5.	Aggarwal [8] 2017	Modified K-Means clustering, SSHLDA, Vector space model	Legal cases dataset	Recall and precision measures	The hybrid proposed algorithm performs better than the original traditional algorithms
6.	Alguliyev [9] 2017	K-Means algorithm	DUC2001, DUC2002 datasets were used	Not applicable	ROUGE value of summaries that were generated through the proposed system was found better when compared with traditional sentence selection systems

(continued)

Table 1 (continued)

S.no	Author	Technique	Dataset	Tool	Discussion
7.	Morchid [10] 2017	LDA model	Wikipedia in three different languages: English, Spanish and French	Mallet Toolkit Proposed Summarization System: ARTEX, BF, BR	Among all the systems used for summarization, BF shows the best result because it selects the first paragraph as the summary and Wikipedia exposes the main idea about a topic in its first paragraph
8.	Guran [11] 2017	FAHP	Turkish documents and human generated extractive summary were used as dataset	A system-based on FAHP and GA	FAHP combining methods perform better than all individual features. Also, FAHP performs equally good as GA and saves time when compared with GA
9.	Kanapala [12] 2017	Survey: sentence scoring methods	Different datasets were used DUC2002, news dataset (English), RST-DTB, Wikipedia (English and Punjabi)	Not applicable	Some more research is needed for comparative analysis of various techniques
10.	Alguliyev [13] 2016	HLO algorithm	Initial population is generated by randomly selecting individuals: RELO (Random exploration learning operator), IKD (Individual Knowledge Database), SKD (Social knowledge database)	HLO algorithm	Optimal summary is generated using HLO algorithm with a termination condition that HLO it terminates when maximum number of generations are met
11.	Nallapati [14] 2016	RNN model	For all models, they used 200 dimensional word2vec vectors	Tesla K40 GPU	Each of proposed model addresses a particular problem in extractive summarization also they suggested a new dataset for multi-document summarization and established benchmark numbers on it

(continued)

Table 1 (continued)

S.no	Author	Technique	Dataset	Tool	Discussion
12.	Shanshan [15] 2016	itextrank model	Online data	Not applicable	Authors proposed a new model iTextRank which was found highly scalable as compared to already present TextRank
13.	Nagwani [16] 2015	Latent drichlet allocation	4000 legal cases from UCI machine learning repository	Java open source technology: MALLET, Hadoop	MapReduce framework results in greater scalability and reduced time. When clustering and semantic similarity both are used simultaneously, summary generated is better
14.	Kinariwala [17] 2015	Fuzzy relational clustering algorithm, K-medoids	FRECCA was applied on Cluster famous quotation dataset	FRECCA	FRECCA performs exceptionally well when compared to k-medoids algorithm and is best suited for generating summary
15.	Babar [18] 2014	Fuzzy logic algorithm	Human-generated summaries were used as Gold standard which is used to compare with summaries generated from proposed method. Online summaries were also used	Not applicable	Use of fuzzy logic in process of summarization improved the quality of summary drastically
16.	Gupta [19] 2014	A graph-based algorithm: weighted minimum vertex cover	Single document summarization was performed using DUC2002 dataset with randomly picked 60 news articles	BIUTEE	Their method of WMVC outperformed the earlier methods that used TE and frequency-based techniques for summarization

(continued)

Table 1 (continued)

S.no	Author	Technique	Dataset	Tool	Discussion
17.	Ferreira [20] 2014	Graph-based techniques	They used three datasets: CNN dataset, Blog summarization dataset and SUMMAC dataset	Not applicable	Several algorithms were discussed in the paper. Among them, com01, com06, com07 were found the best and among these three the best summarization results were given by com01
18.	Ferreira [21] 2013	Survey: text summarization from legal documents	They used three datasets: CNN dataset, Blog summarization dataset and SUMMAC dataset	ROUGE metrices	This paper explains most important strategies found in past 10 years related to text summarization
19.	Erkan [22] 2004	LexRank approach	They used DUC 2003 and 2004 dataset	MEAD summarization toolkit	They have introduced three different methods to compute centrality in similarity graph. The outcome of application of the methods on graphs were quiet effective also they have tried to incorporate more information in LexRank still it results better in most cases. Lastly. they have shown that their methods do not get affected by noisy data
20.	Knight [23] 2002	Noisy channel model	They used 32 pairs of sentences from their parallel corpus and 1035 other pairs of sentences. The second test corpus used was referred as Cmplg corpus: which contained compressed grammatical version of above sentences	Not applicable	They have proposed corpus-based methods for sentence compression using noisy channel framework and decision-based model

(continued)

Table 1 (continued)

S.no	Author	Technique	Dataset	Tool	Discussion
21.	Hovy [24] 1998	SUMMARIST system	They manually created two summaries of almost paragraph length of about 300 words from 2 newspaper articles for the LA Times	Not applicable	Different summaries were generated with previously established systems like SUMMARIST etc., and their quality was examined, It was found that they require some careful and large-scale work in future

5 Discussion

After extensive review of various techniques and existing work done in the field of text summarization, it is found that there is great scope for text summarization in the legal domain.

Obtaining relevant information from a huge corpus of legal data is important for various personnel's like an ordinary citizen, lawyer, scholar, etc. When the document is too long to go through and user is in hurry to get a quick overview then text summarization is the only thing that is needed at rescue.

This is especially significant in the field of law [25]. Summarization of legal documents is an important area for research when it comes to text summarization because a lot of efforts goes in creating an efficient summary of the previous judgments. The court has specialized personnel to generate summaries of judgments in order to underpin their case. Furthermore, many times, the users want to have a look at if any previous evidence relevant to their case is present. Hence automatic text summarization has emerged as an important tool for different stakeholders.

It is also useful for the human summarizers as they can take an overview of important points to be included in summary, legal petitioners often need various important arguments to argue in support of their cases which could be easily be done by text summarization tool as it takes into account the extraction of most crucial sentences from the source document.

Sometimes, there is a requirement for quick perusal of some of legal documents to support a case for which usually court relies on human summarizers, who could attempt to make several mistakes when scanning the documents in hurry, they may miss some relevant information which could be useful for the case. So, automating the process of text summarization would be a great idea for legal domain as it improves efficiency and saves time and human efforts.

This tool is most useful for novice citizen who wants to have a knowledge or reference to legal cases. As we know judgments are available for public use, however, they are too long to absorb for a user who do not have any knowledge about legal system, hence if we can have a tool which can automate the drafting of summaries then user's convenience towards consultative legal information will be significantly enhanced.

The most demanding part in legal text summarization is figuring out the most relevant and informative part at the same time ignoring the unimportant one. The main concerns are:

- Identifying the content that captures the gist of the document to be included in summary.
- Presentation of summary to user in most effective way.
- How to automate the process of extraction from legal text documents as legal text contains certain important and different terminologies from normal text.

6 Conclusion

Automatic Text Summarization is an age-old demand. Now, researchers are getting oriented from extractive summarization to abstractive summarization methods as extractive summaries are coherent and redundant in nature, whereas abstractive summaries produce cohesive and information-rich summaries plus any additional information or comments relevant to the topic are also included in it, therefore, abstractive summaries are better than extractive ones in this criteria.

In this paper, various abstractive and extractive summarization systems are explored. Different text summarization approaches implemented in the recent past are surveyed. Lastly, few research issues that will be addressed as future work certainly has been proposed.

The utmost challenge in text summarization is to condense relevant information from diverse sources whether it is database or internet data or text document. A summarizer should take into account each one of them and generate an effective and cohesive summary.

References

1. Saravanan M, Ravindran B (2010) Identification of rhetorical roles for segmentation and summarization of a legal judgment. Artif Intell Law 18(1):45–76
2. Gupta V, Lehal GS (2010) A survey of text summarization extractive techniques. J Emerg Technol Web Intell 2(3):258–268
3. Khan A, Salim N (2014) A review on abstractive summarization methods. J Theor Appl Inf Technol 59(1):64–72
4. Sinha A, Yadav A, Gahlot A (2018) Extractive text summarization using neural networks. arXiv:1802.10137
5. Khan A, Salim N, Farman H, Khan M, Jan B, Ahmad A, Ahmed I, Paul A (2018) Abstractive text summarization based on improved semantic graph approach. Int J Parallel Program 1–25

6. Ferilli S, Pazienza A (2018) An abstract argumentation-based approach to automatic extractive text summarization. In: Italian research conference on digital libraries, Springer, Cham, pp 57–68

7. Sehgal S, Kumar B, Rampal L, Chaliya A (2018) A modification to graph based approach for extraction based automatic text summarization. In: Progress in advanced computing and intelligent engineering, Springer, Singapore, pp 373–378

8. Aggarwal R, Gupta L (2017) Automatic text summarization. Int J Comput Sci Mob Comput 6(6):158–167

9. Alguliyev RM, Aliguliyev RM, Isazade NR, Abdi A, Idris N (2017) A model for text summarization. Int J Intell Inf Technol (IJIIT) 13:1. https://doi.org/10.4018/ijiit.2017010104. Accessed 25 Sep 2017

10. Morchid M, Torres-Moreno J-M, Dufour R, Ramírez-Rodríguez J, Linarès G (2017) Automatic text summarization approaches to speed up topic model learning process. arXiv:1703.06630 (2017)

11. Güran A, Uysal M, Ekinci Y, Barkan Güran C (2017) An additive FAHP based sentence score function for text summarization. Inf Technol Control 46(1):53–69

12. Kanapala A, Pal S, Pamula R (2017) Text summarization from legal documents: a survey. Artif Intell Rev 1:1–3

13. Alguliyev R, Aliguliyev R, Isazade N (2016) A sentence selection model and HLO algorithm for extractive text summarization. In: 2016 IEEE 10th international conference on application of information and communication technologies (AICT). IEEE, pp 1–4 (2016)

14. Nallapati R, Zhou B, Gulcehre C, Xiang B (2016) Abstractive text summarization using sequence-to-sequence rnns and beyond. arXiv:1602.06023

15. Yu S, Jindian S, Li P, Wang H (2016) Towards high performance text mining: a TextRank-based method for automatic text summarization. Int J Grid High Perform Comput (IJGHPC) 8(2):58–75

16. Nagwani NK (2015) Summarizing large text collection using topic modeling and clustering based on MapReduce framework. J Big Data 2(1):6

17. Kinariwala SA, Kulkarni BM (2015) Text summarization using fuzzy relational clustering algorithm. Int J Sci Res Educ 4370–4378

18. Babar SA, Thorat SA (2014) Improving text summarization using fuzzy logic & latent semantic analysis. Int J Innov Res Adv Eng (IJIRAE) 1(4)

19. Gupta A, Kaur M, Mirkin S, Singh A, Goyal A (2014) Text summarization through entailment-based minimum vertex cover. In: * SEM@ COLING, pp 75–80

20. Ferreira R, Freitas F, de Souza Cabral L, Lins RD, Lima R, Franca G, Simske SJ, Favaro L (2014) A context based text summarization system. In: 2014 11th IAPR international workshop on document analysis systems (DAS). IEEE, pp 66–70

21. Ferreira R, de Souza Cabral L, Lins RD, e Silva GP, Freitas F, Cavalcanti GDC, Lima R, Simske SJ, Favaro L (2013) Assessing sentence scoring techniques for extractive text summarization. Expert Syst Appl 40(14):5755–5764

22. Erkan G, Radev DR (2004) Lexrank: graph-based lexical centrality as salience in text summarization. J Artif Intell Res 22:457–479

23. Knight K, Marcu D (2002) Summarization beyond sentence extraction: a probabilistic approach to sentence compression. Artif Intell 139(1):91–107

24. Hovy E, Lin C-Y (1998) Automated text summarization and the SUMMARIST system. In: Proceedings of a workshop on held at Baltimore, Maryland: October 13–15. Association for Computational Linguistics, pp 197–214

25. Abinaya M (2014) Text summarization using deep belief networks. Master's thesis in information technology, International Institute of Information Technology, Bangalore, India

Prevailing Approaches and PCURE for Data Retrieval from Large Databases

Seema Maitrey, C. K. Jha and Poonam Rana

Abstract Tremendous and exceedingly vast data is collected, nowadays, by every organization which is getting continually increased. It became very difficult to retrieve relevant information from these endlessly rising large group of data. Data mining has emerged to retrieve precious information that gets buried in large databases. Among various functionalities of data mining, clustering became very effective in determining related data. This work is focused on CURE which is one of the most widely used hierarchical clustering techniques of data mining. It started its work by reducing the size of the original database. For that, it made the use of simple random sampling (SRS) technique, followed by partitioning of the reduced database. It also made use of other important techniques but still resulted in a number of short-comings. It is required to eradicate the limitations in the traditional working of CURE clustering. So, this paper avoids the use of sampling and focuses on its enhancement by integrating it with the concept of "Map-Reduce" along with "Corewise Mul-tithreading". This combination is useful for analyzing-searching huge voluminous data by providing the most effective ability of parallel processing, fault tolerance, and load balancing. The proposed approach is parallelization of one of the data mining clustering techniques—CURE and thus named as PCURE (ParallelCURE).

Keywords Data mining · Clustering · CURE · Sampling · Parallelism · M2ing

S. Maitrey (✉) · P. Rana
Department of Computer Science and Engineering, KIET Group of Institutions,
Ghaziabad, India
e-mail: seema.maitrey@kiet.edu

P. Rana
e-mail: poonam.rana@kiet.edu

C. K. Jha
Department of Computer Science, Banasthali Vidyapith, Jaipur, Rajasthan, India
e-mail: ckjha1@gmail.com

© Springer Nature Singapore Pte Ltd. 2019
S. Bhattacharyya et al. (eds.), *International Conference on Innovative Computing and Communications*, Lecture Notes in Networks and Systems 56,
https://doi.org/10.1007/978-981-13-2354-6_52

517

1 Introduction

In the current scenario, the organizations are collating extremely huge amount of data and this data is exponentially growing which tends to get examined inefficiently. To find out valuable information from these continuously growing large data sets has become a challenge for researchers in data mining [1]. Data which converted to big data required data mining analyses in large scale to fulfill its scalability and performance requirements. So, several efficient parallel and concurrent algorithms got into focus, and applied [2]. The huge voluminous data brought many parallel algorithms that used different parallelization techniques into real action. Such techniques were threads, MPI, etc., which produce different performance and usability characteristics. The MPI model which can compute rigorous problems efficiently is very difficult to bring them into the practical use. Data mining is continuously spreading its root in business and in academics over the coming years. Widely used clustering technique, called CURE, produces high-quality clusters in the presence of outliers [3]. Among widely used existing clustering algorithms, serious scalability and accuracy-related problems were encountered whenever they are applied to massive data or huge dimensions. It has been found that CURE has good execution time in presence of large databases [4].

2 Literature Review

A lot of research has been carried out by several researchers on CURE clustering technique. List of some papers on CURE algorithms are described below.

In [5, 6], Sudipto Guha et al. proposed CURE algorithm that showed its efficient execution on large databases. In [7] Qian et al. found the relation of shrinking scheme of CURE and its ability to recognize spherically shaped cluster. In [8], Gediminas et al. provided a new approach to discover clusters from continuously arriving too large amount of dataset and for such cluster formation, it made the use of sampling-based techniques. In [1], a new synthesized data mining algorithm named CA with improving CURE algorithm showed experimental results. Using GA and CURE, M. Kaya, R. Alhajj, in [9], proposed an automated method for mining fuzzy association rules. The process of data mining can be done at a great speed when the sample of the database is used in place of the original database [10]. But this use of sampling is only acceptable if the quality of the mined knowledge is not compromised [11].

The biased result drawn after sampling is required to find out such a technique which can access the appropriate data efficiently from the huge data collection. Therefore, in this paper, an algorithm that is an integration of CURE with the concept of measuring and minimizing (M2ing) is proposed which will offer several advantages against sampling.

3 Problem Formulation

Researchers trusted simple random sampling to be an unbiased approach to the survey, but sample selection bias can occur. Whenever a sample set of the higher population is not abundant enough, the full population presentation is skewed and have requirement of extra sampling techniques. A lot of efforts were put on sampling but still the performance of the algorithm and the desired accuracy or confidence of the sample reflects some gap between them. If a sample of very small size is considered, it may produce many false rules, and thus the performance gets degraded.

Solution:
At the current time, data is accumulating with great speed. This resulted in the enormous size of the database. The huge databases are accessed by a larger group of users. Large databases which are typically associated with Decision Support System and Data Warehouses perform data intensive operations [12]. Thus, the required parallel execution (sometimes called Parallelism) which dynamically reduces response time. The benefits of parallelism lie in the fact that the number of organizations getting increased who are turning to parallel processing technologies. This technology fulfills their desired need by providing them the performance, scalability, and reliability [13].

3.1 Limitations of Random Sampling Algorithm

A lot of research has taken place with the help of sampling. It made the data mining task simple and fast by reducing the database size and executing the query fast in comparison to the accessing and retrieving the entire database [14]. But, it produced the biased result. While picking the data randomly from the database, certain important data can be missed which can lead to the poor result. So, to preserve the good intention of sampling along with improving the quality of the data, the data mining algorithm is integrated with the concept of parallelism. The sampling algorithm and the biased result produced by it are given below as [13, 15]:

```
for i := 0 to n - 1 do READ-NEXT-RECORD(C[i]);
t: = n;                          // t is the total number of records processed yet//
while not eof do                 //process the remaining records1//
start
t := t + 1
d := TRUNC (t x RANDOM( ));      // d is random in the range 0 <=d <= t-1
if.d<n then                      // next record is made a candidate
READ-NEXT-RECORD(C[d])           // substituting one at random
else
SKIP-RECORDS(1)                  //Step to the next record//
end                                                              [13][15]
```

we repeated the step below 3 times..............

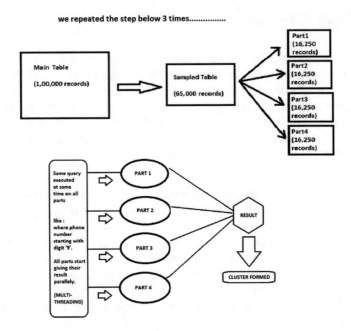

And each time the cluster formed for same data items using same method , was different.

Fig. 1 Cluster formation after using sampled and partitioned database

The limitation of random sampling can be demonstrated by taking a simple and small example which is as follows.

Let the database have 1L records in its main table. It randomly picks 65000 records from the table and split equally into the following 4 partitions, shown below in Fig. 1.

The figure displays the set of records produced for the same query provided each time. It produced the different set of record for the common query. Thus, sampling produced the diverse and controversial result. This can be shown below in Fig. 2.

The figure above produced different result each time for the same query. Thus, generated controversial and biased information. To avoid such a situation, the concept of "Measuring and Minimizing (M2ing)" is proposed in this work.

4 Idea Behind Parallelism

The idea behind parallelism is to break down the task into several parts. Earlier, a single process used to perform all work in a query. It now replaced with many processes which unite or cooperate to accomplish the part of work altogether at the same time. Now, instead of one process doing all work in a query, it results in high performance. It allows the multiprocessor to work simultaneously on several parts

```
run:Phone9                          run:Phone_9                         run:Phone99
'2','XLPTKB','VII','D','9974131341'  '2','XLPTKB','VII','D','9974131341'  '2','XLPTKB','VII','D','9974131341'
'5','JOQTIX','VII','B','9784668440'  '5','JOQTIX','VII','B','9784668440'  '3','XORLAN','VII','E','9982146146'
'13','UKQKTG','VII','E','9873409753' '8','WNEVBZ','III','D','9999409500'  '5','JOQTIX','VII','B','9784668440'
'16','YNQJOY','VI','G','9889180166'  '13','UKQKTG','VII','E','9873409753' '13','UKQKTG','VII','E','9873409753'
'22','RLXIRU','II','D','9990408665'  '17','SVADVO','IV','H','9979180304'  '15','YQGUXS','V','D','9974654216'
'24','SICNPI','I','H','9996885477'   '19','ZARTFQ','V','H','9989016879'   '16','YNQJOY','VI','G','9889180166'
'25','PZZSJJ','VIII','E','9874788116' '22','RLXIRU','II','D','9990408665' '22','RLXIRU','II','D','9990408665'
'40','BDZZUS','VIII','A','9871102269' '25','PZZSJJ','VIII','E','9874788116' '35','RWQYOG','IV','E','9976988061'
'41','OIGSIB','VI','C','9993877617'  '39','CWGIGH','I','A','9791123862'   '39','CWGIGH','I','A','9791123862'
'43','GVHJGO','II','F','9977902553'  '41','OIGSIB','VI','C','9993877617'  '41','OIGSIB','VI','C','9993877617'
'46','HDRFTN','I','A','9795185988'   '46','HDRFTN','I','A','9795185988'   '46','HDRFTN','I','A','9795185988'
'50','GCLIDB','IV','A','9796111347'  '50','GCLIDB','IV','A','9796111347'  '50','GCLIDB','IV','A','9796111347'
'51','UTQKFW','III','A','9787880529' '51','UTQKFW','III','A','9787880529' '51','UTQKFW','III','A','9787880529'
'57','METFFQ','V','E','9784276406'   '57','METFFQ','V','E','9784276406'  '64','HFVTVX','VIII','G','9790333521'
'83','QMLQXC','VII','H','9979276139' '63','JTMTUZ','I','H','9881921933'  '68','WIUFOH','IV','F','9776200048'
'92','TYTDXJ','III','H','9775817079' '68','WIUFOH','IV','F','9776200048' '76','ZOFLDG','I','E','9879067372'
                                     '76','ZOFLDG','I','E','9879067372'  '82','MEYYHR','III','D','9880175752'
                                                                         '83','QMLQXC','VII','H','9979276139'
                                                                         '92','TYTDXJ','III','H','9775817079'
                                                                         '93','YRDRMQ','VI','A','9786079904'
```

Fig. 2 Different results produced each time same query get executed

of a task [16, 17]. Thus, enhance the speed of completion of the task and make the work done faster.

Strategies used in the proper utilization of parallelism in the data mining algorithm are [18, 19]:

i. Independent parallelism: Whole data set is possessed by each process and there is no communication between the different processes.
ii. Task (control) parallelism: Different operations get executed on different partition of data set by each process.
iii. SIMD (Single Program Instruction Data): Same algorithm on different partitions of a data set get executed in parallel by a group of processes and then exchange their partial results.

Here, SIMD strategy is used in the work.

4.1 Parallelism in Data Mining

Very little work on parallel data mining has really been done. The parallel computing seems to be powerful because of the size and complexity of the problems. KDD (knowledge discovery from the database) which is also called as data mining, analyses the huge amount of data, retrieve the useful and interesting information for any organization. Many prior algorithms use the Sampling technique to reduce time but it is also not so effective. Reducing data might result in inaccurate models and can also lack in outlier identification. The parallel data mining algorithm and the high-performance computers can now jointly propose a very effective way to retrieve and access the huge data collection. The large databases can now be aptly handled in a parallel manner. Earlier when there was the practice of using sequential data mining applications that used to take several days or weeks to accomplish their job [20, 21]. At the present time, it is very rare where the sequential access or retrieval is done on large datasets. Parallel processing can return enormous benefits for large databases. Whenever a query is put against a very large database, it can be profoundly improved if the data is kept in partitions and each partitioned is accessed by an individual process.

4.2 PCURE: Parallel CURE Clustering

Clustering algorithms continue to be challenged by ever-increasing dataset sizes. With this work, we have taken a new step towards meeting the emerging computational needs. Specifically, we have demonstrated improved scalability for CURE clustering construction on massive datasets using thread-level parallelism by integrating with M2ing and Multi-Core system. The proposed approach is parallelization of one of the data mining clustering technique—CURE and thus named as PCURE (ParallelCURE).

Finally, we built a PCURE from a larger dataset used. As per our experimental outcomes, it is believed that massively multi-threaded architectures are suitable for massive data set and data mining applications.

5   Description of the PCURE Model

We need an intelligent program which will first find the numbers of cores available and then divide the problem in an adequate amount of threads.

Take the count of threads(t) = count of cores(c), so

Number of threads used to do processing: t * number of cores

The task of multithreading is in-built in M2ing. The M2ing automatically partitions the data and allot them to individual multiple threads which depend on the number of available cores. Finally, the work of research has reached a point where all components required for building a new model for handling large databases are developed. This new model is comprised of the combination of:

CURE (without sampling) + M2ing (in-built Multithreads) + Multi-Core

5.1   Execution Model Designed for PCURE

Initially, the new enhanced model of CURE is implemented in a single machine. It can be extended further across multiple machines as when required with some amendments (where needed). It has given dual benefits, i.e., no data loss along with speed and efficiency while working with large databases. Table 1 and Fig. 3 below depicts the total time elapsed during a search of the required data from the entire database by using the M2ing technique.

Table 1 Time elapsed and records found by using M2ing

Query data	"Nestle", "Moti Mahal", "McDonalds"
Time elapsed	1.537
Overall records	309738

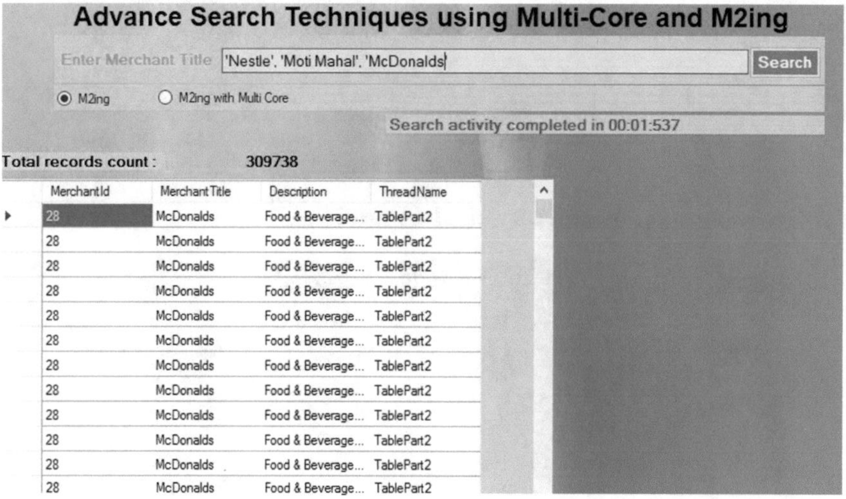

Fig. 3 Records found and time taken using M2ing in original DB

The data in the table above are available from the experiment performed. The screen producing the output is given below in Fig. 3.

The graph below depicts nearly the same time duration as taken by sampling technique. But the sampling acts upon reduced dataset. Whereas, M2ing works upon the complete database. This shows that the M2ing technique on large dataset makes the processing and retrieval faster along with maintaining the quality of the result given below in Fig. 4.

Till now, the experiment was held by three ways—taking the entire database, database with partitioning, and sampling. Finally, experiment by using the M2ing technique. All with their graphical representation are shown combined in one graph. From there, it can be easily manipulated that the traditional approach took more time in comparison to the latest technique of M2ing. But it still can be improved. Comparison of traditional techniques with the new technique of M2ing can be shown in Fig. 5.

Comparing traditional techniques and new M2ing.

The new concept of parallel processing can yield enormous benefits to large databases. Whenever a query is put against a very large database, it can be profoundly improved if the data is kept in partitions and each partition is accessed by an individual process.

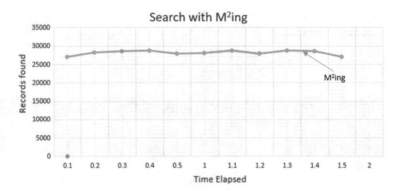

Fig. 4 Advance search from large DB using M2ing technique

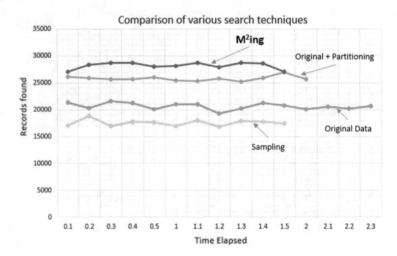

Fig. 5 Various techniques used in clustering and searching from large DBs

Thus, it is found that parallelism can extremely reduce the response time for any query that is broken into sub-queries and each sub-query is assigned to a processor.

6 Future Scope

Data mining algorithms and underlying techniques can be parallelized to make them effective in the analysis of very large data sets. Several parallel strategies, algorithms, techniques, and prototypes have been developed in the recent years. They allow researchers and end users to mine large databases offering scalable performance. Parallel execution of different data mining algorithms and techniques can be integrated to obtain a better model not just to get high performance but also high accuracy. In this work, the CURE is improved by integrating it with the concepts of MapReduce. The working style of MapReduce can be further modified to make it faster in executing huge amount of data.

7 Conclusion

"PCURE for Data Retrieval from Large Databases" presents an organized exploration of several steps involved in a data mining CURE clustering process, provided with both theoretical and practical contributions. It explores the restrictions of the earlier work and mentions the progress of current work. Though sampling is used in the data reduction of large databases, it results in the controversial and biased result. The objective of sampling was novel, i.e., to minimize the database size and retrieve the information out of data very fast. But it gave rise to confusion and sometimes loss of crucial data. This research work combines the good intention of sampling with fast data retrieval and making maximum utilization of available resources by replacing sampling with M2ing. This innovative, improved and enriched CURE will definitely benefit the large transactional databases and data kept in a data warehouse. The enhanced CURE will have the ability to access and process the entire database without any data loss and produce the result very fast and efficiently. The main benefit and enthrallment of this approach are its ease of use.

References

1. Berson A, Smith SJ, Thearling K (1999) Building data mining applications for CRM. McGraw-Hill, New York
2. Sanse K, Sharma M (2015) Clustering methods for big data analysis. Int J Adv Res Comput Eng Technol (IJARCET) 4(3):645. ISSN: 2278-1323
3. Maitrey S, Jha CK (2012) A survey: hierarchical clustering algorithm in data mining. IJESR 2(4) Article No-6/204-221. ISSN 2277-2685
4. Han J, Kamber M (2001) Data mining: concepts and techniques. Morgan Kauffmann, San Francisco, California
5. Guha S, Rastogi R, Shim K (1997) CURE: a clustering algorithm for large databases. Technical report, Bell Laboratories, Murray Hill
6. Guha S, Rastogi R, Shim K (2001) CURE: an efficient clustering algorithm for large databases. Inf Syst 26(1):35–58. Published by Elsevier Science Ltd
7. Qian Y-T, Shi Q-S, Wang Q (2003) CURE-NS: a hierarchical clustering algorithm with new shrinking scheme in IEEE Xplore
8. Adomavicius G, Bockstedt J, Parimi V (2008) Scalable temporal clustering for massive multi-dimensional data streams. In: 2008 workshop on information technologies and systems, WITS 2008—Paris, France, pp 121–126
9. Kaya M, Alhajj R (2005) Genetic algorithm based framework for mining fuzzy association rules. Fuzzy Sets Syst 152:587–601. www.elsevier.com/locate/fss

10. Han J, Kamber M (2006) Data mining: concepts and techniques. In: Gray J (ed) The Morgan Kaufmann series in data management systems, 2nd edn. Morgan Kaufmann Publishers. ISBN 1-55860-901-6
11. John GH, Langley P. Static versus dynamic sampling for data mining. In: From KDD-96 Proceedings. AAAI (www.aaai.org)
12. Maitrey S, Goel N, Tiwari S (2001) Parallelism in CURE clustering technique international conference on issues and challenges in networking, intelligence and computing technologies (ICNICT)
13. Maitrey S, Jha CK (2013) An integrated approach for CURE clustering using map-reduce technique. In: Proceedings of Elsevier. ISBN 978-81-910691-6-3
14. Horton M (2015) What are the disadvantages of using a simple random sample to approximate a larger population? http://www.investopedia.com
15. Steinhaeuser K, Chawla NV (2006) Scalable learning with thread-level parallelism, Master's thesis in 2006. Supported in part by the Defense Advanced Research Projects Agency (DARPA) under Contract No. BNCH3039003
16. Lee K-H, Lee Y-J, Choi H, Chung YD, Moon B (2011) Parallel data processing with MapReduce: a survey, vol 40, no 4. SIGMOD Record
17. Larose DT (2005) Discovering knowledge in data: an introduction to data mining. Int J Distrib Parallel Syst (IJDPS) 1(1). Wiley. ISBN 0-471-66657-2
18. Parsha MK, Pacha S (2013) Recent advances in clustering algorithms: a review. Int J Concept Comput Inf Technol 1(1). ISSN: 2345–9808
19. Fox GC, Bae SH et al (2008) Parallel data mining from multicore to cloudy grids. In: High performance computing and grids workshop
20. Silwattananusarn T, Tuamsuk K (2012) Data mining and its applications for knowledge management: a literature review from 2007 to 2012. Int J Data Min Knowl Manag Process (IJDKP) 2(5)
21. Steinhaeuser K, Chawla NV (2006) Scalable learning with thread-level parallelism. Master's thesis in 2006. Supported in part by the Defense Advanced Research Projects Agency (DARPA) under Contract No. BNCH3039003

A Novel Framework for Automated Energy Meter Reading and Theft Detection

S. M. Adikeshavamurthy, R. Roopalakshmi, K. Swapnalaxmi, P. Apurva and M. S. Sandhya

Abstract Energy crisis is one of the most important issues that the entire world is facing today. The feasible solution for the energy crisis problem needs optimal utilization of available energy. However, the state-of-the-art energy metering systems suffer due to issues such as low battery backup, poor network connectivity, and excessive memory consumption. To overcome these drawbacks, a novel automated energy metering framework is proposed in this paper, which makes use of Microcontroller-based implementation for its operation. Specifically, consumer can get the energy consumption statistics instantly on a LCD screen. Further, whenever any consumer attempts to tamper the energy meter, magnetic sensors get actuated and sends appropriate signals to the microcontroller, which in turn sends theft event messages to the administration side for further processing. Experimental setup and results indicates the good performance of the proposed framework in terms of energy consumption display on LCD screen, which significantly help the customer to monitor their energy consumptions.

Keywords Programmable interface controller (PIC) microcontroller
GSM modem · Electricity theft · Automated meter reading

S. M. Adikeshavamurthy · R. Roopalakshmi (✉) · K. Swapnalaxmi · P. Apurva · M. S. Sandhya
Alvas Institute of Engineering and Technology, Mangalore, India
e-mail: drroopalakshmir@gmail.com

S. M. Adikeshavamurthy
e-mail: as.maroor@yahoo.com

K. Swapnalaxmi
e-mail: swapnabs.bhat@gmail.com

P. Apurva
e-mail: apurvagowda66@gmail.com

M. S. Sandhya
e-mail: sandhyasgowda96@gmail.com

© Springer Nature Singapore Pte Ltd. 2019
S. Bhattacharyya et al. (eds.), *International Conference on Innovative Computing and Communications*, Lecture Notes in Networks and Systems 56,
https://doi.org/10.1007/978-981-13-2354-6_53

527

1 Introduction

A major threat that the world is facing these days is energy emergency. The best solution for energy emergency issue is not just the expansion in vitality creation, yet additionally the successful utilization of accessible vitality. By appropriately checking the energy utilization and maintaining a strategic distance from energy wastage, energy crisis can be diminished to a more remarkable degree. More specifically, currently in India, there are almost 52 million electricity energy meters are interconnected and it is estimated to touch almost 2 billion by 2030 [1].

On the other hand, incorporation of mobile technology into electricity meter system is rapidly advancing, which can decrease the costs. But energy supervising cannot be done efficiently, in the present scenario, because consumers are completely unaware of their energy consumption details. Precisely, consumers get an idea about their consumption precisely, only when the electricity bills are issued. More precisely in India, bills are issued only once in a month or two months. So, the consumers are unaware about their energy consumption during this period of time. In this era of digitalization, no one wants to manually check for the energy consumption by referring to electricity bills of the previous months which is a tedious task. This complex procedure has to be repeated several times in a month to efficiently control the energy usage. Further, if the consumers can check the No. of units consumed regularly, then it is beneficial for the consumer as well as management side to monitor the energy consumption. Based on these aspects, in order to overcome the drawbacks of the manual metering and billing systems, new automated energy metering systems are needed, which can continuously monitor the consumed meter readings. It can also help the management side to detect the tampering of electricity meter and electricity theft.

2 Related Work

In 2011, Landi et al. [2] proposed a framework which processes the main power system quantities and offers probability to deal with the entire power plant. Here, an incorporated Web Server permits to collect the insights of energy utilizations and power quality. This framework enables simple access to the data with the mix of a smart meter and information correspondence capacity and along these lines permits neighborhood and in addition remote access to meter readings. But due to lack of battery backup and excessive memory consumption, this system fails to satisfy the customer level expectations. In [3], an infrastructure called Smart Grid is proposed in which the developing interest of vitality, the limit limitations of vitality administration are talked about. It depends on a vitality meter with low-control microcontroller and the Power Line Communication standards. Koay et al. [4], introduced a system based on Bluetooth technology.

In [4], the authors described that digital meter had already started replacing electromechanical meters in Singapore, where Bluetooth technology is chosen as a probable wireless elucidation to connectivity concern. Khalifa et al. [5] reviewed four communication protocols accessible for automatic meter reading (AMR) gadget and one of them is PLC. Since the information goes through voltage flag, the voltage level conveyed is additionally considered. Despite the fact that PLC gives low upkeep and also great effectiveness, yet it experiences restricted transfer speed and trouble in supporting expansive scale systems. Further, incorporating a few communication protocols with PLC may beat its downsides, for example, WiFi, Micro-control Wireless, and Optical Fiber [6, 7]. Recently, in 2015, Iyer and Rao presented a paper on IoT-based meter reading and theft detection [8], in which the system design eradicates the human engrossment in electricity upkeep. However, in the abovementioned system, if the customer misses to pay the bills on time or trying to tamper the meter, then it automatically alerts the management side which can disconnect the power supply from an autonomous server. This system fails due to need for continuous Internet connection and poor Internet connectivity problems in remote localities.

To outline in the existing literature, less focus is given towards the major issues of automated electricity meter reading, such as battery backup, excessive memory consumption and poor network connectivity. To overcome these drawbacks, this paper proposes a new automated energy metering system which makes use of GSM technology and Microcontrollers for its operation.

3 Proposed Framework

Usage of electromechanical energy meters leads to inaccuracy while taking readings and allows energy theft. In order to overcome these problems, a new automated energy meter is proposed which automatically shows the energy consumption in units and can also detect thefts. The proposed system given in Fig. 1 consists of four modules, namely power supply module, meter reading module, theft detection module, and electricity energy meter, respectively.

The functionalities of each of the modules of the proposed system are explained as follows:

Power Supply Module:

With an end goal of consistently supplying a constant voltage, this unit basically drives the other components of the system and the microcontroller. It consists of a 12 V step-down transformer, bridge rectifier, filter capacitors, and regulators. Initially, a 230 V, 50 Hz load is fed into the system which needs to be regulated to 5 V and 12 V, respectively. Then, the current is passed through a 12 V step-down transformer. The step-down transformer reduces 230 V current to 12 V 50 Hz. AC. The output of the transformer is connected to a bridge rectifier. The rectifier consists of four IN4007 diodes, which is mainly used to convert the AC to pulsating DC. The output current coming from the rectifier comprises of some ripples (noise). The yield of the rectifier

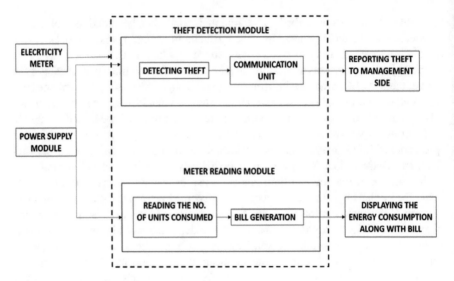

Fig. 1 Block diagram of the proposed automated energy meter reading framework

is directed into the capacitor filter in order to isolate the ripples. The outcome of this will be a varying current and needs to be regulated. For this purpose, the varying current is nourished into 7812 IC and 7805 IC which gives a constant 12 V and 5 V direct current.

Meter Reading Module:

This unit is mainly used to read the number of units consumed by the consumer and display it on a LCD screen. Input of the meter reading unit is connected to microcontroller and the output to energy meter. Here, opto-interrupter is parallelly connected to the LED in the energy meter which indicates consumption of one unit. When a power unit is consumed, the opto-interrupter generates a signal and sends it to the microcontroller. The microcontroller is programmed to keep a track of units consumed and the same information is displayed on LCD screen.

Theft Detection Module:

The prime function of this module is to detect the tampering of electricity meter, disconnect the power supply, and send a message of power theft to the management side. Whenever any consumer tries to interfere with the meter, the theft recognition unit identifies it with the help of magnetic sensors. Then, a signal generated by the sensors is sent to the microcontroller as well as relay circuit. Relay circuit is used to disconnect the power supply. Then, the microcontroller sends a power theft message to the management side via the communication unit comprising of an RS232 interface and GSM modem. In this way, the power theft is detected and alerted to the management side.

4   Experimental Setup and Results

Figure 2 portrays the experimental setup of different components of the system and their respective connections. Figure 3 shows the snapshot of the projected framework. The automated metering system consists on components such as PIC microcontroller, RS232 interface, GSM modem, magnetic sensors, LCD display, meter reading unit, theft detection unit, power supply unit, and an energy meter, respectively. The state-of-the-art readings from the energy meters are collected by meter reading unit and nourished into microcontroller. The energy consumption in No. of units can be observed through an LCD display which is connected to microcontroller. LCD display demonstrates the readings of the energy meters and theft grade. The LCD display is driven by a lithium-ion battery. Crystal oscillator is used to provide electrical signals of precise frequency to the microcontroller by utilizing the vibrating crystal's mechanical resonance. Whenever any consumer tries to tinker the meter, the dimensionality of the magnetic sensors changes. Then, a signal is engendered and sent to the microcontroller indicating theft of energy meter. Microcontroller sends appropriate message to GSM modem via RS232 interface. RS232 interface is used to convert signals shared between a microcontroller and a modem. After receiving the message from microcontroller, GSM modem transmits a theft message to management side. Along with sending a message to the management side, the theft detection unit disconnects the load using a relay circuit.

Figure 4 shows the snapshot of the proposed framework, in which No. of units consumed by the consumer on LCD display as well as the amount calculated depending upon how much units are consumed. For every unit consumed, amount of bill is

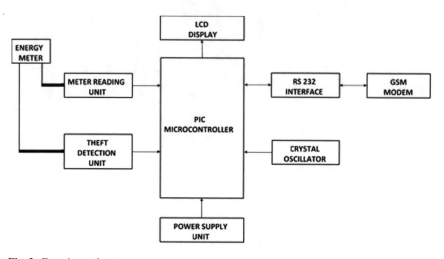

Fig. 2 Experimental setup

Fig. 3 Snapshot of experimental setup of the proposed framework

Fig. 4 Snapshot of results of the proposed framework

calculated. The amount is calculated as follows, Amount of bill (A) = No. of units consumed (n) * price per unit. After calculating the amount, it will be displayed on the LCD screen. This mechanism helps the consumer to efficiently monitor electricity consumption.

5 Conclusion

In this age of energy emergency period, methodologies which help in commendably governing energy utilization and dodging energy depletion are most important. In this paper, PIC microcontroller-based design and implementation of automated energy meter reading system are presented. In this system, the energy consumption is available instantly to the consumer by means of LCD displays, which helps the consumer to know about the energy utilization and hence can monitor it effectively. Further, whenever a theft occurs, load to the energy meter gets disconnected and the management side is alerted by a SMS. The objectives accomplished in this system are ease of accessing energy consumption at consumer end, theft detection is alerted at management end, and disconnection of load whenever a theft occurs. In the future, we can further incorporate a system that alerts the consumer when the energy utilization crosses a specific threshold.

References

1. Langhammer N, Kays R (2012) Performance evaluation of wireless home automation networks in indoor scenarios. IEEE Trans Smart Grid 3(4):2252–2261
2. Landi C, di Ing Dell D, Merola P, Lanniello G (2011) ARM-based energy management system using smart meter and web server. In: Proceedings of IEEE instrumentation and measurement technology conference Binjiang, pp 1–5
3. Garrab A, Bouallegue A, Abdallah FB (2012) A new AMR approach for energy saving in smart grids using smart meter and partial power line communication. In: Proceedings of IEEE first international conference on renewable energies and vehicular technology, pp 263–269
4. Koay BS, Cheah SS, Sng YH, Chong PH, Shum P, Tong YC, Wang XY, Zuo YX, Kuek HW (2003) Design and implementation of Bluetooth energy meter. In: IEEE proceedings of the 4th international joint conference of the ICICS, vol 3, no 7, pp 1474–1477
5. Khalifa T, Naik K, Nayak A (2012) A survey of communication protocols for automatic meter reading applications. IEEE Commun Surv Tutor 3(3):168–182
6. Qingyang L, Dapeng L, Jian W, Ning D (2012) Construction of electric energy data acquiring network integrating several communication methods. In: Proceedings of fourth international conference on multimedia information networking and security, pp 121–124
7. Rafiei M, Eftekhari SM (2012) A practical smart metering using combination of power-line communication (PLC) and WiFi protocols. In: Proceedings of 17th conference on electrical power distribution networks, pp 1–5
8. Iyer D, Rao KA (2015) IoT based electricity energy meter reading, theft detection and disconnection using PLC modem and power optimization. Proc Int J Adv Res Electr Electron Instrum Eng 4(7):6482–6483

Comparative Analysis of Privacy-Preserving Data Mining Techniques

Neetika Bhandari and Payal Pahwa

Abstract Data Mining is the technique used to retrieve useful information, relationships and patterns from huge databases and data warehouses. It is an important phase of the Knowledge Discovery of Databases process which can be applied on Big Data. Mining on Big Data has various concerns which include protecting sensitive data and securing the useful information extracted from unauthorized access. Privacy-Preserving Data Mining (PPDM) aims to protect the sensitive data and information during the mining process. PPDM techniques have gained the attention of the researchers in recent times. In this paper, we have listed them and identified their advantages and limitations along with the algorithms following these techniques.

Keywords Data mining · Knowledge discovery in databases (KDD)
Privacy-preserving data mining (PPDM) · Big Data

1 Introduction

Today in the digital world, we are very data-centric. In the recent years, we have seen that data is ubiquitous and increase in the size of data is tremendous. According to a recent survey published by the Storage Newsletter, April 2017, the total data of the world was 16.1 Zettabytes (10^{21} bytes) in 2016, which is expected to go to 163 Zettabytes by the year 2025. Data is being generated everywhere through various sources and organizations like Healthcare, Biomedical, Defence, Business, Finance, Social media and many more. This data is called Big Data. According to Sagiroglu and Sinanc [1], Big Data represents the huge and massive datasets which

N. Bhandari (✉)
Indira Gandhi Delhi Technical University for Women, Kashmere Gate,
New Delhi, Delhi, India
e-mail: neetika.igdtu@gmail.com

P. Pahwa
Bhagwan Parshuram Institute of Technology, Rohini, Delhi, India
e-mail: pahwapayal@gmail.com

© Springer Nature Singapore Pte Ltd. 2019
S. Bhattacharyya et al. (eds.), *International Conference on Innovative Computing
and Communications*, Lecture Notes in Networks and Systems 56,
https://doi.org/10.1007/978-981-13-2354-6_54

have large, varied and complex structure and face difficulty in storage, analysis and visualization of data. Big Data is thus the data which is huge in size and too complex to be efficiently handled by traditional data management tools and techniques due to its size, velocity and heterogeneous nature.

Extraction of information from this Big Data is a major concern. Data stored in data warehouses and repositories can be used widely to extract useful but unknown information in the form of patterns or rules which can be used for decision-making and analysis purposes. This is called Data Mining [2]. The process used for discovering helpful knowledge or information from data collected from various sources is called Knowledge Discovery in Databases (KDD). KDD involves various steps for Extraction of information which includes Data Cleaning, Data Integration, Data Transformation, Data Mining, Pattern Evaluation and Data Presentation. Like Data Mining, Big Data Mining allows applying analytics on massive data sets to uncover interesting and applicable patterns and knowledge hidden in the Big Data.

There are several challenges associated with Big Data Mining which needs to be addressed. An important concern in big data mining is privacy and security. Big Data includes data from various sources and thus is very critical as it contains sensitive as well as non-sensitive data of individuals and organizations. This data is of great importance and it is required to ensure that it is secure and protected from unauthorized access because it can be misused and exploited. For example, in the IoT era, the recent wearable devices like Apple iWatch, Google Fit, Apple Health Kit can give sensitive information about the user like their location, health condition and financial status on recording and analyzing their daily activities [3]. This data when disclosed can impose threats of misuse on individuals.

Privacy issues of data mining can be handled by Privacy-Preserving Data Mining (PPDM) which aims to protect the critical data and information from being disclosed to untrusted users keeping the utility of data [4]. PPDM techniques aim to lower the risk of misuse of data producing the same results. Many PPDM techniques already exist.

In Sect. 2, we explain Big Data Mining and its needs and challenges. Further, the need of privacy in Big Data Mining is mentioned in Sect. 3 where we introduce PPDM and give a comparative analysis of all the existing PPDM techniques. The conclusion is finally given in Sect. 4.

2 Big Data Mining

Big Data is the huge voluminous data which is collected from various sources and applications and stored in different data warehouses and repositories. It is a challenge to store, manage and use this massive data. It cannot be handled by traditional database management systems because of its size and complex structure. It is required to define new architecture, algorithms, analytics methods and tools to handle this data. The various features/characteristics of Big Data are:

1. **Volume**: Size of data is increasing tremendously that it cannot be handled easily. This data comes from various sources like industry, stock market, social media, healthcare and many more.
2. **Velocity**: Data is streaming in and out at a high speed. It is required to process this data within a specific time limit or else it will become useless and will have no value.
3. **Variety**: Big data is collected from many different sources and in dissimilar and mixed formats. This data is heterogeneous in nature and can include text, audio, video and images thus increasing the complexity to handle it. Data can be:

    (i) **Structured**: Fixed format as in relational DBMS.
    (ii) **Unstructured**: Unknown form or structure. This can be data fetched from any website (say Google search) which includes text, images, video, audio, etc.

4. **Veracity**: This relates to the quality and truthfulness of data. Any uncertainty in data can produce poor quality of information.

Big Data can be used in various applications like business, healthcare, technology, smart devices, IoT and social media. This data is waste to the industry, if it is not converted into useful information. Extraction of useful information as patterns or rules from Big Data is called Big Data Mining. It allows gaining insight into various features and associations which were unknown and undiscovered [5, 6]. These results can be used for critical decision-making and various analysis to make important contributions and benefit the organizations. According to Che et al. [7], various issues and challenges associated with mining big data are:

1. **Variety and Heterogeneity**: Big Data includes mixed data which is very complex to store and process. This data can be structured or unstructured and needs to be transformed into structured format for analysis. It is required to transform data before analysis.
2. **Incompleteness**: Missing values in the data can lead to uncertainty in the analysis. It is important to handle this data by data imputation where missing values are filled by prediction or observation. However, data imputation is difficult to apply on Big Data due to its huge size and complexity.
3. **Scalability**: Data analysis should be able to handle the huge volume of data which is growing rapidly with time.
4. **Timeliness/Speed**: Due to huge size, handling Big Data consumes a lot of time. If it does not handle in a specific time, then the information retrieved may or may not be of value or worth. Thus, the speed of data mining should be fast when handling Big Data.
5. **Privacy and Security**: Big Data often has real-world data like personal data from which any kind of information about the individual can be unearthed. Policies and techniques needs to be defined to manage sharing and use of this data in order to prevent any violation of privacy.
6. **Garbage Mining**: Data in Big Data can become outdated, corrupted, useless and junked over time. This data can cause inconsistencies in the result of mining

process. It is important to remove this data (garbage) by cleaning. Since this data is hidden and has ownership issues, thus it is difficult to remove it.

7. **Accuracy, Trust and Provenance**: The data from various sources may not be trustworthy and accurate. It is required to validate this data and trace its provenance. By tracing the provenance, history and origin of the data can be checked to obtain trust and accuracy of the source.

8. **Interactiveness**: Data mining system should allow prompt interaction with the user to provide feedback of mining results.

3   Privacy-Preserving Data Mining

One of the major concerns of mining Big Data is Privacy and Security. Big Data contains sensitive information which comes from different applications. It is important to protect this data from being disclosed and accessed by unauthorized users as it imposes the threat of exploitation and misuse of data [8]. Big Data can be harnessed to harm user's privacy in various ways. The personal information of users' can reveal the secrets of individuals and organizations which they might not want to be disclosed. Social differentiation of individuals into strata based on their occupation, income, status, etc. can harm their privacy [9]. Information about the individuals can give a deep understanding of their lives on analysis which might not be desired by them.

The whole KDD process has various phases where privacy needs to be incorporated as we need to protect the sensitive data from unapproved access to avoid severe damage to users and organizations. According to Xu et al. [10], privacy concerns of people from various phases of KDD differ. According to the Data Provider who provides data from various sources, it is required that only necessary and required data should be provided to the data collector. The Data Collector collects data from data providers and stores in data warehouses and databases it is required to modify this data to make it secure, not changing its meaning and relevance. The Data Miner use data from data warehouses and databases and mines it to extract useful rules and patterns. For him, it is required to hide these sensitive results from unreliable parties. According to the Decision makers who use the results of data mining for further analysis and decision-making, it is required to check the authenticity of the results obtained from the mining process.

As shown in Fig. 1, there are different sources of data that exist. These sources provide huge data which is further collected and placed in data warehouses and repositories. Data from these data sources are used to extract useful knowledge and information by data mining. The result of data mining in the form of rules or patterns is used by decision makers in organizations for critical analysis.

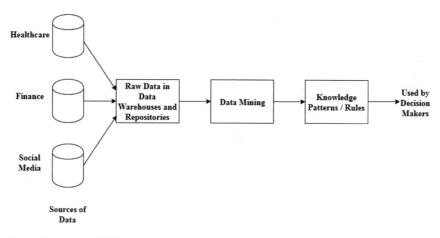

Fig. 1 Illustration of KDD process used for decision-making

While data mining is being done, many challenges need to be addressed to ensure privacy. Privacy can be ensured during data mining by Privacy-preserving Data Mining (PPDM). PPDM is a field of Data Mining used to shield sensitive data or critical information from being used by untrusted unauthorized users [11, 12]. PPDM techniques should be applied for data protection without decreasing the benefits and utility of data. PPDM techniques works at two levels [13]. At one level, the sensitive and critical raw data obtained from the data collector should not be used directly for mining process. At the second level, the critical and sensitive mining results and rules should be excluded from decision-making purposes.

Various PPDM techniques have been proposed which can be classified into different categories based on the following aspects [14, 15]:

1. Data Distribution: These are based on distribution of data and have two types—Centralized PPDM and Distributed PPDM. Distributed PPDM techniques can be based on horizontally partitioned data or vertically partitioned data.
2. Data Modification: These are based upon the method used for modifying data for preserving privacy before mining process. These include Perturbation-based PPDM, Anonymization-based PPDM, Condensation-based PPDM, Randomization-based PPDM and Cryptography-based PPDM.
3. Data Mining Algorithms: These are based upon the data mining technique used for extraction of information. These include Privacy-preserving Association rule mining, Privacy-preserving Clustering and Privacy-preserving Classification.

There are different Privacy-Preserving techniques. These techniques are listed and briefly described in Table 1. Further, their advantages and disadvantages are also identified along with a few algorithms that follow these techniques.

Table 1 Comparison of PPDM techniques

Technique	Brief description	Advantages	Disadvantages	Algorithms
Perturbation-based PPDM (suitable for centralized and distributed)	Original values of data are distorted or replaced with synthetic data. This is done by noise addition	• Simple • Efficient and high accuracy • Reserves statistical information • Treat attributes individually	• Information loss in multi-dimensional data • Need modified Data mining techniques	• Geometric transformation perturbation • Non-negative Matrix Factorization (NMF)
Anonymization-based PPDM (suitable for centralized scenario)	Identification data should be hidden to avoid disclosure of individuals	• Preserves identity of individuals	• Heavy information loss • Prone to linking attacks like background attack and homogeneity attack	• k-anonymity • l-diversity • t-closeness
Randomization-based PPDM (suitable for centralized and distributed scenario)	Scrambling or twisting of data is done in such a way that it makes it difficult to check whether it is correct or incorrect	• Simple • Efficient • Does not need knowledge of distribution of data	• Treats all records equally including outliers • Not suitable for multiple attributes • High information loss	• Randomized response forest for privacy-preserving classification
Condensation-based PPDM (suitable for centralized scenario)	Data is constructed into clusters or groups based on constraints. This condensed data is used to generate clusters	• More efficient • Preserve statistical information of data • Works on synthetic data • Provides additional layer of protection	• High information loss • Affects data mining results	• Condensation approach to PPDM
Cryptography-based PPDM (suitable for distributed scenario)	When multiple parties collaborate to compute results from their own inputs not knowing each other's data	• Works with multiple parties	• Not suitable for too large number of parties • Fails to protect output on computation	• SMC-based protocols like secure sum, secure set union • SMC-based homomorphic encryption

4 Conclusion

Data is very important in today's world. Various organizations have huge data from which useful information can be pulled out using the KDD process. This useful information can be further used by the organization to make important decisions. Thus, data plays a crucial role in decision-making. It is important to keep this data and information secure during the mining process. PPDM helps in keeping the data and information secure during the mining process. Various PPDM techniques exist already and each has its own advantages and limitations. Also, various algorithms have been proposed which use these techniques. We have identified the advantages and disadvantages of these techniques and also listed a few algorithms which use them along with a brief description of each technique.

References

1. Sagiroglu S, Sinanc D (2013) Big data: a review. IEEE
2. Chen MS, Han J, Yu PS (1996) Data mining: an overview from a database perspective. IEEE Trans
3. Bertino E (2015) Big data—security and privacy. IEEE
4. Jin X, Wah BW, Cheng X, Wang Y (2015) Significance and challenges of big data research. Elsevier (2015)
5. Tyagi AK, Priya R, Rajeshwari A (2015) Mining big data to predicting future. Int J Eng Res Appl
6. Wu X, Zhu X, Wu GQ, Ding W (2014) Data mining with big data. IEEE Trans Knowl Data Eng
7. Che D, Safran M, Peng Z (2013) From big data to big data mining: challenges, issues and opportunities. Springer
8. Subaira AS, Gayathri R, Sindhujaa N (2016) Security issues and challenges in big data analysis. Int J Adv Res Comput Sci Softw Eng
9. Bardi M, Xianwei Z, Shuai L, Fuhong L (2014) Big data security and privacy: a review. China Communication
10. Xu L, Jiang C, Wang J, Yuan J, Ren Y (2014) Information security in big data: privacy and data mining. IEEE
11. Natarajan R, Sugumar R, Mahendran M, Anbahagan K (2012) A survey on privacy preserving data mining. Int J Adv Res Comput Commun Eng
12. Senosi A, Sibiya G (2017) Classification and evaluation of privacy preserving data mining: a review. In: IEEE Africon 2017 proceedings
13. Divanis AG, Verykios VS (2009) An overview of privacy preserving data mining. ACM Cross-roads 15
14. Malik MB, Ghazi MA, Ali R (2012) Privacy preserving data mining techniques: current scenario and future prospects. In: IEEE international conference on computer and communication technology
15. Vaghashia H, Ganatra A (2015) A survey: privacy preservation techniques in data mining. Int J Comput Appl

Author Index

Printed in the United States
By Bookmasters